Terrors of the Table

The Curious History of Nutrition

WALTER GRATZER

OXFORD
UNIVERSITY PRESS

Great Clarendon Street, Oxford OX2 6DP

Oxford University Press is a department of the University of Oxford.
It furthers the University's objective of excellence in research, scholarship,
and education by publishing worldwide in

Oxford New York

Auckland Cape Town Dar es Salaam Hong Kong Karachi
Kuala Lumpur Madrid Melbourne Mexico City Nairobi
New Delhi Shanghai Taipei Toronto

With offices in

Argentina Austria Brazil Chile Czech Republic France Greece
Guatemala Hungary Italy Japan Poland Portugal Singapore
South Korea Switzerland Thailand Turkey Ukraine Vietnam

Oxford is a registered trade mark of Oxford University Press
in the UK and in certain other countries

Published in the United States
by Oxford University Press Inc., New York

First published 2005

British Library Cataloguing in Publication Data
Data available

Library of Congress Cataloging in Publication Data
Data available

Typeset by Footnote Graphics Limited
Printed in Great Britain
on acid-free paper by
Clays Ltd., St. Ives plc

ISBN 0-19-280661-0 978-0-19-280661-1

1 3 5 7 9 10 8 6 4 2

Contents

Picture acknowledgements

Harriette Chick: The Wellcome Trust; ***The Virgin and Child*, 1509, by Hans Bergmaier:** Germanisches Nationalmuseum, Nuremberg. Lauros/Giraudon/*www*.bridgeman.co.uk; **James Lind:** akg-images; **Unfinished portrait of Sir Gilbert Blane by Sir Martin Shee:** The Royal College of Physicians of London; **Andreas Vesalius:** © TopFoto.co.uk; **Herman Boerhaave, *c.* 1723, by Aerty de Gelder:** © Topfoto.co.uk; **Advert for Rose's Lime Juice Cordial:** Courtesy of Cadbury Schweppes plc; **Lavoisier and his wife, 1788, by Jacques-Louis David:** The Metropolitan Museum of Art, New York. © Bettmann/Corbis; **"Scientific Researches! . . .", 1802, by James Gillray:** The Royal Institution, London/www.bridgeman.co.uk; **Frederick Accum:** © Hulton-Deutsch Collection/Corbis; **Justus von Liebig, c. 1865:** © Austrian Archives/Corbis; **Liebig's Laboratory at Giessen:** © Bettmann/Corbis; **Paris slaughterhouse, 1874:** Mary Evans Picture Library; **Advert for Oxo, 1918:** © TopFoto.co.uk; **John Kellogg:** © Corbis; **Advert for Kelloggs, 1934:** Mary Evans Picture Library; **Pigeon before and after vitamin treatment:** C. Funk, 1922, from *Beriberi, White Rice and Vitamin B*, by Kenneth J. Carpenter, 2000. Courtesy of the University of California Press; **Advert for Pinkham's Vegetable Compound:** © Bettmann/Corbis; **Advert for bile beans:** Courtesy of the Advertising Archives

The publisher and author apologize for any errors or omissions in the above list. If contacted they will be pleased to rectify these at the earliest opportunity.

Introduction

'I look upon it, that he who does not mind his belly will hardly mind anything else.' Thus Samuel Johnson, giving vent between mouthfuls to an opinion that could serve as a text for our incontinent times. Dr Johnson was fortunate, for good food was abundant in England in his day, his digestion was sound, and he was little vexed by dietary admonitions. The study of nutrition – of what happens to one's dinner after it has entered the gullet – was by then a well-established topic in physiology, and indeed Johnson would have been familiar with the illuminating writings on the subject by the great Dutch physician and savant, Hermann Boerhaave, whom he greatly admired. But the science of food scarcely impinged on the eating habits of the epicures, let alone the poor. A half-century after Johnson's death, Jean Anthelme Brillat-Savarin published (anonymously) his classic treatise, *La Physiologie du Goût*, a captivating, discursive paean to the pleasures of the table. (If our primal ancestors ruined themselves for an apple, Brillat-Savarin asks at one point, what might they not have dared for a truffled turkey?) But of physiology, as we understand it, there was none: the closest approach was probably the observation that the more abundant the papillae on one's tongue the more acute one's perception and appreciation of taste.

But the history of nutritional science is full of fascination and drama, for it encompasses every virtue, defect, and foible of human nature. Scholarly rigour offset by pig-headed perversity, noble principle by shameless roguery, self-sacrifice by opportunism, all these are on display. Reason contends with superstition, mountebanks prosper, and savants quarrel among themselves. The pageant stretches back two millennia and more, but it has never appeared more gaudy and extravagant than today. The story begins, more or less, with the doctrine of humours – those elusive essences that were held to define men's natures and determine what they should eat and drink, and what avoid, if they were to stay healthy in body and mind. The ancients began to formulate precepts for a good life around 600 BC. A Greek sage, Alcmaeon of Croton, announced to the world that not only too little food but also too much was injurious to health. The medieval physicians and philosophers, like their Greek and Roman predecessors, denounced over-indulgence and inveighed against the dangers of obesity. Their declamations, like Alcmaeon's modest insight, eerily prefigure the admonitions of learned bodies and governments today, who have made it a mantra for our time.

A new era dawned during the 16th and 17th centuries, when the great Andreas Vesalius and his followers gradually put paid to the humours. It remains a mystery that physicians should have clung for so many centuries to the petrified doctrines handed down in ancient texts. The only historical parallel was the universal practice of bleeding, which was followed in ancient Asia, India, and the Arab world before reaching Europe; there, although rooted originally in the humoral theory, it reached its zenith much later, in the early 19th century, and caused the death of innumerable patients.

It was early in the 17th century that savants first began to concern themselves seriously with the processes of digestion, and through the history of nutritional science from then on there march a succession of remarkable personages. The greatest of all, arguably, was Antoine Laurent Lavoisier, whose life was terminated on the scaffold during the French Revolution. It was he who introduced the rigours of quantitative analysis into chemistry, and applied that science to the study of the mechanisms by which food generates the body's energy. His high-spirited widow, who had shared in his work, married the egregious American, Benjamin Thomson, Count Rumford of the Holy Roman Empire, spy, soldier of fortune, physicist, and inventor of Rumford soup, who contributed much to the problems of feeding the poor in a war-torn Europe. Throughout the 18th and 19th centuries progress came from unexpected quarters: an obscure naval surgeon carried out a momentous experiment on shipboard and could bring himself to believe the results; a gunshot made a fist-sized hole in a man's stomach that opened a window on the function of the gastric juices; an Alpine ascent supplanted a theory that was leading science down a cul-de-sac; the caprice of a new cook in an Indonesian hospital threw a sudden light on the malnutrition that was ravaging the population. Impending starvation in the wake of the Franco-Prussian war stimulated a programme of research by the two pre-eminent physiologists of the 19th century. They uncovered the principles of the chemical transformations of foods, on which most later work was based (but it brought the greatest of all the savants into savage conflict with the opponents of vivisection and destroyed his marriage).

But where ideas about nutrition were concerned, the paladins of progress were always up against the forces of reaction. The celebrated chemist, Justus von Liebig, pugnacious, domineering and powerful, had formulated the rule that foods contained only three nutritious elements, two that furnished the energy for life and one that replenished the body's tissues. He would not entertain the suspicion of other less influential figures that trace substances – vitamins as they came to be known – might be equally important for sustaining life. And so there appeared on the market a variety of foods, often promoted by Liebig himself, most notably baby feeding formulas, devised on the best scientific principles. They were devoid of vitamins, and it was the babies of the better off who mainly suffered from this disastrous regime. But when vitamins finally arrived, they were pounced on and strenuously promoted by the food industry, and greeted with near-hysterical enthusiasm by a gullible public. The vitamin craze ushered in an unprecedented outbreak of quackery, which eclipsed the preceding fads of Sylvester Graham alias Doctor Sawdust, the Kellogg brothers, Horace Fletcher, the Great Masticator, and their like.

And so to our own era of superabundance – in the countries of the north, that is to say. The culture of excess has brought with it the blessings of obesity, diabetes, heart disease, and a host of other ills as yet only foreshadowed. Our food now is polluted as never before – not, as in Victorian times by enterprising scoundrels who watered the milk, recycled the tea, added strychnine to beer to make it more bitter, and coloured children's sweets with copper arsenite or lead chromate – but by huge industrial interests, protected most often by governments. We are fed hormones, antibiotics, and pesticides, and the poorest, especially, are crammed with concoctions made up of chemically enhanced sugars and fats. The fear of cholesterol (though latterly, to be sure, only the 'bad cholesterol') has, in our society, supplanted the devil, as the roaring lion who walketh about, seeking whom he may devour. And it is as if the theory of the humours had returned in the form of the contending diets that pursue each other through the marketplace. Perhaps in time the wheel of history will turn and allow us or our descendants to recapture the lost innocence that allowed the eupeptic cleric, the Reverend Sydney Smith to proclaim:

> Serenely full, the epicure would say,
> Fate cannot harm me, I have dined today.

I have tried, so far as I could, to tell the story with a minimum of technical jargon, but the underlying science is briefly set out in the Appendix. I have also given at the end of each chapter suggestions for further reading for those who would like to know more. References are included in the text only to passages of direct quotations and to especially difficult or controversial topics.

I have had much help and encouragement from a number of friends while writing this book. I am indebted most of all to Professor Bill Bynum, who read the manuscript, detected a series of errors and infelicities, and made many valuable suggestions for its improvement. Peter Brown gave me much advice and directed me to useful sources. Alan Weeds put me right on matters of biochemistry. I am grateful also to my successive editors at Oxford University Press: to Michael Rodgers, who stimulated the idea in the first place, to Abbie Headon, Marsha Filion, James Thompson, and especially to Latha Menon, whose perceptive criticisms were greatly needed, and to Michael Tiernan, whose copy-editing eliminated no small number of gaffes and inaccuracies.

1

The Ravages of War

Harriette Chick and the children of Vienna

On an autumn day in 1919 Dr Harriette Chick from the Lister Institute in London, and her colleague, Dr Elsie Dalyell, arrived by train in Vienna, and were greeted by Professor Clemens von Pirquet of Vienna University. The Great War was over, the Habsburg monarchy had fallen, and the City of Dreams was dilapidated, its population in the grip of famine and disease.

Harriette Chick was no stranger to Vienna. Daughter of a lace merchant, she had graduated from University College London, and gone on to advanced study in her chosen field of bacteriology at the universities of Munich and more especially Vienna, which was then, and remained for another three decades, one of the world's pre-eminent centres of medical science. In 1905 she had joined the Lister Institute, where her interests took a new direction: it was nutrition that now engaged her attention. In 1914 her male colleagues went off to the war, some to forward dressing-stations on the Western Front, others to base hospitals. For the women it was an opportunity to show that they could run the research activities of the Institute without any help from the men. Harriette Chick, in particular, took charge of a broad programme of nutritional research germane to the needs of the military and of the undernourished population at home. The existence of trace substances inseparable from a life-sustaining diet had been generally recognized. The menace of scurvy still hung over the Navy and, in hostile climes, the Army. Another deadly malnutrition disease, beriberi, was also making its effects felt among the soldiers in Mesopotamia. The team of women researchers at the Lister devised prophylactic measures to avert the spread of these scourges. Another of Harriette Chick's concerns was the study of rickets, and experiments with animals convinced her that this too was a disease of malnutrition. But von Pirquet and the rest of the illustrious Viennese medical establishment were in no doubt that they were dealing with an infectious disease, rampaging through the hunger-enfeebled population.

It was a dismal prospect that the two Englishwomen encountered in Vienna. The paediatric wards of von Pirquet's clinic were choked with infants and babies.

All were in the throes of infantile scurvy, and many were dying. In the streets old people hobbled, barely able to support themselves. The bent posture, the muscle pains, the ulcerated eyes, all were signs of ricketsial disease. What then was rickets?

The English disease

Rickets has stalked the human race throughout recorded history. Herodotus, writing in the 5th century BC, recounts what he observed on a visit to the battle-field close to the city of Pelusium, where a hundred years earlier the Persian tyrant Cambyses had defeated the Egyptians in the course of his bid to conquer Africa. The field was littered with the remains of the soldiers who had died, and Herodotus found that the skulls of the Persians were so thin that they cracked when struck by a pebble. Those of the Egyptians, on the other hand, would break only under the impact of a hefty stone. The local people, when questioned, opined that their Egyptian craniums were strengthened by their custom of going bareheaded in the sun, whereas the Persians wore turbans. This is, as will emerge, a highly plausible explanation.

But the manifestations of rickets were first explicitly described in the 17th century. The disease was marked in growing children by bowed legs, a curved spine, enlarged, rather square heads, swollen joints and abdomen, and bad teeth. It was, according to one observer, writing in around 1620, a malady 'wherein the *Head* waxeth too *great*, whil'st the *Legs* and *lower parts* wain too *little*'.[1] Early in the 17th century rickets was one of the afflictions recorded in the Bills of Mortality of the City of London, compiled by the 'searchers'. These were mainly old women, charged with inspecting each week's crop of corpses and recording the causes of death. The first precise description of the disease by an observant physician appears to have been published in 1646. It occurs in a doctoral thesis presented by Daniel Whistler of Oxford to one of the great centres of medical learning of the day, the University of Leiden. In 1650 an independent and much more detailed account, under the title *De Rachitide*, was published by the Regius Professor of Physick at Cambridge University, Francis Glisson.

The characteristic signs of rickets appear in northern European paintings depicting the Christ child – an indication of how commonplace the condition was throughout the region; so it was probably known as the English disease only because it was mainly English physicians who had written about it. The etymology of rickets is uncertain, but the term appears to have been in common circulation in the 17th century, although the physicians opted for the more learned-sounding *rachitis*. This affectation aroused the ire of the celebrated herbalist and physician Nicholas Culpeper, who wrote of 'children that have the Rickets, or rather (as the Colledge of Physicians will have it) the *Rachites*, for which name for the Disease they have (in a particular Treatise lately set forth by them) Learnedly Disputed, and put forth to publick view, that the World may see they took much pains to little purpose.'[2]

The incidence of rickets fluctuated widely during the next two centuries with the prosperity of the population, changes in taste for different foods, and the quality of harvests. Early attempts at treatment by diet were based on Galen's ancient theory of humours (Chapter 3). One writer averred that it could be cured by 'by cauterizing the *Vein* behind the Ear'. The swollen abdomens of rachitic children gave rise to the belief that the liver was somehow implicated, and livers of various animals were used in treatment, probably with some success. Some luckless infants were strapped into a straitjacket of leather and iron for months on end. Sometimes the condition cleared up, often not.

With the coming of the industrial revolution and the growth of the city slums, poverty, poor diet, and the soot-laden air which filtered out the ultraviolet part of the sun's spectrum ensured that rickets became endemic in the English population. In the 19th and early 20th centuries bandy legs were commonly associated with slum dwelling. It was only when the Boer War was going badly enough to force the Army into an urgent recruiting drive that the deplorable state of the nation's health started to cause concern in official circles. Only a year before, in 1901, the Quaker philanthropist Seebohm Rowntree had written a report on the condition of the poor in one British city, York;[3] most of them, adults and children alike, he discovered, subsisted on little but white bread and tea. Rowntree's revelations met with nothing but apathy. But when the head of the Army Medical Service, the DGMS, announced that no less than 40% of volunteers (and many more in industrial areas) had to be rejected on medical grounds, because of deformities, rotten teeth, doubtful hearts, and other infirmities, the Government reacted with alarm. Its first act, characteristically, was to drop the minimum height for foot-soldiers to five feet (just over one and a half metres, down three inches, or 7.5 cm, from the limit set in 1883). Eventually, though, a committee was set up (the medical profession in the shape of the Royal Colleges of Physicians and of Surgeons having declared such an investigation otiose) and confirmed all that Seebohm Rowntree had said. Even so, its recommendations had more to say about such matters as living and working conditions, sanitation, and alcoholism than about diet, and little was done. White bread, the staple food, had, as will be seen, minimal vitamin and mineral content, so children, especially, suffered from malnutrition. A study in 1895 revealed that 83% of children in the Bethnal Green district of London had to make do with bread as their only solid food in 17 out of 21 of what passed each week for meals. The poor, both children and adults, were perpetually hungry. One consequence was that the mothers' milk was deficient in nutritive value, while dairy milk was in very short supply and also of inferior quality. Similar conditions prevailed among the poor in other countries. In the Columbus Hill district of New York, for instance, a report in 1917 by two paediatricians disclosed that 90% of black children were rachitic. When a group of babies from this milieu, aged between 4 and 12 months, were fed cod liver oil they matured normally, while a control group, fed at home in the usual manner, developed all the symptoms of severe rickets.

Hunger and the Great War

By the time the Great War engulfed Europe and conscription was introduced to meet the insatiable demands of the 'meat-grinders' on the Western Front, little had changed: 40% of the men of military age were still found unfit for service. The other armies of Europe had similar problems of manpower, as well as provisioning. The calamitous rout of the Italians at Caporetto in 1917 was attributed by investigators to the collapse of the soldiers' resilience and morale, following the reduction of an already inadequate food ration. The Second Army, under its luckless commander, General Cadorna, who was blamed for its misfortunes, had been fighting in the mountains for two years in appalling conditions. By the time the Austrians attacked, the men had been subsisting for seven months on a calorie intake that left them miserably hungry and enfeebled.

In Britain malnutrition was already widespread after one year of war, and food was becoming ever scarcer as the U-boat blockade in the Atlantic took effect. As early as 1898, a leading physicist, Sir William Crookes, had given warning of the dangers: millions, he had told the British Association for the Advancement of Science, were being spent on protecting the coastline, on commerce, ships, explosives, guns, and men, and yet 'we omit to take necessary precautions to supply ourselves with the first and supremely important munition of war – food'. But little was done, and by 1916 there was real alarm. The Government had already asked the Royal Society to set up a committee to consider nutritional and related problems engendered by the War. It advised, for instance, against stopping imports of citrus and other fruits, which the responsible authorities viewed as luxuries. The committee took evidence from Harriette Chick and her colleagues at the Lister Institute, who formulated a schedule of nutritional needs of adults and infants, and made recommendations for best use of the existing resources. An almost immediate result was a large improvement in flour yields from wheat, and a nutritive 'war bread' appeared in the shops. A Minister of Food was appointed: Lord Davenport had grown rich in the food industry and was not disposed to take scientific advice. He laid down what he believed to be the minimal requirements of bread, meat, and sugar and appealed to the patriotism of the populace, which he hoped would comply with a system of voluntary rationing. His designated norms were insufficient to sustain health and strength, but in the spring of 1917 ships were being sunk at such a rate that food reserves had dwindled to an estimated three weeks' worth. The situation was aggravated by the confiscation of much of the richest calorie source of all, animal fat, as a source of glycerol (glycerine) for the manufacture of the explosive, nitroglycerine. For Lord Davenport the strain was too great; his health failed and he resigned. His successor took a much firmer grip. Meat farming was discouraged on grounds of inefficient energy yield per acre of land, and as the year wore on rationing was progressively introduced, allied to price controls. These measures were successful, or at least equitable, and with the introduction of convoys shipping losses diminished.

The British population, at all events, did better than the German. For one thing, British scientists were at this time better informed about nutrition, and especially the recent discoveries of vitamins, than their German confrères. The Germans, moreover, seem to have failed even to consider the problems of feeding the citizenry, believing evidently that their agriculture could be made self-supporting. A mere year into the war, in the winter of 1915, flour had to be rationed to 200 g (seven ounces) per person per day. Meat and dairy fats became scarce, prices rose, profiteers flourished, and the signs of malnutrition appeared in the urban population. A year later the situation had worsened and the country seemed weeks from collapse. Relief came in the nick of time from Field Marshal Mackensen's lightning campaign of conquest in Romania, with its rich agrarian resources. But, in time, mutinous unrest at home, aggravated by renewed food shortages, allied to a deteriorating military situation in the east, more than events on the Western Front, ensured the terminal collapse of Germany, and with it that of Austro-Hungary. The demoralized and debilitated population laid the blame on a mysterious Jewish financial conspiracy and the seeds of National Socialism were planted.

It is time now to return to Harriette Chick and Elsie Dalyell, taking stock of the grim situation in Vienna in 1919. The armistice had not put a stop to the hunger and malnutrition ravaging the populations of the Central Powers. Not only humanitarian concerns, but also fear that revolutions might erupt, concentrated the minds in the Western regimes. So it was that within a few months of the War's end the Lister Institute and the Medical Research Committee (the precursor of the Medical Research Council, founded the following year), prompted by the Government, ordained a mission to Vienna. Its purpose was to ascertain whether the effects, observed in animals, of 'accessory food factor' (that is to say, vitamin) deficiencies had a parallel in humans, especially infants, and whether their afflictions could be cured in the ways that had been found to work in the animals.

The doctors of Vienna had for five years been isolated from the advance of nutritional research, and in any case had had other preoccupations. They knew nothing of the recent studies in Britain and the United States (Chapter 9). The mission's host, Clemens Freiherr (or Baron) Pirquet von Cesenatico, was director of the University Kinderklinik, a man of great distinction and authority. A scion of the Lower Austrian aristocracy, he had been destined for the Church but turned to the study of philosophy and eventually to medicine. He chose paediatrics as his speciality and his researches into immunology – he developed vaccination procedures and coined the term 'allergy' – brought him early recognition, so that in 1908, at the age of 34, he was elected the first occupant of the chair of paediatrics at Johns Hopkins University in Baltimore. He stayed only one year in America before returning to Europe as Professor of Paediatrics in Breslau, and two years after that he moved back to Vienna, where he remained for the rest of his life (dying in 1929 in a suicide pact with his wife).

Von Pirquet received the two Englishwomen with courtesy but did not conceal his belief that, in seeking a nutritional cause and cure for the rickets that was rampant in his clinic, they were wasting their time. He adhered to (and had acted on) the firm belief that it was an infectious disease which attacked children with a susceptible predisposition. This was the generally accepted wisdom in Vienna as elsewhere, but there was one man, at least, in the Austrian medical hierarchy who was willing to listen. Here is how Dame Harriette Chick recalled the events, when in 1974, in her hundredth year, she mounted the podium to deliver the British Nutrition Foundation's annual Prize Lecture:

> Vitamin deficiency was widespread and our message of the cause of its various diseases was received with polite incredulity. But there was one man who believed, Professor K. E. Wenkebach, the well-known heart specialist, Director of the First Medical Clinic of the university and a very influential medical man.[4]

Wenkebach was a respected and popular figure, celebrated as a lecturer: *Beim Herrn Professor Wenkebach/ Da bleiben alle Bänke wach*, ran the medical students' doggerel, (that is to say, 'at Herr Professor Wenkebach's lectures the students at all the benches stay awake'). Wenkebach was Dutch and he had read about the work of his compatriot, Christiaan Eijkman (Chapter 8), who while a medical officer in the Dutch East Indies some thirty years earlier had identified the cause of beriberi as the lack of something in the diet that was needed in only minute amounts. (The term 'vitamin' had not yet been coined, nor had the concept taken root.) Rice was a good source of the mysterious substance, which, however, was lost when the grains were polished. The work was published in a Dutch journal and excited little interest at the time. Wenkebach invited Harriette Chick and her companion to his house, where he assembled a group of his colleagues, for an impromptu lecture and discussion. The interest of the doctors was aroused, and soon thereafter a formal lecture was arranged before the medical association (*Gesellschaft der Ärtzte*).

But Harriette Chick and Elsie Dalyell had not come to Vienna just to talk. They had already busied themselves in the wards of von Pirquet's clinic. Finding a child with acute (and deadly) infantile scurvy, they prepared a vitamin C extract from lemon juice. Infants cannot tolerate large amounts of acid, so the juice was carefully neutralized and concentrated to a small volume under vacuum. The child quickly recovered. Next they saw an adult patient with one of the manifestations of vitamin D deficiency (the cause of rickets): keratomalacia is ulceration of the cornea, and if not quickly treated leads to blindness. Butterfat was procured from the Quakers – the Relief Mission of the Society of Friends, active in Austria – and the patient's sight was saved. These feats had already been bruited around Vienna, and at the lecture to the medical faculty there was much enthusiasm, especially, as Harriette Chick later recalled, on the part of the younger doctors, for many had been prisoners of war in Russia and had themselves experienced the consequences of vitamin deficiencies, notably night-blindness.

The observations in Vienna on the effects of starvation, and especially of vitamin D deficiency, caused by the dearth of dairy products and animal fats generally, affirmed that what had been gleaned from animal experiments indeed pertained also to humans. The growth of many of the Viennese infants was severely retarded, and the elderly poor were martyrs to 'hunger osteomalacia', a softening of the bones accompanied by debilitating muscle pains and betrayed by a stooped posture and a 'waddling gait'. This distressing condition was rapidly alleviated by cod liver oil. It was enough to convince the sceptical von Pirquet. He wrote in the preface to a report on the work of the English mission:

> When Dr Chick and her colleagues began their work in my Klinik in 1919, I had little expectation that it would lead to results of much practical value. At that time I was of the opinion that vitamin deficiency in our ordinary diet was a very exceptional occurrence, as for instance in case of infantile scurvy (Müller–Barlow disease). With regard to the aetiology of rickets I held the view that it was an infectious disease, widely prevalent in this part of Europe, producing severe symptoms only in the case of children who possessed special susceptibility as a result of an inherited tendency, of a faulty diet or of defective general hygiene. I imagined rickets to be a disease comparable to some extent with tuberculosis.[5]

In Vienna most children did indeed give a positive reaction in the tuberculin test, which von Pirquet himself had developed, though in the majority symptoms of the disease never emerged. Von Pirquet now conceded that the inferences Harriette Chick had drawn from her animal experiments were fully vindicated.

In June of 1920 the little team returned to London, but von Pirquet wanted them back to develop a programme for the elimination of rickets in his city. He put his proposal to the Lister Institute and the Medical Research Council, which referred it to its Accessory Food Factor Committee. The Committee deliberated and acceded to von Pirquet's request. Soon Harriette Chick and a small band of doctors were ensconced in the Kinderklinik. Rickets, the 'English disease' of earlier epochs, is a defect of bone and joint formation. The infant's bones and joints are too soft to support the weight of the body and they deform. To devise a diet that would prevent rickets enough patients were needed to ensure that firm conclusions could be drawn from trials. In addition to the 20 beds that he made available in the Kinderklinik, von Pirquet secured another 60 or so at the American children's charity clinic in Meidling, a southern suburb of Vienna. A grant was secured from the International Red Cross and with the help of the Quaker mission a large hut was converted to meet the needs of the study. As a rigorous criterion for diagnosing rickets in its first stages Harriette Chick chose X-ray images of the bone structure, and von Pirquet made one of his hospital radiologists available. The babies were divided into two groups, according to what had by then become the recognized procedure for clinical trials. One group was to receive von Pirquet's standard diet, enriched with supplements containing vitamin D; the other, the control group, the same diet without supplements. Von Pirquet's diet was based on milk from cattle kept in stalls, which was low on fat and deficient, therefore, in the fat-soluble vitamin D. The more fortunate group had the additional benefit of full-cream milk powder and cod liver oil. The control group soon started to show the signs of rickets, while the properly fed babies remained healthy. The control babies where at once treated with cod liver oil and in some cases ultraviolet radiation from a mercury lamp (which, like sunlight, stimulates the body to make its own vitamin D) and soon recovered. A similar controlled experiment was performed on adult patients with hunger osteomalacia, again with wholly conclusive results. In 1922 Harriette Chick and her team went home, their mission accomplished. They had saved many lives and alleviated much human misery, for their prescriptions were followed throughout war-ravaged Europe. 'The crucial experiment', Professor von Pirquet wrote in his preface to the report on the visit, 'was thus successfully made. The British workers succeeded with the accuracy of a laboratory experiment, in a city where rickets is extremely prevalent, in maintaining a large number of artificially fed babies free from the disease, and further, in the same wards, they were invariably successful in healing children admitted with rickets already developed.'[6]

The Second World War

In 1919 the Allied Commander-in-Chief, Marshal Foch, ruminating on the wreckage that the Versailles Treaty was leaving behind, opined that what it had achieved was not peace but a 20-year truce. And in 1939 war duly returned to Europe. What lessons, then, had been learned in the intervening two decades? Enlightenment came slowly. In Britain, even in the mid-thirties, much of the nation was still undernourished. Indeed in some ways the standard of health seemed worse than in 1914. In 1935 no less than 60% of would-be recruits for the Army did not measure up to the required, and none too stringent, minimum standards.

Subsidized milk had been made available to children in 1927, and in 1934 the Milk Act reduced the price to a halfpenny for one-third of a pint for children in primary schools. It now seems scarcely conceivable that this was still more than some parents could afford. By then a majestic figure had appeared on the scene: John Boyd Orr, a Scotsman, born in 1880, one of seven children.[7] He trained as a teacher, and from an early age developed a consuming preoccupation with the state of the Scottish poor. Dissatisfied with the life of a primary school teacher, Boyd Orr enrolled as a medical student at Glasgow University, and having qualified, opted for a career in research. His chosen field was nutrition, but before he had got properly started the Great War began, and Boyd Orr joined up as a medical officer. He was twice decorated for heroism under fire, transferred to the Navy, and by the War's end had been allocated to investigate problems of military nutrition. Demobilized, he returned to Scotland as director of the Rowett Institute, and embarked on an arduous programme of running the laboratory, raising money, and pursuing research into the nutritional problems of the poor in Britain and the deprived in underdeveloped countries. Thanks to his urging, an Act of Parliament was passed to distribute food to pregnant women, a measure that reduced maternal mortality in childbirth by 75%. In 1936 Boyd Orr published a famous report, entitled *Food, Health and Income*, in which the degree of malnutrition in Britain was exposed and analysed. To say that it galvanized the Government would be an overstatement, but it did goad it into sluggish action. (Boyd Orr, by then a Lord, lived to the age of 90, having served as a Member of Parliament and Director-General of the United Nations' Food and Agricultural Organization – a function he found burdensome and superfluous – founded the Emergency Food Council to grapple with the post-war problems of food distribution, and had been rewarded by the Nobel Peace Prize.)

Cod liver oil, the miracle food

Bones are made of a form of calcium phosphate, but attempts, when this was discovered, to treat rickets with calcium- and phosphorus-containing foods, were a failure. A great advance came in the 19th century, when a French doctor, Armand Trousseau, declared that, while calcium deficiency was indeed the proximal cause of rickets, lack of sunshine was a contributory factor, and that cod liver oil, but not vegetable oils, could effect a cure. Moreover, he made the connection between rickets and osteomalacia. But Trousseau's insight fell, for the most part, on deaf ears (and he made one false inference, in imputing the beneficial properties of cod liver oil to its high iodine content). Trousseau's name at least survives in the textbooks in 'Trousseau's sign', a low calcium level in the blood, still used in diagnosis of rickets.

Then, in 1889 a surgeon, John Bland-Sutton, reported that he had given cod liver oil, together with crushed bones and milk, to lion cubs born in the London Zoo, and whereas all previous litters had died of what was unmistakably rickets, all those he had treated survived. In 1905 came the definitive proof: Edward Mellanby (later the enlightened First Secretary – the head – of the Medical Research Council) showed that puppies succumbed to rickets if reared on a diet devoid of animal fats, but thrived when the diet was supplemented with egg yolks or cod liver oil (but, again, not vegetable oils). The solitary caveat was that these measures were effective only if the diet also afforded a sufficiency of calcium and phosphorus for bone growth. At this stage the water was muddied by a Glasgow physiologist, who asserted that puppies on a restricted diet were spared rickets if allowed plenty of fresh air and exercise. The observation was real enough, but its meaning was misinterpreted: exposure of the skin to sunlight stimulates the body to synthesize its own vitamin D. Not long thereafter, a German researcher, seeking evidence for a beneficial effect of sunlight, noted that the symptoms of rickets could be partly assuaged by irradiation with ultraviolet light from a mercury vapour lamp (a dangerous proceeding for other reasons, as later transpired). His paper on the subject received little attention, but its message was not lost on Harriette Chick, who, as we have seen, treated some of the Viennese children in this manner.

The virtues of cod liver oil were by this time widely recognized, and its popularity grew when one Charles Fox, in Scarborough in the North of England, discovered that when fresh cod livers were steamed the exuded oil was largely (though not entirely) free of the rancid fishy taste that had made earlier preparations so repulsive. Soon extraction plants were set up in Newfoundland and in Scandinavia, and their product was much sought after as a cure-all for rheumatic and other conditions. It was to be some time before the active factor, calciferol, or vitamin D, was isolated (Chapter 9), but an important step was the demonstration by one of the founders of nutritional biochemistry, Elmer V. McCollum of Johns Hopkins University, that the factor was not the same as the ingredient of milk and butter which averted blindness (xerophthalmia). McCollum showed that cod liver oil and butterfat, heated to the temperature of boiling water, retained their antirachitic activity in young rats, while no longer preventing blindness. This was an early indication that different 'accessory factors' lurked in the common foods: McCollum had discriminated between vitamins A and D. Later analyses showed that cod liver was not unique, for halibut liver contained twenty times more vitamin D (and is now more widely used as a source), and tuna liver some twenty times more yet. The vitamin D content of butter and egg yolk is, by comparison, minute.

A lesson learned: the nutritionists go to war again

In the years between the two World Wars the nutritional scientists were busy. The lessons of the First World War had been digested, and in Britain, especially, the nutritionists saw it as their mission to ensure that the problems of malnutrition, whether in war or peace, were surmounted. The most thorough and penetrating investigations were undertaken by a remarkable partnership of two exceptional, strong-minded individualists, R. A. McCance and Elsie Widdowson.[8] Robert McCance was an Irishman, born in 1898, one of the generation of young men who went directly from school to war. He joined the Royal Naval Air Service and flew an observation plane from the turret of one of the ships of the Grand Fleet's Second Battle Cruiser squadron, based at Scapa Flow. McCance, against all probability, survived the war unscathed, and in 1919 applied for a position with the Irish Ministry of Agriculture. But he was ill-qualified for the work and so was advised to study for a diploma in agriculture at Cambridge. By the time he arrived there, after a six-month interval for reflection on a farm in County Antrim, he had decided that the social upheavals and impending revolution in Ireland made a career there an unattractive proposition. He therefore resolved instead to read for the Natural Sciences Tripos. His enthusiasm for science was kindled, and on graduating he joined the laboratory of the professor of biochemistry, Frederick Gowland Hopkins, who was then working at the very frontiers of nutritional research on 'accessory food factors' (and of whom more anon). In 1926 McCance was awarded the PhD degree, but decided that he must qualify in medicine, and so enrolled at King's College Hospital in London. With the restless energy that so impressed all who encountered him in later life, McCance found enough spare time to develop a research project on the carbohydrate content of cooked fruits and vegetables. The aim was to develop a diet for diabetics, for effective insulin preparations were not yet available and it was only a starvation regime that could delay the worst symptoms of the disease and eventual death. McCance's work, which demanded exacting chemical analyses, was made the subject of a Medical Research Council report.

McCance remained in the Biochemical and Diabetic Department at King's College Hospital, extending his analyses to a wide variety of foods. It was in the basement kitchens, where he would go to collect joints of cooked meat for analysis, that one day in 1933 he first encountered Elsie Widdowson – 'a momentous meeting' he called it, for it was the start of a collaboration that endured for more than 60 years. Elsie Widdowson was a Londoner, eight years McCance's junior.[9] At school she was influenced by her chemistry teacher and went to study, not in a women's college as was customary, but at Imperial College where she was one of only 3 women in an intake of 100. She completed her degree course in only two years and, intent on research, joined the biochemistry laboratory of a noted protein chemist, known to all as Sammy Schryver. There she took part in a programme to separate and identify the amino acids – the constituents (20 in number) of proteins. After a year of this, she applied for, and got, a position funded by the Department of Scientific and Industrial Research (the DSIR), in the laboratory of a leading plant physiologist, Helen Archbold Porter, to study the synthesis of carbohydrates (sugars, starches, and cellulose) in apples. After three years, during which she devised methods of separating and analysing these constituents, collected apples of different varieties from the Kentish orchards, and found out how the carbohydrate constituents came and went during ripening, she presented her PhD thesis. By then she had had enough of apples and wanted to move on to animal physiology. But research positions were not easily come by at the time, especially for women, and the best she could manage was an interim year in the Courtauld Institute of Biochemistry at the Middlesex Hospital. Having spent the year applying unsuccessfully for all the jobs that came up, she was advised by the Director of the Courtauld to moderate her ambitions, and try for something in the less than glamorous field of dietetics. So she went to study for a one-year diploma at King's College of Household and Social Science. The course took her to the kitchens of King's College Hospital to learn the trade of cooking in bulk and to the encounter with Robert McCance.

After their first conversation in the basement kitchens McCance invited Elsie

Widdowson to his modest laboratory and told her about his work. She, with nearly four years of carbohydrate analyses behind her, at once pointed out some inherent errors in his analytical techniques. McCance forthwith proposed a collaboration, and applied for and received a grant from the Medical Research Council for a broader programme of food analysis. Elsie Widdowson nevertheless took time to complete her diploma, and later owned that what she had learned in the course about practical nutrition stood her in good stead in the years to come. She also spent time in the kitchens at another London hospital, St Bartholomew's, studying dietary requirements, especially of diabetics. What was needed, she decided, was a comprehensive survey of the compositions of all common foods, raw and cooked, and she and McCance set about making the analyses. The labour took them four years, and their monograph, *The Chemical Composition of Foods*, first published in 1940, instantly became the international standard for the computation of dietary requirements.

In 1938 McCance and Widdowson moved their little operation to the medical school in Cambridge. The Munich crisis had created an atmosphere of uncertainty and apprehension. War might be imminent, and food shortages could be expected, so the dietetic requirements of the population needed to be defined. Widdowson and McCance had by then entered into studies on human subjects, especially themselves. This work had begun with the discovery that in diabetic coma the metabolism went grossly awry, for example in respect of salt elimination. McCance got healthy volunteers to consume a salt-free diet for several days. They were then made to sweat in a hot-air tank for two hours each day for two weeks. They lay on rubberized sheets in the hot, dry atmosphere. Their water loss was determined by weighing them before and after the treatment, and the salt loss by hosing them down, together with their sheet, and analysing the amount of sodium chloride recovered. Kidney and other functions were then measured.

This did not satisfy McCance, who tried another method of eliminating salt from the urine. He had learned from the intrepid, indeed reckless, physiologist J. B. S. Haldane the technique of over-breathing. Haldane, who had been inured to self-experimentation as a small boy by his physiologist father, courted death on many occasions, apparently with some relish – in the laboratory, on the battlefield in the First World War (which he confessed to having hugely enjoyed), and most of all while working on underwater physiology for the Admiralty during the Second World War.*[10]

Over-breathing consists of breathing in and out as deeply and rapidly as possible for 45 minutes. This eliminates carbon dioxide from the body to such an extent that the blood becomes alkaline. McCance performed this action in order

* Haldane remarked that these high-pressure experiments were really quite safe; true, they could lead to a perforated eardrum, but that was a small matter, for while it might leave you a little deaf, it would enable you to exhale tobacco smoke through your ear, which was, he opined, 'a social accomplishment'.

to see what happened to his salt excretion. Elsie Widdowson stood by in case of untoward consequences, and left the room only when the experiment was over and McCance was recovering. Returning, she found him unconscious and blue because the alkalinity of his blood had interrupted the signal to the respiratory centre of his brain, and he had stopped breathing. She ran to fetch help, but when she got back he was starting to recover once more. His first words were 'Give me the bottle', and a urine sample was quickly produced. This was only one of the occasions on which McCance might have died or suffered serious damage, but the results clarified the relation between salt and water loss and probably saved lives in the North African desert campaign.

When war came, McCance and Widdowson embarked on their famous 'experimental study of rationing', encouraged by the Medical Research Council. The idea was to establish to what extent the home-grown food supplies would suffice to support healthy life in adults and children. There would be a dearth of meat and dairy products, and possible vitamin and mineral deficiencies. Together with a resolute group of volunteers, Widdowson and McCance lived for three months on a restricted intake of meat, cheese, and sugar or jam, far below what was then generally regarded as adequate or tolerable, but with unlimited quantities of potatoes and bread (which remained unrationed and plentiful throughout the War, as an additional two million acres of the British countryside were brought under the plough). They then subjected themselves to arduous fitness tests in the Lake District. McCance cycled from Cambridge to the Lakes against a stiff north wind on snowy and icy roads: it took him two and a half days. He and his volunteers then climbed the peaks with ropes and rucksacks full of bricks or ice-picks on their backs. This self-imposed penance, which lasted intermittently for nine months, led to the important conclusion that the minimal diet, carefully thought out, was indeed sufficient to maintain a high level of fitness, but for one particular: the dairy food did not yield enough calcium for health, and so addition of calcium phosphate to flour for baking bread was made statutory. (There was a predictable outcry from the 'natural food' faddists, who foresaw terrible consequences, and a book published at the time, with the title, *The Calcium Bread Scandal*, warned of an epidemic of kidney stones and hardened blood vessels.) The statute remained in force long after the War.

The outcome of the study guided the formulation of nutritional norms when food rationing was introduced. Many other nutritionists participated in advising the newly established Ministry of Food, which was far better organized to meet the impending crisis than its counterpart in the First World War. Chief among them was the Professor of Biochemistry at University College London, Sir Jack Drummond. It was he who ensured that all mothers of young children received milk, fresh orange juice, and cod liver oil, delivered to the doorstep. Sugar consumption was reduced, meat was severely rationed, and fish and whale-meat were promoted. The 'Dig for Victory' campaign was initiated to encourage the cultivation of vegetables and fruit, and within a year nearly a million allotments had been created. The incidence of heart disease fell precipitously, and the nation became healthier than at any period in its history, before or since.*

McCance for his part, like Haldane and many more physiologists, did other war-related work, studying among other issues the problems of survival at sea. One tangible outcome was the discovery that a supply of sugar or boiled sweets could avert dehydration. The committee over which he presided also found, less surprisingly, that it was fat rather than thin people who best maintained their body temperature on immersion in cold water. McCance never let up. In old age he was still working and attending meetings, and on one such occasion, when he was late for a lecture in Glasgow, the organizers received word that he was at that moment pedalling past Newcastle.

When the War ended, food shortages in Britain did not altogether abate, and indeed rationing became in some respects more severe. One problem was caused by the traditional British preference for white over brown bread, for there was evidence that white flour was deficient in 'accessory factors' required for health, and especially normal development of growing children. In 1946 Sir Edward Mellanby, head of the Medical Research Council, called a meeting of his nutritionists to design a standard post-war loaf that would provide the necessary elements. Should the nutritive value of bread be enhanced by lacing it with iron compounds and B vitamins? A population on which this proposition could be tested was needed. Elsie Widdowson was in attendance, on leave from Germany where she had been sent to advise on the peculiar and difficult problems of feeding the starving survivors in the concentration camps (Chapter 8). Mellanby, she later reminisced, took her aside and instructed her thus: 'There must be many hungry children in Germany. You go and find out the truth of all this.' She returned to Germany, and Harriette Chick's mercy mission of a quarter of a century earlier was re-enacted. With a colleague she toured the country in search of a suitable orphanage, where the merits of different types of bread could be tested. They found what they needed in Duisburg – an institution full of malnourished

* Drummond had had a distinguished career in academic research. Among his many achievements was the purification of vitamin A. He was an exuberant lecturer, who once coated a haddock with white phosphorus to demonstrate the bioluminescence of rotting fish, but when the lights were turned out the subterfuge was at once discovered, for the lecturer's hands glowed more brightly than the fish. Drummond left the academic scene after the War to head the research laboratories of Boots the Chemist in Nottingham, and in 1952, at the age of 61, met a grisly end. He, and his wife and ten-year-old daughter, touring in Provence, pitched their tent in a field near the small village of Lurs. There, the next morning, their bodies were discovered. Drummond and his wife had been shot and their daughter clubbed to death. The case became a *cause célèbre* in France and Britain. A 75-year-old peasant farmer, Gaston Dominici, was convicted of the murder on the basis of little evidence, except a confession, later withdrawn. Conspiracy theories blossomed around the episode, and most recently a historian published his reasons for believing that Drummond was acting as a spy – for whom is not revealed. He had visited Lurs on several previous occasions, and had shown interest in an insecticide factory nearby, which could have been suspected, in the prevailing paranoid climate of the Cold War, of producing nerve gases.

and underweight children. These were divided into five groups, each fed with a different kind of bread as the main energy source – wholemeal, partially white, completely white, and white bread with added B vitamins or iron. All contained added calcium carbonate. The result was that all groups did equally well, and Elsie Widdowson brought five little girls to Cambridge to present them to a meeting of the British Medical Association. Other studies on infant nutrition followed, which improved the health of German children, and it was not until 1949 that Elsie Widdowson left Germany for the last time.

Back in England, she resumed her earlier work on body composition, doing fat analyses on stillborn fetuses and on maturing animals, especially pigs. (She once brought, in the boot of her car, a dead seal, discovered on a beach in the Orkney Islands, to Cambridge, where she was by then head of Infant Nutrition.) She analysed infant milk formula products, and discovered that in Holland babies not breast-fed were being given a milk substitute containing maize oil. The content of unsaturated fatty acids led to a vastly increased deposition of body fat. She was called on to advise on infant nutrition in Britain and among the impoverished people of Africa. Elsie Widdowson retired from her position in 1973, and set up a new laboratory at Addenbrookes Hospital, from which she retired for the second time 15 years later. But her work did not end there, and at the age of 84 she was still busy dissecting and measuring, while looking after her mother who lived to be 107. Lucid, modest, and good-humoured, Elsie Widdowson continued to write and lecture to the end of her life: she died aged 93, having outlived her inseparable colleague and friend of six decades, Robert McCance, by seven years.

2

The Scurvy Wars

The history of scurvy and its eventual conquest embodies all the best and the worst elements of human nature. There is horror and heroism, wisdom and acumen wrestling with ignorance and obscurantism, greed and vanity; and there is self-sacrifice, vying with a heedless indifference to suffering and death. Many more sailors in the world's navies were lost to scurvy than in battle, and more explorers died of it than from any other cause. Scurvy, also written scurvie, skyrvie, or scorbie in old English manuscripts, was known to the Greeks and Romans, and was given the name *Purpura nautica*, although a spurious distinction was generally made between 'sea scurvy' and 'land scurvy'. Diagnosis was often obscured by other conditions that prevailed in circumstances of deprivation, especially venereal diseases, to which sailors were held to be especially prone, ague, and 'the bloody flux'. It was not, in fact, until the 17th century that a clear distinction was made between scurvy and syphilis, which had some symptoms in common.

Scurvy is a hideous disease. It is marked by bloody patches under the skin (whence the name 'blacklegs' for scurvy in Scotland), extreme lassitude, constipation so persistent that it often required the attentions of a surgeon, joint pains, softening of the muscles, so that an indentation made in calf or biceps with a finger remains for minutes or more, and loosening of the teeth and rotting of the gums, which made chewing an agony and in time an impossibility. To add to these afflictions, the sufferers emit an intolerable stench of putrefaction. When scurvy appeared among the crusaders (most often during Lent when no meat was eaten) the best the barber-surgeons could do was to cut away the necrotic gums from around the teeth so that their patients might eat.

The most curious aspect of scurvy, as it appears in early histories, is the extent to which cures or preventatives were known, and yet were disregarded. The great navigators of the 15th and 16th centuries were dogged by the disease. More than half of Vasco da Gama's crews died of it on his expeditions to the Indies. Some made transient recoveries after eating oranges following a landfall, and in 1510 Cabral reported that on his way down the west coast of India scorbutic sailors had been cured by citrus fruits. Other fruits, such as a wild pineapple, were also found efficacious by the Spanish apothecaries in South America, who urged that every opportunity should be taken to reprovision ships with fresh fruit, especially lemons and sour oranges. Similar observations were made by Dutch seafarers in the 16th century, but sources of fruit were often elusive, and problems of storage on board were hard to surmount.

The belief nevertheless persisted in official quarters that scurvy was caused by the damp air of the oceans or the foul air below decks, and in the British Navy there was a widely held belief that it was lazy or sluggish sailors who most readily succumbed. Others again held that scurvy was a contagion, transmitted by crowding. Sir Richard Hawkins described in meticulous detail the scurvy that had broken out during his South Seas expedition in 1593, and also the cure effected by citrus fruits, yet he did not include them in his provision lists. By the beginning of the 17th century reports of cures by citrus and other fruits were frequent. In 1607, for instance, a well-known savant, Sir Hugh Platt, published a book on seamen's food, which commended lemon juice as an infallible prophylactic and cure. Sauerkraut was also mentioned, and Jacques Cartier recorded that while nearly all his sailors had been killed or rendered incapable of work by scurvy on his voyage to America, the survivors had been made whole again by an infusion of a tree bark prepared for them by local Indians. (Some of the men, he further reported, were by the same means also cured of 'the French pox', which had plagued them for years.) Spruce beer was an infusion of leaves, often effective, as, in northern climates especially, was the weed *Cochlearia officinalis*, known as scurvy grass or sometimes spoonwort. The legend ran that its virtues had been discovered by a group of sailors put ashore to die in Greenland, when their condition was judged to be hopeless.

When the East India Company, which began to trade at this time, sent four ships from Woolwich to the east, scurvy seriously afflicted all but one, on which the captain had laid in a provision of citrus fruit. The Dutch East India Company kept scurvy at bay by the same means, and also planted fruit trees on the Cape of Good Hope, where its ships stopped to reprovision. Even one of the most illustrious scholars of the era, the Hon. Robert Boyle (see Chapter 4) made public his conviction that scurvy would always yield to an infusion of fresh lemons in barley water. The success of such measures made little impression on the doctors, however, perhaps because of the obdurate adherence of the profession to Galen's theory of humours (Chapter 3). This precluded any notion that a disease could be caused by the absence of something vital, rather than the presence of something noxious. According to learned opinion, scurvy was brought on, if not by bad air, then by too much salt, for the sailors' daily diet included an allowance of extremely salty preserved meat, so hard that the men sometimes carved figures out of it. (Besides this the main staple was ship's biscuit, baked from a dough of flour and water, and dried to an iron hardness which often made the product impossible to break, until the weevils got at it and made it friable. Where there were no weevils the biscuit could be stored for 50 years or more. But in the ships the damp atmosphere allowed weevils and maggots to multiply, and many of the sailors found the infested biscuit so repellent that they could eat it only in the dark. Most often they soaked it in water and mixed the resulting paste with their salted meat and a little vinegar. This confection was called lobscouse or skillygolee.)

Whatever the agent of scurvy, it was an excess of 'black bile', or 'corruptions of the humours' that it was thought to provoke. Then in 1617 there appeared a prescient and influential book, *The Chirurgeon's Mate* by John Woodall, a ship's surgeon, who became the first Surgeon-General of the East India Company and later surgeon at St Bartholomew's Hospital in London. Woodall was a rational man, and he wondered why scurvy, classified according to the doctrine of humours as a 'cold' disease, should yield to treatment by a 'cold' remedy like an acid fruit juice, prescribed in the canon for 'hot fevers' (see Chapter 3). He came close, indeed, to questioning the validity of the whole system. Woodall favoured fresh food and plenty of alcohol, but the sovereign cure for scurvy was lemon juice, or failing that, lime or orange juice or tamarind pulp. Failing even those, the ancient remedy of water acidulated with 'elixir of vitriol', that is to say sulphuric acid, he hoped, might serve. This of course was wholly inert. Woodall's book alerted the more receptive surgeons and captains to the value of citrus fruits.

James Lind and the controlled experiment

Even the sluggish Navy Board in London, which appeared to have taken the view that sailors could be more easily replaced than cosseted, was dismayed when in 1744 Commodore Anson returned from his voyage of circumnavigation, begun four years earlier. True, he had brought back a mass of captured Spanish treasure, but his fleet of seven ships had been scattered by the tempests around Cape Horn: one had been wrecked and two had turned back, so that, after further vicissitudes, Anson had reached Spithead with only one ship. Some 1400 of the nearly 2000 men who sailed were dead, four killed in battle, most of the rest by scurvy. It had not helped that many of the men had been scarcely fit to serve from the outset. The press-gangs had evidently met with limited success, and Anson had been compelled to recruit boys of 10 or even less, with a schoolmaster, as required by law. In place of a contingent of Royal Marines the Navy Board sent on board 500 veterans – Chelsea Pensioners, many of them infirm. Most of those who could run deserted before the squadron sailed, and Anson had to reject many of the barely mobile remainder. Those who did sail all perished, and were probably the first victims of scurvy. Anson was fêted as a hero, which indeed he was, and his high reputation helped him over the ensuing years to wield a powerful influence for reform in the Admiralty. It was probably his patronage that enabled James Lind to carry out his celebrated experiments and to gain some publicity for the results.

Lind was born in 1716 into a respected Edinburgh family. The Scottish Enlightenment had made the city a place of intellectual and academic ferment, not least in science and medicine. The Medical School opened its doors in 1726, and was organized on modern lines, although it was to be some years before Latin as the language of the lectures gradually gave way to English. Theses, too, had to be written in Latin. The faculty had strong contacts with the famous medical school in Leyden, presided over by the great Hermann Boerhaave, and it was one of Boerhaave's former students, George Langlands, to whom in 1731 Lind was apprenticed under the prevalent system of the time. Lind's interest in scurvy may have been kindled at this early stage, for Boerhaave had considered the disease and thought it could be prevented by the consumption of fruit: he set the greatest store by crab apples.

In 1739 Lind left home, armed with his apprenticeship certificate, but no university qualification, and headed south to join the Royal Navy (in which his elder brother was already serving) as a surgeon's mate. Ship's surgeon was a less than glamorous calling, low in the world's esteem and commensurately rewarded, and not the most obvious choice of career for a promising young man of good family. The Navy was being steadily enlarged, for the threat of war was always present and the hugely profitable and expanding East India trade was leading the demand for ever more escorts. But conditions, except for the officers, who were generally well-provided for and socially sought after, were woeful. 'No man will be a sailor', Dr Johnson declared, 'who has contrivance enough to get himself into a jail; for being in a ship is being in a jail, with the chance of being drown'd.' The food was often already rotten when dishonest victualling agents brought it on board, and soon infested with vermin, and the water was stagnant and foul. The sailors' quarters were cramped and devoid of all ventilation. Gastrointestinal disorders were endemic, as were venereal diseases, and these between them generated much of the surgeons' work. Accidents, however, also took their toll, and, in tropical climes especially, yellow fever and typhoid. But scurvy was the greatest scourge. The persistence of the belief that it was caused by bad air and the miasma of the lower decks, allied perhaps to unwholesome food and the gross saltiness of the meat, is perhaps not surprising, considering especially that scurvy on land broke out mainly where men were closely confined – in jails, asylums, and barracks. Lind never quite managed to distance himself from this theory. On his ship, as on others, the decks were swabbed down with dilute vinegar, and tar and tobacco were burned to neutralize the noxious air.

Lind spent his first eight years in the Navy at sea, almost without interruption. There were voyages down the coast of Africa, to the Caribbean, and to the Mediterranean, but the longest periods without touching land were generally on patrol in the English Channel. Throughout this time he made observations and took punctilious notes. He had the opportunity to judge the effects of quack nostrums, widely peddled at the time. The Admiralty directed that 'Doctor

James's Fever Powder', a dire concoction containing the toxic antimony,* should be tried out on the men of the Channel Fleet. This brew was being marketed with the aid of an advertising promotion of extraordinary scope. It was aimed even at children, for the father of the heroine in *Little Goody Two-Shoes* was taken ill 'in a place where Dr James' Fever Powder was not to be found', and so died. Horace Walpole swore by it: if his house were on fire, he declared, it would be Dr James's Powder that he would rush to save. It was tried on George III during his bouts of madness, and Oliver Goldsmith was thought to have died of its effects. Dr James's Fever Powder and another antimony-based panacea, the 'Pill and Drop', devised by a notorious charlatan called Joshua Ward, were thought to have taken many sailors' lives. Anson's expedition, news of which appalled Lind, had been disastrously provisioned by the Admiralty, which had put its faith in the advice of a society doctor, formerly physician to the Navy at Greenwich, William Cockburn. Edinburgh-trained, he nevertheless relied on a quack remedy, his 'Electuary', to achieve wealth and position. The Electuary was also sold to the Navy as a cure for dysentery and like complaints, thanks to the support of a prominent admiral, Sir Cloudisley (or Cloudesley) Shovell. (Shovell was a sea-dog of a traditional stripe, who met a spectacular end in 1707, while returning with his squadron from an abortive attack on the French naval base of Toulon. The story went that a sailor with better navigational skills than the Admiral tried to warn Shovell that they were on course for the rocks of the Scilly Isles. Affronted by the presumption of a common seaman, Shovell ordered the man to be hanged. Retribution followed swiftly, when the ship struck the rocks and some 700 men perished, the egregious admiral among them.) Cockburn (who was eventually to be buried in Westminster Abbey) was wedded to the view that scurvy was caused by the abidingly work-shy nature of sailors, but he conceded that it was treatable. Unfortunately his prescription consisted not of citrus juice, but of vinegar. The Navy Board was evidently still unsure of their man, for they sought a second opinion from one of the most prominent doctors of the day, Physician to the King, Richard Mead, who advised that 'elixir of vitriol' (dilute sulphuric acid) should be preferred to vinegar, but was to be mixed with alcohol, sugar, and herbs. For good measure the Navy Board chose to add Ward's deadly 'Pill and Drop' to the menu. Citrus fruit was not recommended, but rather condemned as 'the commonest cause of fevers and obstructions of the vital organs'.

Five years after Anson's return Dr Mead published a book, *Discourse on the Scurvy*, in which he alluded with scepticism, indeed sarcasm, to accounts of

* Antimony was used in medicines from ancient times. A famous monograph on its virtues, *The Triumphal Chariot of Antimony*, was published in the 17th century under the pseudonym of Basil Valentine. Antimony was chiefly known for its laxative properties, and is still favoured in homeopathic medicine. It was (and perhaps still is) used in parts of France in the form of a large tablet of the metal itself for the treatment of constipation. The trace that dissolves during its passage through the gastrointestinal tract generally produced the desired effect. The tablet would be recovered and would serve a family for generations.

sailors *in extremis* cured by citrus juice. From the evidence, he insisted, it was 'very plain that this malady is a kind of corruption of the blood'. This doctrine seems still to have been tacitly accepted by most of the medical establishment. Yet the value of citrus and other fruits must have been widely discussed, though many even of its upholders despaired of the prospects of preserving anything fresh against putrefaction in dank holds. This was undoubtedly one of the reasons why a majority continued to look to quack nostrums for salvation. All the same, in about 1740 Admiral Vernon, known to the sailors as Old Grog, from the coat of grosgrain cloth that he always donned in foul weather, introduced the modification to the rum ration, which still bears his name: the rum was diluted with water and a shot of lemon juice. But Vernon was in a small minority. (It was not until 1795 that the Admiralty made grog the official sailors' daily restorative.) The tide of apathy turned only when William Pitt took command of the direction of war, perceived the deficiencies in the way in which the Navy was organized, and appointed Anson First Lord of the Admiralty. With the fearful effects of scurvy on his famous expedition still fresh in his mind, and troubled by the failure of the Admiralty and its responsible body, the Sick and Hurt Board, to act, Anson made the scurvy problem a priority.

In 1746 James Lind was posted to HMS *Salisbury* as surgeon. This was a new ship, a fourth-rate man-o'-war, with a complement of some 350 men, under an intelligent and liberal captain, George Edgcumbe, who was to become an Admiral, a Member of Parliament, and finally the first Earl of Mount-Edgcumbe. Lind thought it outrageous that naval doctors, who knew the disease at first-hand, had not been encouraged to undertake a systematic examination of all the contradictory observations and theories about scurvy. He may have been stimulated by a report which emanated early in 1747 from John Hammond, a surgeon on another naval vessel, the *St. George*. During three cruises, Hammond stated, 19 deaths in every 20 had been the result of scurvy. Sir William Cockburn's doctrine, that the disease was brought on by laziness of the sailors who contracted it, Hammond condemned as 'a vulgar error'. Scurvy stemmed, he believed, from inadequacy of the diet, much salt, and bad water. The patent medicines of Joshua Ward and others were a fraud. Tests of possible remedies, he concluded, had to be undertaken.

Lind approached the problem with a relatively open mind. He thought that damp air on the voyages and poor hygiene must be important factors, and that 'lack of exercise and melancholy humour' were no help either. He doubted whether fresh fruit in general and vegetables entered into the equation, and yet the curative properties of citrus fruits were beyond doubt. But Lind knew that proof would be demanded, and he undertook to obtain it. His famous declaration ran: 'I shall propose nothing dictated from theory, but shall confirm all by experience and facts, the surest and most unerring guides.' The experiment that he performed on the *Salisbury* in 1747 is generally regarded as the first controlled clinical trial of all time. That is to say, the experimental subjects are divided into separate groups, all treated in exactly the same manner except that one receives the substance under test, whereas another, the control group, does not. Lind's protocol was in fact more complex, for he chose to compare simultaneously the outcomes of several treatments, each to be administered to a different group of subjects as a supplement to the standard diet.

It was the *Salisbury*'s second cruise. Severe scurvy broke out, as it had on the first cruise, affecting 80 of the sailors. This was what Lind had been waiting for. He now selected 12 men, all in a fairly advanced stage of illness, and as closely matched as possible for their symptoms, and divided them into six groups of two. All were accommodated together and for 14 days received the same basic diet. Breakfast was gruel sweetened with sugar, the midday meal mutton broth, a pudding of undisclosed nature, or sweetened boiled biscuit, and supper consisted of barley with raisins, rice and currents, sago with wine, 'or the like'. But each of the six groups received a different additive, thus: *group 1* – 1 quart cider daily; *group 2* – 25 drops 'elixir of vitriol' (sulphuric acid), presumably diluted in water, three times a day, and gargling with the same; *group 3* – two spoonfuls of vinegar before each meal, some also added to the food and also used as a gargle; *group 4* – a half-pint of sea water daily; *group 5* – two oranges and one lemon per day, but this was continued for only six days because the supply ran out; *group 6* – an 'electuary' or paste, the 'bigness of a nutmeg', three times a day. It contained a curious mixture of ingredients, all with supposed curative properties, namely garlic, mustard seed, 'balsam of Peru' from a native tree, dried horseradish, and 'gum myrrh'. To drink, the men received barley water, acidulated with tamarind pulp, and sometimes 'cream of tartar' (potassium hydrogen tartrate) was added as a mild purgative. A further, more loosely regulated control group received only an analgesic, a laxative and cough syrup, but the observations on these were not included in Lind's report.

The results were as conclusive as could be conceived. After six days the two men on citrus fruit were free of symptoms and fit for duty, and were put to work nursing their less fortunate companions. The two men on cider showed some improvement when the two weeks were up. As to the remainder, the treatments had done them little or no good. The grand and indisputable conclusion was 'that oranges and lemons were the most effectual remedy for this distemper at sea'. The following year the Treaty of Aix-la-Chapelle ended the War of the Austrian Succession after 17 bloody years, and Lind left the Navy and returned to Edinburgh. There he resumed his studies, was in due course awarded his doctoral degree for a thesis with the title, *De morbis veneris localibus*, which treated of syphilitic lesions, and was licensed to practice medicine. It was not until 1753 that his *magnum opus* appeared in print, under the title, *A Treatise on the Scurvy, in Three Parts, Containing an Inquiry into the Nature, Causes and Cure of that Disease, Together with a Critical and Chronological View of what has been published on the subject*. It was written in English rather than Latin, ran to 400 pages and was dedicated to Lord Anson.

Lind's *Treatise* was a monumental, if muddled, work. The ancients (Hippocrates, in particular), he argued, knew of scurvy, and there was but one kind: sea scurvy and land scurvy were one and the same, in fact the best description of the disease had been published by a doctor with the Hungarian army in the recent war. Lind surveyed the entire literature on the subject and found much of it to be nonsensical. But his beliefs concerning the aetiology of the disease did not deviate greatly from much of what had gone before, and indeed harked back to Galen. Cold air and moisture, Lind conjectured, aggravated by lack of exercise and a discontented state of mind, caused the pores of the skin to become obstructed. 'Undesirable humours' were consequently denied their normal route of escape – the usual cause, Lind thought, of 'putrid diseases' – and 'acrimony and corruption' ensued. Lind knew that the weight of excreta did not balance that of food ingested, and so supposed, as did those of his contemporaries who thought about such matters, that either sweat or (in cold weather) 'insensible perspiration' must make up the deficit. Pure, dry air, moreover, was essential for all bodily functions; moisture, Lind supposed, 'weakens the spring and elasticity of the air, making it unfit for the many salutary purposes obtained by respiration'. Worst of all, it prevented the formation of blood from chyle (the milky fluid found in lymphatic glands and the thoracic duct, which emulsifies fatty matter in food during digestion). This process was an article of belief among the physiologists of the time (Chapter 5), and Lind further hazarded that the transformation was impeded in states of physical idleness, which led to 'gross' chyle. An excess of fluid in the circulation would result, the pressure of which would express blood into the skin, whence the bloody patches on the legs of scurvy victims.

Lind thought that warm dry air, circulated from the ship's galley, might help to keep scurvy at bay. In this he was anticipated by the Reverend Stephen Hales, the learned cleric who had made remarkable contributions to plant and animal physiology and to chemistry. Hales had invented a hand-operated pump to circulate air through ships, and, perhaps in response to Lind's *Treatise*, the Admiralty caused these to be installed in many of His Majesty's vessels around 1756. As to treatments of scurvy, Lind was in no doubt that the electuaries and pills, containing antimony and mercury compounds, were useless and dangerous, but that bleeding – the panacea for most conditions at the time and for long afterwards – could help. Drinking acids, whether vinegar or dilute sulphuric acid, might be salutary, a conclusion rationalized in terms of the current concepts of the digestive processes; mustard or anything to promote sweating was good, but fresh fruit and vegetables were generally redundant. He cited the works of a Leyden scholar, Johannes Bachstrom, who had stated in 1734 not only that land and sea scurvy were the same malady, but also that it stemmed from a want of fresh vegetable matter, 'which is alone the primary cause of the disease'. In dismissing Bachstrom's conclusion Lind made an exception of course for citrus fruits. And here, he urged the use of 'rob' of oranges and lemons, an extract that he thought would avert the problems of spoilage during long voyages. 'Rob' was prepared by squeezing the juice, letting the sediment settle, placing an open vessel containing the clear liquid into a bath of boiling water and letting it evaporate over a period of hours until it attained the consistency of syrup when cooled. It was then bottled and corked. This was generally sterile and would keep, sometimes for years. It could be added to water, rum, or other drinks. This process, alas, destroyed the ascorbic acid – the vitamin C – which is the active constituent of citrus fruits.

Lind revised his *Treatise*, which appeared in a new edition in 1772 and was translated into several languages. He went on to publish two further ponderous works, *An Essay on Preserving the Health of Seamen in the Royal Navy*, and *Essay on Disease Incidental to Europeans in Hot Climates*. But it was not until 10 years after he had left the Navy that Lind made his deepest impact on the welfare of British seafarers: in 1758 Anson, the First Lord, appointed him physician to the Haslar Naval Hospital in Portsmouth. This eventually accommodated 2000 patients, many of them just returned from cruises, and especially blockade duty off the French coast, with severe scurvy. Anson, in his anxiety to limit the incursion of scurvy into the Western Squadron, which was charged with blockading Brest and other French ports, took Lind's advice and sent out provisioning ships carrying fresh fruit and live animals. This measure proved a great success, and remained Admiralty policy thereafter, but on long cruises the fruit still rotted and scurvy remained a problem. The 'rob', on which he conducted careful trials, was a sad disappointment to Lind, who finally conceded that fresh fruit was best. Yet he remained wedded to the causative link between scurvy and damp, unwholesome air, clogged pores, and putrefaction, and on this he mainly based his regime of treatment. Lind remained in charge of the hospital until his retirement in 1783 at 67 on a modest pension from the Admiralty. He was succeeded at Haslar by his son. By then he had suffered another bitter experience, for while at Haslar he had invented a simple and effective shipboard still for producing drinking water from sea water, but the credit for this was shamelessly stolen from him. First a French naval doctor constructed an identical device and claimed it as his own invention, and then a young naval surgeon, Charles Irving, made some trivial and indeed deleterious modifications to Lind's design, which he then derogated in petitioning Parliament for a reward. Irving's still was adjudged a success in trials at sea and he was awarded a prize of £2000 from the Admiralty, Lind having been forced to sit on the committee that endorsed it. Poor Lind received no recognition, honours, or awards from his masters at the Admiralty and died, largely forgotten, in 1794 at Gosport, a few miles from Haslar.

No end to nostrums

Neither Lind's *Treatise* nor the published evidence of ship's surgeons and captains put a stop to the debate about the efficacy or otherwise of citrus fruit, and whether sulphuric acid or vinegar might not be as good, if not better. The waters were muddied by many ambitious practitioners, such as David MacBride, a Dublin surgeon, and his brother John, a naval captain, who promoted malted barley as a cure and prophylactic. This, they asserted, could be stored for years and, when used to prepare an infusion called 'sweet wort', would cure scurvy. Trials on scorbutic sailors, one of them by Lind at Haslar, produced no positive results. In 1768 Captain Cook took along a generous supply of malted barley (40 bushels, or some 320 gallons, packed in hogsheads), quantities of portable soup (a hard, gummy cake made by drying down a concentrate of meat, offal, and vegetables), sauerkraut, and a little rob of lemon, when he set out on his South Sea expedition of scientific exploration. His ship, HMS *Endeavour*, remained largely free of scurvy and Cook's surgeon, William Perry, put this down to the action of the sweet wort, but in fact the *Endeavour* was never away from land for long, and so was able to take supplies of fresh fruit on board quite regularly.

The pattern was repeated during Cook's second great expedition in 1772, when he was charged by the Board of Longitude with the task of evaluating John Harrison's latest model of chronometer, and also instructed to look out for a new Southern Continent. There were no deaths from scurvy on his ship, the *Resolution* (although one was recorded on the accompanying vessel, HMS *Adventure*), and Cook put this down to the virtues of the sweet wort, although a series of other remedies, real or supposed, were administered to the men. This time they included, besides elixir of vitriol and sauerkraut, a 'carrot marmalade', prepared according to the recipe of 'Baron Storch of Berlin', by evaporating carrot juice and rob. When he made his report on the voyage Cook was lauded for finding the key to the elimination of scurvy. His faith in the sweet wort was endorsed by one of the most elevated doctors of the day, Sir John Pringle, who had already added to the obfuscation that surrounded the scurvy problem by recommending sugar as a preventative, for he knew that it enhanced fermentation processes, thought at the time to oppose putrefaction. For his espousal of the useless malt wort Pringle was honoured by the Royal Society, which awarded him its highest distinction, the Copley Medal, and soon after elected him its President. Pringle had served as an army doctor and done much to improve the conditions of soldiers. He had witnessed suffering and death from scurvy in the field, but had consistently drawn the wrong conclusions. His theories had also, in 1770, contributed to a tragic episode, when one of his own protégés, a highly esteemed young doctor named William Stark, became one of the earliest martyrs to the noble tradition of self-experimentation. Stark tried a wide variety of different diets in order to determine which would induce and which prevent scurvy. Fresh fruit and vegetables did not feature. Eventually he began a regime of cheese and honey, which, if his patron's sugar theory had been correct, should have averted the disease. But when he became severely scorbutic Pringle instructed him only to avoid salt and apparently never even thought to administer a spoonful or two of lemon juice. Stark's condition grew worse and he apparently contracted an infection of the kind to which scorbutic subjects were very susceptible. He was bled, which probably hastened his end, and he died after eight months of relentless self-inflicted suffering.

The debate continued. In 1776 Dr James Badenoch reported that he had restored a severely scorbutic patient to perfect health by administering nine gallons of sweet wort and one quart of lemon juice. He scorned the notion that the lemon juice might have had anything much to do with it. It is as if one had inferred from the intoxicating effect of gin and tonic that the tonic water must be the active component. Some, Badenoch wrote, might suppose that the lemon had had some part in the cure, 'but the contrary is well known to every practitioner who hath tried that remedy in the cure of scurvy at sea, the utmost to be expected from it being only to mitigate or to prevent the increase of that disease'. Other supposed specifics against scurvy, in use for centuries, were salop or salope – dried orchid root, often used as a thickener for soups – and portable soup. Sea-water purgatives were also recommended by many ship's surgeons, as was the inevitable bleeding.

The soda-water cure

The theory, so passionately espoused by Sir John Pringle, that fermentation was the mechanism by which nature kept putrefaction in check, cast a long shadow. If scurvy was a 'putrefactive' disease (as for instance the malodorous decay of the gums was taken to imply), it ought, Pringle kept insisting, be prevented or reversed by the elements of fermentation. The notion was taken up by David MacBride, who, in a publication of 1764 entitled *Experimental Essays*, made a daring and fatuous intellectual leap: putrefaction was accompanied by the emission of 'fixed air' (carbon dioxide), and therefore the loss of this gas must be its cause. To reverse the process, then, it was necessary only to put back the fixed air.

Now gases at this time – the 'factitious airs' of the new chemistry developed by Humphry Davy – were much in fashion, and were widely believed to have medicinal attributes, for did not the recently discovered oxygen support life better than mere air? The properties and origins of 'fixed air' were being investigated by the two famous 'pneumatic chemists', Joseph Black of Edinburgh and Joseph Priestley in Birmingham. One of Priestley's experiments was to bubble his fixed air into water, thereby generating soda water, or as he termed it, 'artificial Pyrmont water', after the fashionable spa. He too wondered whether it might not cure scurvy, and indeed the 'mephitic waters' of the spas were believed to expel all manner of noxious humours from the body. In 1772 Priestley presented a paper on the various new 'airs' before the Royal Society in London. On the strength of this he was invited to dinner by the Duke of Northumberland, who had procured a bottle of water, distilled from sea water by the egregious Irving. This the Duke decanted for his guests, who would have experienced the flat, rather metallic taste characteristic of distilled water. Priestley thereupon announced that he could restore the 'briskness' of the water with his fixed air. The Duke and the whole company applauded the idea and, thus encouraged, Priestley set up an apparatus in his lodgings to generate fixed air from chalk and acid, and bubbled the issuing gas through London water (water of good quality from the New River Head in Islington). An enthusiastic report about the resulting soda water was passed to the Earl of Sandwich, who had succeeded Anson as First Lord of the Admiralty (and was also an inventor in his own right, for he had conceived of the sandwich, a breakthrough that allowed him to satisfy his hunger without losing valuable time at the gaming tables). Sandwich demanded an evaluation of Dr Priestley's proposal 'for rendering salt water fresh' by the Royal College of Physicians and a demonstration of the process. This ceremony was duly enacted before a distinguished audience that included Benjamin Franklin and the famous botanist, Sir Joseph Banks, who had sailed with Cook on his first voyage. The College report pronounced Priestley's water wholesome: the treatment had 'communicated no noxious qualities to the water'. Lord Sandwich thereupon directed that the efficacy of Priestley's soda water as a scurvy cure should be tried out by Cook on his second expedition. Instructions for preparation of the health-giving water were drawn up by Priestley and sent to Cook, and an apparatus was constructed on the *Resolution*, but there is no evidence that it was ever put to use. Cook was able to take on fresh provisions and water at frequent intervals, and there was, as we have seen, no significant incidence of scurvy on his ship during the three-year voyage. It needs scarcely to be added that soda water is without effect on scurvy or on any other disease.

Lemons and limes

As the 18th century drew to a close sailors (and many landsmen as well) continued to suffer and die from scurvy. The next aberration was the oxygen theory, provoked by the experiments that Lavoisier, especially, had been conducting in Paris (Chapter 4). Because oxygen – Priestley's 'dephlogisticated air' – was now known to be inseparable from life, the idea took hold that scurvy was a consequence of oxygen-starvation in the unventilated lower decks. It was first given coherent expression by another Edinburgh doctor and former surgeon's mate, who had briefly served under Lind at Haslar, Thomas Trotter. In 1783 the young Trotter had sailed as surgeon on a slave ship, but, appalled at the conditions of the slaves on the long voyage from the African coast to the Caribbean plantations, had signed off on his return and gone back to the Medical School in Edinburgh. Yet he had not wasted his time on the slaver. He had been persuaded that the stinking air in the hold was indeed responsible for the scurvy, but he also noted that when the slaves were given guavas to eat they showed a preference for the sour, unripe fruit. Was this perhaps an inborn defence against scurvy? A controlled experiment might show. Trotter found nine scorbutic slaves and divided them into three equal groups. To one he gave ripe, to another unripe guavas, and to the third lime juice; all but the first group quickly recovered. Three years later in Edinburgh Trotter published his findings in his *Observations on the Scurvy with a Review of the Theories Lately Advanced on that Disease*. By the time the enlarged second edition appeared in 1792, Trotter had rejoined the Navy and was doing his best to ensure that his ship was properly provisioned, to the extent of going out to buy lemons himself in the market in Portsmouth.

Trotter presently published his theory that as 'vital air, or what is more properly called *oxygene*, is a component principle of acid fruits, we have reason to conclude that this is the quality which they restore to the human body in scurvy'. This shows that Trotter must have read, or at least been aware of Lavoisier's recent work, for *oxygène* was Lavoisier's coinage. It was, moreover, a misnomer, which implied that, as Lavoisier erroneously believed, it was the characteristic principle of acids. But why other acids, such as elixir of vitriol and vinegar, should have been ineffective appeared a paradox. Trotter's suggestion was that the weaker acids of fruits yielded up their oxygen most easily.

The next man to seize the idea and run with it was Thomas Beddoes. He was a distinguished chemist and doctor, who had done much research into the chemical and physiological properties of the 'factitious airs', notably nitrous oxide, or laughing gas. At his Pneumatic Institution in Bristol he treated various diseases with the gases and ministered to the local *beau monde*. Humphry Davy began his illustrious career under Beddoes's guidance. Beddoes owned that he too had pondered the likely relation between scurvy and oxygen deficiency, and concluded that the disease developed with the 'gradual abstraction of oxygen from the whole system'. But he disagreed with Trotter that other acids were useless, and thought that tartaric acid might serve. Its low cost would also commend it to the Admiralty. He further noted, more cogently, that the Inuit, who lived in a climate often associated with scurvy and subsisted on a carnivorous diet, did not suffer from the disease. Meat, he inferred, must therefore be loaded with oxygen. Tartaric acid of course was useless, as was the continued use of vinegar. And reports by Trotter that he had effected cures with citric acid, which he had crystallized from lemon juice, proved a delusion. This marked the demise of the oxygen theory.

Enter Gilbert Blane

In 1780 the Royal Navy was fully engaged in the war with France. That year the Channel Fleet was ravaged by scurvy, and was saved from disaster only by the even worse state of the French ships. The following year the West Indies Fleet under Admiral Rodney, Commander of the West India Station, was similarly afflicted. Rodney had brought with him on his flagship his Scottish doctor, an MD graduate of Glasgow Medical School, called Gilbert Blane. Blane owed his position to influential patrons: he had carried letters of introduction from his

Glasgow professor, William Cullen, to the celebrated surgeon and anatomist, John Hunter in London, and it was Hunter who recommended the young man to his friend, Admiral Rodney. The Admiral, who suffered from gout, must have found Blane's ministrations very satisfactory, for he appointed him to the special position of Physician to the Fleet. It was in the Caribbean that Blane had his first sight of scurvy. After barely 12 months he and Rodney were back in London and they presented a strongly worded report to the Admiralty. During the last year, it noted, some 1600 sailors had died, most of them from scurvy and especially infections arising out of it, and only 60 from enemy action. Blane was clear that scurvy was not itself infectious, and he thought that sufferers should be treated onboard ship, as long as good hygiene prevailed, rather than in hospital, where they were far more likely to succumb to 'flux and fever'. Finally, fresh vegetables and fruits, especially citrus, were a certain preventative and cure for scurvy. In the Caribbean, vessels should be provided for the purpose of collecting and delivering fruit from the islands.

After a short stay in London Rodney and Blane returned to their station, where they remained until the American War of Independence ended two years later. Blane had been present at six naval engagements, and had distinguished himself by his martial ardour in an action in which several of the ship's officers were killed. On his return to England he left the Navy and with Rodney's help secured a position as physician to St Thomas's Hospital in London. His report had done little to change opinions in the Admiralty. A letter from an East India Company surgeon, recounting his experiences with lemon and lime juice, prepared from good-quality fruit and casked, fared no better. It was passed by their Lordships to the Sick and Hurt Board, which duly reported that, rob of lemon and orange having proved unsatisfactory, there was no reason for further trials with the juice. 'Portable broth, wort, sour krout etc', the Board proclaimed, 'are much more efficacious in the cure and prevention of the scurvy.' Yet, when scurvy struck Admiral Hood's Mediterranean Fleet in 1793 his physician had vessels sent to gather citrus fruits from the land, and administered the juice to the scorbutic sailors, with dramatic results. From his position in London Blane now lobbied, through a friend on the Admiralty Board, for supplies of citrus fruits to be put on ships heading out to the East India station. The first ship thus provisioned was HMS *Suffolk*. Every sailor was given a daily dose of lemon juice with sugar in his grog ration, and at the end of 19 weeks at sea not a single case of scurvy had been recorded. Blane made the same recommendations in his work, *Observations on the Diseases of Seamen*, in which he reiterated James Lind's conclusions. Other naval surgeons weighed in, mainly in support of citrus juice. Finally the Board of the Sick and Wounded Sailors (as it was now called) agreed, at least if a request came from a commanding admiral, to sanction the necessary provisions. The change of heart came in the year after the death of Lind. Nelson (who believed that salt was the true cause of scurvy and would use none himself) was one who secured an issue of 30,000 gallons of lemon juice when he set out in

1804 on his 18-month blockade of Toulon, and received funds to take on board another 20,000 gallons in the Mediterranean. The fruit was most often put into barrels, which were then filled with sea water and stoppered. This probably prevented decay induced by bacteria or moulds. By the time of Trafalgar scurvy had at last been largely banished from the Royal Navy.

Gilbert Blane, meanwhile, had achieved the social standing that, as an evident snob, he had always coveted. Soon after starting at St Thomas's Hospital he had re-established contact with the Duke of Clarence, whom he had known as a midshipman in Rodney's fleet, and who was destined to become William IV. Clarence secured for Blane the additional office of household physician to the Prince of Wales. In 1788 Blane, like Pringle before him, was invited to deliver the Croonian Lecture to the Royal Society. In 1794 the First Lord of the Admiralty, Earl Spencer, appointed Blane one of the three Commissioners, in fact Chairman, of the Board of the Sick and Wounded Sailors. Two years later he was head of the Navy Medical Board. After his retirement from medical practice, and by then a baronet, Blane continued to write on scurvy and the health and welfare of seamen. He died, laden with years and honours, in 1834.

Disenchantment

By the early 19th century scurvy had disappeared from the British and other navies, but, astonishing to relate, the argument about its cause and cure came once again to the surface as the century unfolded. Its resurgence arose mainly out of the unhappy experiences of members of Arctic expeditions, especially those sent out by the British Navy to search for the Northwest Passage, to get to the North Pole, or merely to make scientific observations. Many of these men fell victim to scurvy, despite carrying liberal amounts of lemon or lime juice. The issue came to a head in 1875, when two ships, the *Alert* and the *Discovery*, under the command of Sir George Nares, headed north, lavishly provisioned and with an abundance of lemon juice. In 1853 Sir James Ross's expedition had returned from the Arctic three years and six months after leaving England, largely free of scurvy. In the summer of 1850 Ross's ship, the *Investigator*, had been trapped in the ice. Throughout the following winter and spring, despite punishing sledging expeditions towards the North Pole, there had been no scurvy. By the autumn of 1851 concern grew about an impending shortage of provisions; the food allowance was cut by a third, and the lemon juice ration by a half, and yet the crew remained healthy throughout the winter and spring. In May the medical officer reported the first case of scurvy. It was now 27 months since the *Investigator* had left home, and another 15 months were to elapse before rescue arrived. In all there were three deaths from scurvy.

Why then did Nares's men fare so much worse, when they were better provisioned and all other circumstances were very similar? The first case of scurvy occurred in May 1875, only seven months into the expedition. By the time the sledging parties went out in April 1876, scurvy was rampant, and there were three deaths among the sledge crews. Discouraged, Nares brought his ship home in October of that year. Disparate experiences multiplied, medical opinion was more than ever divided, and obfuscation reigned once again. Nansen's Arctic expedition of 1893–96 had been scurvy free, and when he discovered that Nansen had carefully sterilized all his tinned food, the Inspector-General of the Royal Navy's medical services inferred that scurvy must be caused by tainted food, that it was nothing more than 'ptomaine poisoning'. A committee of the Royal Society under no less a luminary than Lord Lister, the veteran hero of surgical asepsis, was assembled to consider yet again the origins of the malady. 'It would appear', it concluded after prolonged deliberation, 'that neither lime juice nor fresh vegetables either prevent scurvy or cure it, and that it is not the absence of these which is the cause of the disease, but that scurvy is a disease produced through the eating of tainted food.'[1] 'The doctrine of anti-scorbutics', wrote one highly placed English doctor, 'is untenable.' It was, he declared. 'the last remnant of an obsolete physiology', comparable in his view to the phlogiston theory in chemistry. The senior surgeon on Captain Scott's first voyage to the Arctic in 1901 asserted flatly that 'there is no antiscorbutic activity in any food or drug'. It was only, he concluded, a question of maintaining the purity of the food, but despite all his efforts scurvy quickly broke out; Ernest Shackleton, for one, then serving with Scott as a junior officer, was seriously affected and had to be sent home to recover. Food poisoning was not the only theory. The conviction had persisted in some quarters that fruit and vegetables were unnecessary, because the Inuit got along perfectly well without them for most of the year, and relied largely on fresh, commonly raw or undercooked, meat and blubber for their nutrition. A sufficiency of raw meat can indeed provide the necessary vitamin C, and this was the probable reason why Nansen's men thrived. A theory that achieved wide currency was that a defective diet caused the blood to become more acid, demanding treatment with alkali. The potassium content of potatoes was, by some *non sequitur*, thought to account for their beneficial effect in averting scurvy. The most outlandish notion of all, propagated as late as 1911 by the chief medical officer of a shipping line, was that scurvy is carried by a parasite of shipboard cockroaches.

The mystery was resolved by none other than Harriette Chick and her team of researchers at the Lister Institute, who in 1918 published their definitive study in the medical journal, *The Lancet*, based on both experimental and historical evidence.[2] Here is how the story unfolded: Harriette Chick's colleague, Alice Henderson Smith, noticed that no consistent distinction had ever been made between limes and lemons – a confusion that undoubtedly cost many lives. The lime juice then issued to the Navy and the Army, she wrote, came from the sour lime of the West Indies, which had been available in Lind's day. The sour lime does not grow in Europe, and the juice was therefore largely replaced by that of Spanish lemons, with perhaps a little from the Spanish sweet lime. In 1796 the

Napoleonic Wars restricted the supply from Spain, and most of what reached the Navy at that stage came from Portugal; but then in 1803 a cheap source of lemons became available from Sicily and Malta, abundant enough for the juice, always referred to as 'lime juice', to be provided prophylactically on all ships. Within the decade scurvy was eliminated from the Navy. But around the middle of the 19th century complaints were heard about the quality of the lemons from Malta, and the Admiralty reverted to shipments from the West Indies, where new plantations had been established to meet European demand. The fruit was shipped to the Navy yard at Deptford on the Thames in London, and squeezed there. Dr Henderson Smith noted that the expeditions of Ross and Nares had been so similar in provisioning as to constitute a controlled experiment like that of Lind. The difference was that Ross's men had received Mediterranean lemon juice and Nares's West Indian lime juice. The West Indian limes, then, were the culprit.[3] This conclusion was triumphantly borne out by Harriette Chick's experimental study on guinea pigs and monkeys, which strikingly revealed that lime juice, preserved in the customary way with an admixture of rum or with salicylic acid and alcohol, rapidly lost its antiscorbutic activity. Even when fresh, it was far less effective than lemon juice (by a factor of four). Lemon juice retained its activity much longer, and could be kept for extended periods without deterioration, especially if sulphite (still used today) was added as a preservative. (As we now know, it prevents the oxidation of vitamin C.)

At some stage in the 19th century, American seamen began apparently to notice that their British counterparts were regularly dosed with lime (or lemon) juice, and would refer to their ships as 'lime-juicers'. Later the term 'limey' came into use among American sailors, and then among Americans generally, to denote (most often in an unflattering spirit) any Englishman, much as the French would speak of 'les rosbifs'. James Lind would probably have been delighted.

Lime juice for all

In 1846 Edward Sturge, scion of a prosperous Birmingham merchant family, convinced evidently that seafarers the world over would soon all be drinking lime juice, bought the island of Montserrat in the West Indies. There he caused lime trees to be planted in huge numbers, having in mind the demands not only of the seafarers, but also members of the British public, who might welcome the juice as the basis of a refreshing hot-weather drink. Sturge negotiated the sale of his entire output to a firm of manufacturing chemists in Liverpool by the name of Evans and Co. They clarified and bottled the liquid and made a success of the business. Other, smaller companies also started up at about the same time, and then in 1865 a Scot, Lauchlan Rose, who had inherited the family firm at Leith, decided to enter the same business. Lime juice was at that time supplied to practically all naval and merchant vessels. To make it more palatable, Rose sweetened the juice with sugar for public consumption, while that in the large jars destined for ships was laced with rum as a (largely ineffective) preservative. It was Rose, however, who discovered that fumigating casks with the smoke of burning sulphur was much more effective, and a little later he took out a patent on sulphur dioxide (transformed into sulphurous acid when dissolved in water, and sodium sulphite when then neutralized) as an excellent preservative. (Harriette Chick, who in her great study during the First World War recommended its use, acknowledged Rose and Co. as the indispensable source of all her experimental samples.) Rose's Lime Juice Cordial became the first soft drink to be marketed in Britain, and endures to this day, along with Rose's lime marmalade, its popularity still undimmed.

Land scurvy

Scurvy is of course the same on land and on sea, but it took some centuries for this fact to gain acceptance. It had always been a presence in times of famine, and in the more insalubrious corners of communities – in prisons, asylums, in labourers' camps, and especially in the army. The incidence was high in the poverty-stricken regions of Eastern and Northern Europe, and pockets persisted long after the disease at sea had been largely overcome in the early years of the 19th century. A notorious outbreak occurred, for instance, in 1823 in the Millbank prison in London. This institution was modern and efficiently run under an enlightened regime. It had been built in response to the concerns of social reformers in Parliament, whose motto had been that the felons were sentenced to be detained but not to suffer disease or eventual death. But social reactionaries had complained that the inmates were coddled and that their food was better than that of many honest working folk. Forced to yield to such pressure, the prison authorities had curtailed the rations, largely eliminating meat and reducing the staple of potatoes. By 1825 half the prisoners were scorbutic and a large proportion were suffering from infections, especially dysentery. After much debate, doctors called upon for advice directed that the meat ration should be restored and all inmates given three oranges each day. The scurvy rapidly disappeared, but it was too late to stem the dysentery.

It was also found that in many county jails, and especially in military 'glasshouses', scurvy frequently erupted. A perceptive doctor noted that its incidence was correlated with the absence of potatoes, which were served in large quantities in some prisons and not at all in others. Dr Baly published his observations, and received intelligence from two navy surgeons that potatoes had indeed proved effective in combating scurvy at sea when supplies of citrus fruits had run out. Despite some claims to the contrary, it became clear that the potatoes should be eaten raw, but no doubt some of their content of vitamin C was left if they were not cooked for too long. The potato (*potatl*) was a native of the New World and treated with suspicion when it was introduced into Europe (Chapter 3). Its cultivation as a crop began in the early 18th century in northern Europe and the

British Isles, and very soon outbreaks of scurvy in the poor parts of Scotland and Ireland ceased. Communities soon became so dependent on the potato that the potato blight, which struck in 1845, caused an estimated million deaths from starvation in Ireland alone. In Scotland the consequences were less cataclysmic, but even so, the economic effects of the failing potato harvests led to a deterioration in institutional diets beyond the scarcity of potatoes. Scurvy appeared in prisons and among such groups of workers as the navvies, constructing roads and canals. These manifestations caused dismay in the medical profession. Misdiagnoses were frequent, a distinction was still commonly made between 'sea-scurvy' and 'land-scurvy', and theories about the origin and treatment of the condition proliferated. Some focused on fresh vegetables, some on meat, others on milk, and others again on minerals. In 1848 the potato harvest returned to its normal level, and scurvy petered out.

Scurvy also severely afflicted the 'Forty Niners' in California, for when the gold rush began in 1848 the journey west took many months by sea and even longer by land, when the river steamers and railways could get only as far as Kansas City, and the route by wagon across the Rocky Mountains was slow and arduous. It was at this time that citrus fruit imports into California were followed by massive plantings. Five years later came the catastrophic outbreaks of scurvy among the troops in the futile and notoriously mismanaged Crimean War. The army doctors demanded that consignments of citrus fruit be sent out from the Mediterranean, but all fresh food was scarce from the start and, especially in the winter, supplies by sea and land were sparse and unreliable. Many scurvy cases were treated (along with cholera and dysentery, to which scurvy appeared to predispose) by Florence Nightingale's staff at the hospital in Scutari. The Turks were not spared, but the French suffered most acutely, probably because their doctors had no faith in the efficacy of citrus fruits. This state of mind also helped scurvy to flourish during the siege of Paris in 1870, where most doctors seemed to adhere to the view that the disease was due to damp and fatigue, or was a contagion brought back by soldiers from the Crimea. How, they asked, could a wholesome diet of white bread, horsemeat, and wine, which was available to the defenders for a time at least, be inadequate for the preservation of health? Fresh fruit and vegetables might help to assuage scurvy in the same way that cinchona bark mitigated the effects of malaria, but their lack, by the same token, could scarcely be the cause. This misapprehension cost many lives and persisted into the First World War.

One of the moguls of medicine in Britain was Professor Colonel Sir Almroth Wright, model for Bernard Shaw's Sir Colenso Ridgeon in *The Doctor's Dilemma*, and founder of the Inoculation Department at St Mary's Hospital in London. (It was there that Alexander Fleming first observed the death of bacterial colonies, where the *Penicillium* spore, entering through the open window, had settled on the Petri dish.) Wright was an imperious man, with an unshakeable faith in his infallibility. He had devised a method of measuring the acidity of small blood samples, and had found that the blood of six scorbutic soldiers, shipped home after the siege of Ladysmith during the Boer War, was a little more acidic than that of most healthy people. Scurvy, he at once inferred, was probably a condition of 'acid intoxication'. This was forthwith adopted as the new dogma by the British Army Medical School, to which Wright was attached. Wright was troubled by no doubts: 'Nothing in pathology is to my mind more certainly established than that the essential essence of scurvy is to be found in a diminished alkalinity of the blood.'[4] It should accordingly be treated with salts (buffer salts) that reduced the acidity, and 'lime juice ought to be eschewed as containing citric acid'. With experts like this, what need for charlatans?*

During The Great War scurvy asserted itself among badly fed prisoners-of-war and in the Russian Army, and also frequently in troop ships. A highly publicized and poignant episode was the decimation of the garrison at the siege of Kut el-Amara in 1916. Major-General Charles Townshend had been sent on an ill-conceived mission to break the stalemate in the Mesopotamian campaign. He was to march from Basra through marshy, fever-ridden country, and attack Baghdad. His mixed force of British and Indian soldiers was confronted by superior and well-led Turkish troops, and his grossly overextended supply-lines were eventually cut. He chose to make his stand at the village of Kut, and the garrison held out for 146 days, up till then the longest siege in history. Food supplies ran out and the British soldiers survived by eating their horses, but this was not an option for the Indians, who died in their hundreds of scurvy and the ensuing infections.

Infantile scurvy occurred, often with fatal results, in slums and in institutions such as orphanages, when the diets were poorly planned. It was aggravated by the heat-sterilization of milk and by proprietary baby foods if babies were not breast-fed, nor was the condition always correctly diagnosed. Some doctors believed that the distinctive symptoms of infantile scurvy were complications of rickets. Infantile scurvy was common in the years of privation during and immediately after The Great War. If orange juice was available the symptoms rapidly vanished.

It is distressing to relate that scurvy has appeared anew in recent years, not, as the adherents of Linus Pauling's vitamin C regime (Chapter 9) have loudly asserted, in the developed world, but in the poor countries of the South, worst of all during the endemic famines in the Horn of Africa. Children are especially vulnerable. With vitamin C so easily obtained, and the cure so sure, this is a barely conceivable scandal.

* The *Oxford English Dictionary* of 1933 described scurvy as a disease 'induced by exposure and by a too-liberal diet of salted foods'.

3

In the Beginning

The fumbling start of nutritional theorizing

Although some of the earliest civilizations had rules about diet, and the curative virtues of different foods, the study of nutrition arguably began late in the sixth century BC. It was about this time that the Greek philosophers first pondered such fundamental issues as what happens to food during digestion, and how the tissues of the body are assembled. Chief among these scholars was probably Alcmaeon, a member of the Pythagorean school in Croton, a Greek colony in the south of Italy. He laid it down that good health could be achieved and maintained only through an equilibrium between what went in and what came out, and that an imbalance, reflected by obesity or emaciation, led to disease. Put more pithily, as by Nerissa in *The Merchant of Venice*, 'they are as sick that surfeit with too much as they that starve with nothing'. The Hippocratic school, which flourished a century later, was also much taken up with obesity – which suggests that over-indulgence may have been a common failing in southern Europe at the time.

Hippocrates of Cos was born, probably around 460 BC, on the Aegean island of that name, and was said to have lived to the age of 90. Nothing is known of his life, but writings, in particular the *Corpus Hippocraticum*, are (insecurely) attributed to him. This work extols the virtues of moderation, and lays stress on the need for regular exercise. The body uses the substance of foods to repair worn parts and to generate energy. The digestive system is likened to two workmen sawing a log: one pushes, the other pulls, that is to say what is driven in is, in some form, released. The amount of food required is a function of the constitution, of the amount of energy expended, and of the season of the year. The treatise offers much advice on slimming: obese people must exercise hard on an empty stomach, taking only a little diluted and moderately cold wine as restorative, and afterwards eat while still out of breath. The food should be rich in nature and well spiced, so that the appetite is soon sated. One meal a day will suffice, bathing should be eschewed, and one should sleep on a hard surface and move about in the nude. Hippocrates gave advice to couples desiring either a girl or a boy – if the former, the father should confine his diet to cold, watery foods, if the latter, then the opposite. As to women after they had conceived, they should consume only mild foods – fish, especially – and desist from strongly flavoured vegetables such as onions and garlic.

Salt was considered important, and various unpleasant conditions were said to arise from its deficiency. Its supposed expectorant property was valued, and a saline solution mixed with vinegar was used as an emetic to treat the sick. So too was inhalation of water vapour from a hot salt solution, which presumably carried some of the salt into the nose and mouth. For 'diseases of the spleen' there was nothing better than a mixture of two parts cow's milk to one part salt solution, taken first thing in the morning on an empty stomach. There were also specific prescriptions for various afflictions, involving the use of pepper and other spices.

Aulus Celsus was a Roman doctor who lived around the end of the millennium and wrote copiously. He had firm ideas about nutrition and developed in particular the concept of 'strong' and 'weak' foods, the former being the more useful in the treatment of illnesses.[1] Bread was the strongest and most nutritious of victuals, and wheat made the best and strongest bread. Below that there was a comprehensive scale of meats, fishes, vegetables, and fruits, arranged in order of strength. Older animals and vegetables, for instance, were stronger than young ones, and there were distinctions between domestic animals reared in different circumstances, between wild and domesticated species, and between meats variously prepared. Other odd theories flourished in Rome. One, apparently taken seriously somewhat before Celsus (during the second century BC), was propounded by no less an authority than the great statesman Cato the Elder (Marcus Porcius Cato). Cabbage, he thought, was a sovereign panacea (a theme that recurs throughout much of later history). Eating cabbage or, perhaps even better, drinking the urine of a cabbage-eater, would look after digestive diseases, ulcers, warts, and inebriation. Celsus was more moderate in prescribing foods and drugs for all the diseases prevalent at the time, and his treatise won him many followers, but he was eclipsed by the arrival of Galen.

Galen of Pergamum

It was Galen of Pergamum, 'the Prince of Physicians', who first formulated a coherent (if misconceived) theory of nutrition. Claudius Galenus was born in about AD 130 into a prosperous family. Pergamum was a city near the Aegean coast of what is now Turkey, and at the time part of the Roman Empire. It was a noted centre of learning, famous for its libraries and academies, and Galen's father was an educated man with interests in architecture, astronomy, and mathematics. At an early age the son entered the Aesclepium, the sanctuary and school of the god of medicine, as a *therapeutes*, or attendant. There he would have learned the elements of medicine, as laid down by Hippocrates, whose system, based on the theory of the bodily humours, he embraced and adhered to ever after. On the death of his father Galen left Pergamum to pursue his education in the major centres of medical learning, Smyrna (now Izmir), Corinth, and Alexandria. On his return to Pergamum in 157, he secured the highly desirable appointment of physician to the gladiators. This gave him new insights into human anatomy and physiology at a time in which dissection of human corpses was frowned on; he later wrote that wounds are windows into the body. But after four years of this his wider ambitions took him to Rome itself.

In Rome Galen quickly established himself as a savant and philosopher, for he held that philosophy was a necessary part of a doctor's education. His anatomical demonstrations and lectures attracted wide attention and brought him eventually to the notice of the intellectually inclined emperor, Marcus Aurelius. Galen in the end served as personal doctor to him and to the three emperors who followed. He wrote ceaselessly and was one of the earliest proponents of vivisection, performed mostly on Barbary apes, dogs, and pigs (all treated with what seems to the modern mind callous indifference to prolonged suffering). It was said that he employed 20 or more scribes to catch his every word, and there is no doubt that he was a man of invincible confidence and self-esteem. His influence permeated medical thought for the next millennium and a half, until the theory of humours and much of Galenic physiology was overturned by the work of Vesalius and his successors. Galen died in Rome in about 205.

Galen taught that life derived from *pneuma* – the cosmic breath. When this entered the body in the form of air it underwent transformations in the brain, the heart, and the liver. The first turned it into animal spirit, responsible for motion and the senses, the second into vital spirit, which heated the blood, and the last into natural spirit, which drove nutrition, digestion, and growth. Digestion – the conversion of inanimate matter into body constituents – was driven by the 'innate' animal heat, and was akin to cooking. Blood was formed in the liver and dispensed to the arterial and venous systems. These were separate body components which never mixed, but the venous blood was transformed by a 'vital spirit', derived from the *pneuma* in the inspired air, in the left cardiac ventricle into bright crimson arterial blood.

Galen's view of bodily composition and function enlarged on that of Hippocrates and Aristotle. All things, living and inert, were made up of admixtures of the four elements – earth, air, fire, and water – respectively dry, cold, hot, and moist. A person's 'complexion', or character, could be defined in terms of proportions of these elements. So in *Julius Caesar* Mark Antony says of the dead Brutus: 'His life was gentle, and the elements/ So mixed in him that Nature might stand up/ And say to all the world, "This was a man."' Complexions, then, were combinations of four extreme states, *sanguine, phlegmatic, choleric*, and *melancholic*. These were associated with pairs of elementary attributes, namely: hot and moist, cold and moist, hot and dry, cold and dry. Each type in turn represented one of the four humours, which were the bodily counterparts of the four elements – respectively *blood, phlegm, green (or yellow) bile, and black bile.* One's character could be tilted towards another complexion by circumstances, especially sickness. The same elements, with the same characteristics, were combined in all foodstuffs. Fruits, for instance, were cold and moist, and were best (in Galen's view) avoided altogether, for they could give rise to diarrhoea and fevers. They could, on the other hand, be used to combat excess of choler. Galen assured his followers that his father had lived to 100 by avoiding fruit all his life. In actuality, denying fruit to children probably contributed in a major way to malnutrition and infant mortality throughout many centuries. Moreover, nursing mothers and wet-nurses were also enjoined to eschew fruit, and this again would have contributed to vitamin deficiencies in mothers and children. Many diseases, Galen thought, were the result of a bad diet or simply over-indulgence, and to contract gout, kidney stones, or especially arthritis was a scandal (prefiguring Samuel Butler's imaginary society in *Erehwon*, in which sickness is considered a crime).

The principles laid down by Galen were disseminated throughout Europe by many centres of medical learning, of which the most famous was the School of Salerno in Italy. It flourished during much of the first and second millennia, having benefited greatly from the arrival early in its history of a mysterious learned traveller, known only as Constantine the African. It was he who brought the Arabic versions of Galen's works and those of other authors to Salerno and translated them into Latin. The School's famous tract, the *Regimen Sanitatis Salernitatum*, first appeared late in the 11th century. It defined the qualities of foodstuffs and the nature of diseases, warned, for instance, against the fearful danger of restraining intestinal wind and many other hazardous practices, and was regarded with almost religious reverence. The *Regimen* enlarged on Galen's prescriptions. Foodstuffs were described in terms of the intensity of the four qualities, measured in three degrees. Ginger, for instance, was wet in the first degree and hot in the third, while beef was both hot and dry in the first degree. A satisfactory meal, like the ideal temperament, should represent a balance of all four characters; it should be intermediate between moist and dry, and between hot and cold. Different foods were suitable for the sick and for the healthy. Foods that engendered black bile, among which were apples, pears, peaches, milk, cheese, and the meat of calves, hares, goats, and deer, were deadly for the sick.[2]

The School was run by Benedictine monks, and reports of their skills were brought back by Crusaders, who stopped off in southern Italy to have their wounds and diseases treated on their journey back from the Holy Land. Bleeding, with its long history, was reintroduced at about this time, to restore the balance of humours in the sick. This legacy to the physic of the next eight centuries is witness to the School's enduring influence. The *Regimen*, in fact, was still in use among the physicians of Europe into the 17th century.

The Galenic wisdom had by then spread to the Middle East by way of Persia, brought there by Christian schismatics fleeing from persecution in Italy, and translated into Syriac, the preferred language of scholars in the region. Then, after the Arab conquest of Persia, the works were translated into Arabic. The system was well ensconced in Byzantium by the beginning of the first millennium, and elaborate nutritional schemes, notably the Dietary Calendar, thought to have been compiled by one Hierophilus the Sophist, were widely circulated. The Calendar set out what foods should be eaten at different times of year to combat the ill-humours of the season. So in January the dangers of 'sweet phlegm' must be averted by small doses of high-grade, aromatic wines, sipped slowly. During the next three hours no food was to be consumed, and thereafter a specified diet was prescribed. Another important Byzantine encyclopaedic work, with the title *De Alimentis*, listed foods possessing a series of attributes. There were foods that generate good humours, that are digestible, that are indigestible, that retard the digestion, that move the bowels, provoke phlegm, produce black bile, wind, heat, excrement, and many more. Some of these are recommended for warming, for cooling, for slimming. There are variations on the Galenic theme: the least digestible of foods are the meats of billy-goats, bulls, and rams; mulberries, blackberries, and the like hurt the head, while wood-pigeons (though not their wings) engender excrement.

The Galenic doctrine endured with little change throughout the Middle Ages, although the categories of quality were elaborated in various ways by some schools (like the calibrated degrees of hotness and moistness defined by the School of Salerno). But the principles were immutable. So young children were of their nature phlegmatic, thus cold and moist, and therefore the diet of milk (likewise cold and moist) had to be tempered with dunked bread and a little poultry meat. The provenance of the milk, which was held to derive from blood, could influence the disposition that a baby would acquire. Wet-nurses therefore must come from good racial stock, and should be of good character and sober. (It was not explained why the character of a wet-nurse, but not that of a cow or goat, would transmit itself to the child.) Should a trace of blood enter the milk, the baby would grow up a murderer. In adolescence children grew sanguine or choleric, and were to be given cold and moist food, including 'gross' or red meats. The old revert to the state of babyhood, becoming cold, though dry, and so must be fed mainly hot and moist foods. Prescriptions existed for the feeding of the sick or the socially maladapted: milk was good for the melancholic or choleric, and hard cheeses (which were categorized as hot and dry) for the phlegmatic, and so on. Moreover, the bodily functions were driven by one or other of the elements, appetite by green bile, digestion by blood, storage of digested food by black bile, and evacuation by phlegm.

Humoral theories, it should be said, also arose in many other parts of the world. They may indeed have had their roots in China around the 6th century BC, whence emerged the concept of the *Ch'i*, the 'subtle wind', in essence the *pneuma*. This engendered six elements, reduced in later writings to five. They were fire, water, earth, wood, and metal. These were linked to the *yin*, or feminine character, which was cold, moist, and condensed, or to the *yang*, or masculine – hot, dry, and expanded. Ho the Physician, writing in the 6th century BC, divided diseases into different classes, each caused by a deficiency of one element. Diets were indicated, according to the affinities of foods with the elements. The natures of foods corresponded loosely with those laid down by Galen, so some were 'hot' (meats, blood, ginger, and hot spices), others 'cold', notably green vegetables. Diarrhoea and dysentery – the causes of much infant mortality – were 'cold' maladies, for which green vegetables were forbidden.

In India rice was hot and sugar cold, while in Malaya most vegetables were cold. In Persian medicine beef, starchy vegetables, and cereals were all 'hot', while mutton, sugar, pulses, and melons were cold. There is in general little correspondence with Galen's classification.

After Galen

Galen's teachings, as is apparent, held sway for the ensuing millennium more or less unchanged. They were in essence reiterated by even the most illustrious scholars over the centuries, such as the great Arabic polymath Avicenna (Ibn Sina), born in Asia Minor near Bokhara in about 970 or 980. He was a philosopher, poet, astronomer, mathematician, and physician, and disseminated an essentially Galenic vision of the human body throughout the Middle East and beyond.

Galen was regarded as infallible, and any departure from his teaching as heresy. Anatomists whose observations deviated from Galen's might conclude that the human body had undergone subtle changes since the master's day. A break came with Leonardo da Vinci, a man not given to uncritical acceptance of other men's conjectures. He performed many dissections, and drew conclusions about the physiological functions of the organs, which diverged from those of Galen and were in many respects well in advance of what had gone before. But Leonardo's anatomical work had little known influence because he published none of it, and it came to light only much later from studies of his notebooks.

The earliest public criticisms of Galen were probably contained in the writings of Paracelsus, otherwise Theophrastus Phillipus Aureolus Bombastus von Hohenheim (*c*.1493–1541), sometimes called 'the Luther of Medicine',[3] a truculent and disagreeable character who alienated everyone he encountered. A Swiss son of a doctor, he left home at the age of 14 and migrated between the European and Arabian academies, absorbing the teachings of medicine and alchemy. He served for some years as a surgeon in the mercenary armies in Europe, and eventually returned to Switzerland in 1527, where he was invited to Basel to treat the celebrated scholar, Johannes Frobenius. He was appointed to the post of city doctor, and, under the name Paracelsus ('greater than Celsus', the first century Roman physician), taught at Basel University, where, as a prelude to his lectures, he ceremonially burned a volume of Avicenna's most admired work. Frobenius died and Paracelsus, caught up in an ensuing lawsuit, took flight.

Paracelsus was the leading figure in the movement known as the iatrochemists – effectively proto-pharmacologists, who practised alchemy with the object of discovering substances to cure diseases and promote longevity – and he also dabbled in the alchemy of transmutation of elements. He declared that matter was made up of three elements – not the earth, air, fire, and water of Greek philosophers, but the principles defined by the Arabic alchemists. These he called sulphur, mercury, and salt. They represented all that was combustible or mutable, like real sulphur; that which was hard, stable, and resistant to fire, like common salt; and metals, which differed from everything else. The elements revealed themselves when vegetable matter was heated, for seeds or spices, such as nutmegs and cloves, gave off a 'spirituous' vapour, produced an aromatic oil, and left a residue of ash. Aroma and flavour derived from the element, mercury, while the sulphur, which was moist, sweet, and cohesive, appeared in oils and fats. Living matter, and in particular the organs of the body, were pervaded by vital forces, or *archei*, reminiscent of Galen's *pneuma*. These were somehow involved in the processes of digestion, whereby the nutritious constituents of foods were separated out from the excreted inert matter. Most importantly, he rejected the theory of humours. Diseases, he insisted, did not arise from any form of global imbalance, but each had its seat in one or other organ. Therefore bleeding was pointless (and alas for the innumerable victims over the next four centuries that the healing profession did not take note). Paracelsus made contributions to the practice of *materia medica*, which before his time was based almost exclusively on plants and plant extracts, by introducing inorganic (mineral) substances. Thus, amongst his other innovations was the use of mercury to treat syphilis.* Paracelsus's doctrines had considerable influence on other European thinkers, notably the much deeper Johannes Baptista van Helmont in Holland (Chapter 4) and Thomas Willis in England.

But the writings of many physicians before the 18th century still revealed an adherence to the old doctrines, based on the four humours. Consider, for instance, the Rev. Dr Thomas Muffett (alias Muffet, Moffett, or Moufet), who lived from 1553 to 1604 and set out the principles of a good diet in his treatise, *Health's Improvement, or Rules comprising and discovering the Nature, Method or Manner of Preparing all sorts of Food*, whence the following passage:

> Touching the difference of meats in substance: some are of thin and light substance, engendering thin and fine blood, fit for fine complexions, idle citizens, tender persons, and such as are on recovery out of some great sickness: as chicken peepers, rabbet suckers, young pheasants, partridge, heath-poulse, godwits, all small birds

* This therapy, which was felt by many sufferers to be worse than the disease, remained in use until the 20th century. 'Twenty minutes with Venus, twenty years with Mercury', the old saw ran.

being young, all little fishes of the river, the wings and livers of hens, cockchickens and partridges, eggs warm out of the hen's belly, etc. Others are more gross, tough, and hard, agreeing chiefly to country persons and hard labourers: but secondarily to all that be strong of nature, given by trade or use to much exercise, and accustomed to feed upon them: as poudred beife, bacon, goose, swan, satfish, ling, tunnis, salt salmon, cucumbers, turneps, beans, hard peaze, hard cheese, brown and rye bread, etc. But meats of a middle substance are generally the best, and most properly to be called meats; engendering neither too fine nor too gross bloud.[4]

Muffett was a fashionable English doctor and a noted entomologist, who wrote learned treatises on silkworms and on spiders. (He believed spiders to possess great medicinal properties, and his name is now commemorated in the nursery rhyme about his daughter Patience, the Little Miss Muffet who was frightened by a spider.)

Milk remained food fit only for young children and the very old and sick. It was forbidden to the young and healthy, for their character was choleric. It would damage their constitution in a variety of ways, from inducing agues and rheums to causing cramps and convulsions, on account of 'repletion'. It might also generate stones. Babies thrived on the stuff, but, as in earlier eras, the character of the woman from whom it came transmitted itself to the child, so that she had to be chosen with special care. The same held true for the old and sick, who were commonly suckled. Here is what Dr Muffett had to say about one of the beneficiaries, who was supposed to have subsisted throughout his declining years on wet nurses' outputs. Dr Cajus was the founder of Gonville and Caius College in Cambridge.

What made Dr. Cajus in his last sickness so peevish and so full of frets in Cambridge, when he suckt one woman (whom I spare to name) froward of conditions of bad diet; and contrariwise so quiet and well when he suckt another of contrary disposition? Verily the diversity of their milk and conditions, which being contrary to the other, wrought also in him that sucked them contrary effects.[5]

Scholars during the 16th century were in fact thought generally to have a weak, 'melancholic' disposition and to be less healthy and long-lived than husbandmen. The denizens of the universities were expected to be celibate, and in an influential work published in 1584, *The Haven of Health* by Thomas Cogan, were urged to eat purslane, for 'it representeth the rage of Venus; wherefore it is much to be used of students [that is scholars] that will live honestly unmarried'. Cogan goes on to lay down the best diets and meal-times for students in Oxford houses. He and others also enunciated various rules about foods: those that are easily digestible should be taken before those that are harder to digest, for otherwise the digestible foods might putrefy in the gut in the time that it takes to process the tougher kind. Then, moist foods must be eaten before dry, and helped down with frequent small draughts of liquid (milk being good, since it is 'made of bloud twice concocted'). And also, foods that are 'leuse and slipperie' are to be taken before those that are hard and compacted, for the latter might otherwise move down the digestive tract prematurely.

Andrew Boorde, another authority of the time, disapproved of water, which 'is not holesome, sole by it selfe', being 'colde, slowe and slacke of digestion' but 'ale for an Englysshman is a natural drynke'. For a Dutchman, on the other hand, beer (which was then ale without the hops) was most suitable.[6]

The 16th century, in England and France especially, saw the rise of a refined cuisine. Sauces, scholars of the period tell us, were devised on Paracelsian principles. Fats and oils (sulphurous) served to bind forms of mercury and salt, such as meat concentrates – the genesis of stocks – wines and vinegar. Herbs also came into wide use at this time. Sugar, on the other hand, was thought a danger to the constitution, for it generated acid in the gut, strong as *aqua fortis* (or nitric acid). Its use in the kitchen was therefore eliminated, except in the hands of specialized cooks (the precursors of the Victorian pastry chefs) who, in the houses of the rich at least, prepared sweets in a separate kitchen. The dishes placed before the rich at table during this time would not have appeared particularly alien to the gourmets of today.

The rich and the poor at table

Theories about food were all very well at times and in places of abundance, but for prehistoric man life was hard, or, as Hobbes had it, 'solitary, brutish, and short'. Few survived much beyond 20. The majority, even if they received sufficient protein, probably fell victim to malnutrition, especially vitamin deficiencies, and to putrefied food, others to the hazards of the hunter's accident-ridden life. The diet in Europe was primarily meat and the roots and seeds of some plants. In warmer climates, in America and the Near East for instance, circumstances were more favourable because of the more nutritious root vegetables that were available. Grains came into use for preparing breads in Neolithic times. Flour ground from wheat, maize, or millet was made into a paste with water and baked in much the same way that flat breads still are. The merits of yeast were discovered in Egypt around 2000 BC. Meat was scarce. Around the Mediterranean olive oil probably did much to maintain health, but the diet was in general monotonous. In Athens it improved as the city-state became more prosperous; its wealthy citizens were appalled at the sustenance of the Spartans and their notorious 'black broth' of pork stock, salt, and vinegar. In imperial Rome the rich feasted, while the poor subsisted on a dreary regime consisting mainly of bread and a porridge made from millet. The food of the better-off was strongly flavoured with a highly prized sauce, called *garum* or *liquamen*, prepared by allowing the liquid exuded from fermenting fish to putrefy for weeks or months. The other ubiquitous flavouring was derived from the pungent plant *asafoetida*.

The food of the European peasant in the Middle Ages and after afforded little variety. Vegetables were, on the whole, despised, and fruit was available only in season. Yet fresh meat was scarce for much of the year, because in the winter feeding of animals became problematical. Most were therefore slaughtered around the year's end and the meat was salted. Bread became a staple, and the wheat was also turned into frumenty, a thick milky substance exuded by the grain when it was soaked in water. Vitamin deficiencies were probably prevalent in the winter and early spring. Much of the population would have suffered from more or less serious scurvy through a large part of the year, as well as vitamin A deficiency (see Chapter 8), revealing itself in skin lesions, night blindness, and susceptibility to infections. (Night blindness was a recognized symptom even in the time of Hippocrates, and the recommended remedy then and later was liver.) With prolonged vitamin A deprivation more serious eye conditions could supervene, leading often to permanent impairment of vision or total blindness. Carotene (the precursor of vitamin A) came from eggs and from milk, though less in the winter when such cows as had not been slaughtered were kept in stalls where they had no access to fresh grass and were fed mainly on dried hay. Poultry meat would not have helped much, but the liver and kidneys of game, when available, could have provided some of the vitamin.

Town-dwellers were generally worse off than the peasants, especially in winter. Animals were kept in even the most densely populated districts of cities like London and Paris. The streets and rivers were rendered insalubrious by the rotting waste from domestic animals and from the abattoirs, which discharged their blood and offal without restraint. Pigs, which were allowed to roam the streets freely, were the nearest approach that there was to street-cleaners. Vermin fed on the refuse and spread diseases. In winter, when there was a dearth of fresh vegetables (as of animal fodder), and meat and fish were all salted (or in the Scandinavian lands, dried), diseases of malnutrition would undoubtedly have been rife. In some years famines swept through Europe, especially when, after a wet summer, the harvests failed or the wheat crop was destroyed by the fungus known as 'black rust'. At such times thousands died of starvation, and especially of protein deficiency. Pellagra (Chapter 8) was another common manifestation, often probably mistaken for leprosy.

After a warm, wet spring especially, ergotism was an additional hazard. Ergot is a fungus, *Claviceps purpurea*, which grows on rye and secretes a powerful hallucinogen, closely related to lysergic acid diamide, familiar in our time as LSD.[7] Ergotism resulted when infected bread was eaten, and took one of two forms, gangrenous or convulsive, both of them exceedingly unpleasant. Gangrenous ergotism, known in the past as St Anthony's Fire or sacred fire (*ignis sacer*), made itself felt by an itching and burning skin, followed by dry gangrene of the extremities through failure of the circulation. The loss of fingers and toes commonly ensued. The convulsive form, most common in northern Europe, attacked the nervous system and led to seizures resembling epileptic fits, and to hallucinations. It was sometimes fatal. Historians have conjectured that convulsive ergotism was responsible for the notorious denunciations of supposed witches in the town of Salem in Massachusetts in 1692. It was not until the mid-19th century that ergot was recognized as a parasitic fungus.

The rich were shielded from many of the travails of the poor, but had problems of their own. It may also have been partly a consequence of vitamin A deficiency, but more probably of calcium and phosphorus, that they so often fell prey to kidney and bladder stones. The rough bread and huge quantities of meat that they consumed (for they scorned vegetables) would not have provided enough of these elements. Much ingenuity went into the cuisine of the day. There were complex sauces and many varieties of meat and fish dishes, but the feasts of the rich, judged by the surviving menus and kitchen manuals, were conspicuously devoid of vegetables and fruit. It would have been the unhealthiest form of over-indulgence.

The potato comes to Europe[8]

The deep suspicion that the peasants harboured of any unfamiliar food can perhaps be attributed to the prevailing beliefs connecting diet (according to disposition) with a variety of ill-defined ailments. So when the potato appeared in Europe it met with hostility. It was brought to Spain from Peru by the *conquistadores*. Initially grown in Seville on a modest scale, it was apparently fed mainly to the sick in the Hospital de la Sangre. From Spain it slowly migrated into Italy and thence to other parts of Europe. It arrived in England in 1586, some 15 years after its introduction in Spain, brought apparently by Francis Drake, who had gathered some specimens on the South American coastal lands. An English physician discovered that the new roots had wonderful curative properties: they were nutritious, cured consumption and 'fluxes of the bowels', and 'increase seed and provoke lust, causing fruitfulness in both sexes'. The peasants remained unconvinced, and the success of the new root was no greater in Germany. In Burgundy potatoes were condemned as a cause of leprosy, and elsewhere in Europe they were blamed for other diseases. This was unfortunate, for the potato has remarkable virtues. It is sufficiently rich in vitamin C to serve as an excellent antiscorbutic, and it is an undemanding and abundant crop, affording nourishment for animals as well as people. It was introduced into Ireland by the English landlords soon after its arrival in England, but only around 1663 was it harvested in significant amounts. Soon thereafter it became the preferred, indeed almost the only crop that the peasants chose to grow. Until the potato famine, caused by a fungus, in the mid-19th century it sufficed to sustain the population. Very gradually it came to be accepted throughout Europe, as will be seen later.

Maize and other New World gifts

Maize (or, as it became known in the United States, corn) was the other great culinary bequest from the New World to Europe following the Spanish conquest. Columbus brought back specimens in 1493 from his first expedition, and it was planted in the south of Spain. This was the Caribbean maize, which was a staple crop (co-existing with manioc, or cassava) of the Indians, and its flour, like that of the manioc root, was used to make a gruel and bake into cakes, often with other foods. It was supposed that maize had spread to other European countries from the original Spanish planting, but recent genetic studies of different maize strains suggest otherwise. The likelihood is that it was only with the introduction of North American strains that maize became established in northern and eastern Europe. It is thought to have been discovered by Giovanni Verrazano in the course of his travels up the east coast in 1524, to have been brought back by him to Italy and to have spread to France, Hungary, and elsewhere from there. Its popularity – for it is a hardy and easily cultivated crop – had untoward effects. It did not offer the broad nutritional benefits of the potato, and the reliance that impoverished peasant populations placed on it led to the spread of pellagra (Chapter 8).

Other foods also entered Europe from the Americas at this time, but none of comparable importance. Cocoa beans were much prized by the Mexican Indians, who made them into a bitter drink. Columbus presented some of the beans at the court of Ferdinand and Isabella, but it was only when an unknown gastronome thought to sweeten it with sugar that its popularity grew, and it became the fashion in the prosperous cities of Europe. Such indeed was its popularity among the French aristocracy by the 17th century that it featured in Madame de Sévigné's amiable gossip: the Marquise de Coëtlogon, she reported, was so addicted to chocolate that she had, as a result, been brought to bed of a baby boy who was 'black as the devil'. Solid chocolate for eating evolved only later.

4

Dawn of the Scientific Age: the Road to the Scaffold

Groping towards the light

The gradual eclipse of the Galenic doctrine, the authority of which remained unquestioned for 15 centuries, was hastened by the great Andreas Vesalius, seen as the founder of modern medicine. He was born in Brussels in 1514, and came from a long line of doctors. His father had served as apothecary to the Holy Roman Emperor Maximillian and to his grandson, Charles V. Vesalius was admired for his deep knowledge and elegant use of the classical languages, which he absorbed in his studies at the University of Louvain (Leuven). He went on to study medicine at the Sorbonne, returning to Louvain in 1536 when war broke out between France and the Holy Roman Empire, with which he was identified. But soon, having apparently quarrelled with the medical faculty, he moved to Padua, and it was at the university there that the bulk of his great work was done. He taught in the medical school, and illustrated his lectures with demonstrations of dissections. Initially a follower of Galen, Vesalius had the independence of mind to question the master's teachings, and by about 1540 he had begun to formulate his doubts. In 1543 he published one of the seminal works in the history of science. Entitled *De Humanis Corpore Fabrica* (The Fabric of the Human Body), it was written, like all learned works of the period, in Latin, but not in the rather debased macaronic Latin of the European physicians and scientists: Vesalius expressed himself in the crystalline prose of Cicero and the great classics, for he held that precision of thought demanded precision of language. The *Fabrica* was the first comprehensive textbook of anatomy, splendidly illustrated, it is thought by students from Titian's workshop in Venice. That same year Vesalius left Padua and academic work to succeed his father as physician to the household of Charles V, serving also as an army surgeon during the Habsburg campaign. He acquired a brilliant reputation as a physician and surgeon, but left the imperial court for private practice in Brussels, having by then published a second and enhanced edition of the *Fabrica*. In 1564 Vesalius accepted a call to succeed the great anatomist Gabriele Fallopio in Padua, but first set off on a mysterious journey to the Holy Land, possibly a pilgrimage, or to collect medicinal plants. He died on the return journey, at the age of 50.

Vesalius envisaged the body as a machine, and the stomach as a workshop and store. It mixes and compacts the food and drink delivered to it and impels it into the intestine. The juice is extruded from the mass, passes through the venous system into the liver and is used to make blood. The waste matter is carried away with the bile to be excreted. The blood is distributed around the body by the great vein which Harvey later showed to be the vessel that returns it to the heart. So Vesalius had not altogether shaken off the Galenic view of the liver as the central agent of blood supply.

Hard on Vesalius's heels came another son of Brussels. Johannes Baptista (or Johan Baptista, or Jan Baptist) van Helmont was born in 1579 into the Flemish nobility, and was, like Vesalius, educated at the University of Louvain. Opting, after flirtations with other disciplines, for medicine, he was awarded his doctorate in 1599 and set out on the scholar's tour round the European centres of learning. It was 10 years before he returned to his country, settling near Brussels, in the village of Vilvoorde. There he remained, experimenting and practising medicine, until his death in 1644.

Van Helmont's philosophy was a strange mishmash (not unusual for the epoch) of fearless and incisive reasoning, mysticism, and religiosity. He was a resourceful and ingenious experimenter, whose chemical studies led him to the discovery of several gases – carbon monoxide, carbon dioxide, sulphur dioxide, and hydrogen chloride, which he recognized as all being distinct from air. For this state of matter he coined the word 'gas', from 'chaos' – thereby drawing a parallel some centuries ahead of its time. There were, he further observed, two classes of gas: *gas sylvestre* was one that would not burn or support combustion, and *gas pingue*, a combustible gas. He reduced Galen's four and Paracelsus's three fundamental elements to two – air and water – rejecting fire as devoid of substance and earth as reducible to water. But he did not dissent from the teachings of the alchemists, and indeed thought that he had effected the transmutation of mercury into gold with the aid of a chunk of the philosopher's stone.

But van Helmont's most remarkable achievement was what is generally seen as the first ever quantitative experiment. It was predicated on his belief that plants are composed of water. He planted a willow seedling, weighing 2.3 kg, in a pot containing 91 kg of soil, weighed dry, and waited for five years while it grew into a substantial tree with a weight of 77 kg. Weighing the soil, he discovered that it had scarcely changed its weight. From this experiment he concluded that the tree did indeed derive its mass from the water received during growth, and that the soil had not significantly contributed. This was excellent as far as it went, but van Helmont must have been aware that other factors entered into the growth of plants. One of his critics was Robert Boyle ('son of the Earl of Cork and father of modern chemistry', as he was sometimes known). Boyle was indeed the 7th son and 14th child of the Earl, born in 1627 in Lismore Castle in Ireland. He was a remarkable polymath, a member of the club of leading London intellectuals calling themselves the 'Invisible College', the nucleus of what was to emerge as the Royal Society. Boyle, whose name is enshrined in Boyle's Law, which defines the relation between the pressure of a gas and its volume, contributed to many areas of science, but especially to chemistry, of which his book, *The Sceptical Chymist*, of 1661 is a landmark. He examined the effects of depriving animals of air, but was deterred from pursuing physiology any further by 'the tenderness of his nature'. Boyle's criticism of van Helmont's experiment was the following:

> several plants, that thrive not well without rain water, are not yet nourish'd by it alone, since when corn in the field, and fruit-trees in the orchards have consum'd the saline and sulphureous juices of the earth, they will not prosper there, how much rain soever falls upon the land, till the ground by dung or otherwise be supplied again with such assimilable juices.[1]

Plants, in other words, were composed of something other than just water. Boyle was undoubtedly driven in this and many similar disquisitions by his fierce rejection of the 'spagyrists' – those who embraced Paracelsus's doctrine of three elements – and no doubt of van Helmont who had gone a step further in reducing them to two.

Van Helmont equated nutrition and bodily functions with fermentation, which he saw as a chemical process like the formation of alcohol during the transformation of grape juice into wine, rejecting, therefore, Galen's hypothesis of innate heat (the fire, derived from the *pneuma*, which smouldered perpetually in the heart and was distributed around the body by the blood). The digestive system, van Helmont inferred by chemical and anatomical reasoning, converts dead material – the food – into living tissues with the aid of six 'ferments' (or as we would now say, enzymes, a remarkably prescient vision).[2] The site of the first ferment (acid) is the stomach, the second the duodenum, which receives the chyle (the milky suspension of fatty globules produced by the first digestive process) from the stomach and adds the supposedly alkaline bile to absorb its acidity. Then the chyle enters the venous system and passes into the liver, where the nutritious part is used to form blood and the remainder is converted by another ferment into faecal waste. In the fourth fermentation the dull red venous blood has reached the heart, where it is converted into the bright crimson arterial blood. A fifth fermentation imbues this with a vital spirit, which enables it to nourish the organs by way of a sixth and final local fermentation. Van Helmont was evidently ignorant of Harvey's discovery of the circulation of the blood, or mistrusted it, for in this part of his scheme he hews to the Galenic furrow. Diseases, van Helmont thought, arise from chemical imbalances. So, for example, the gastric acid was in some way responsible for the first stage of digestion, but an excess of acidity would cause derangements, for the power of the bile would be insufficient to neutralize the chyle on its emergence from the stomach.

Indigestion should therefore be treated with an alkaline agent, and this was typical of his general chemical (or iatrochemical) view of therapeutic method.

Up to this point van Helmont's theories are perfectly rational and, at many points, remarkably far-sighted, but unfortunately he conflates them with a muddled admixture of Galenic and mystical elements. There is, he asserts, a kind of orchestral conductor, residing in the stomach, who coordinates the bodily activities. This is in essence Paracelsus's *archeus*, but in van Helmont's scheme the spirit supervises an army of subsidiary *archei*, each responsible for the functioning of one or other organ. A disease is a malfunction caused by a chemical insult to an *archeus*, which has to be coaxed back to its normal state by chemical means, that is to say, a suitable drug. There are, however, further layers of complexity, in particular a 'sensitive soul' to which the governing *archeus* is subject. This sensitive soul supplanted the 'immortal mind' at the time of the Fall, thereby depriving man of his immortality. The entire corpus of van Helmont's work was edited and published after his death by his even more wildly inventive son, Franciscus Mercurius van Helmont. His medical treatise appeared as *Ortus Medicus* – the Fount of Medicine.

In his later years van Helmont concentrated increasingly on medical practice, and his therapeutic procedures seem to have grown progressively more bizarre as he sought a panacea that would do for medicine what the philosopher's stone was meant to accomplish in alchemy. It was magical healing that he now sought. To cure a sick patient, he advised, one should fill an eggshell with a sample of his still warm blood, and lay it under a broody hen. One must then mix the blood with some meat and feed it to a hungry dog or pig, and the patient's sickness would pass into the animal. He equated this drawing-out of the disease with the power of magnets to procure miracles, and published a treatise on the subject. But this drove him into deep trouble, for it attracted the attention of the Church. Worse yet, he had, even though he saw himself as a devout Catholic, firmly rejected the power of religion to effect cures. His vexations started in the medical faculty of the University of Louvain, which issued a warning to desist from such impiety. Then, in 1623 the Spanish Inquisition, which was very active in the Low Countries, turned its cold eye on his writings. Two years later his works were impounded and in 1626 the Holy Office called him to account. He was accused of heresy, arrogance, and trafficking with Lutherans and Calvinists. These were deadly charges, and after an investigation lasting until 1630 he was forced to recant. For his punishment, as 'an adherent of the monstrous superstition of the Paracelsian school', he was condemned to house arrest. The wrath of the Holy Office was still not slaked, and in 1634 van Helmont was thrown into prison. He was released after a year, but the sentence of house arrest was reimposed, this time under more severe conditions, and it was only in 1642 that he was adjudged to have expiated his guilt. His treatment by the Inquisition had nevertheless, by the standards of the time, been exceptionally lenient, surmisedly because of the influence of his and his wife's families, both connected to the Flemish ruling caste. Van Helmont died at his home in Vilvoorde, apparently of pleurisy, in 1644.

Chemists and mechanists

The study of mechanics, brought to the forefront of science by Galileo, sent its ripples into medicine and physiology. Descartes was probably a prime mover with his purely mechanistic view of living organisms. An important figure in the late 16th and early 17th centuries was Santorio Santorio, also known as Sanctorius. He lived from 1561 to 1638, and was much influenced by his friendship with Galileo. In 1611 he was made professor at the University of Padua, where he engaged in much ingenious experimentation, involving often the construction of new kinds of instruments. But his fame rested largely on his quantitative measurements of body weight under different conditions. He tried to relate what went in with what emerged, and so recorded over some period the amount of food that he consumed, his own weight, and the weight of what he excreted. It did not add up, and Sanctorius therefore conceived an elusive essence, called 'insensible perspiration', which accounted for the deficit in body weight. This experiment had considerable influence.

William Harvey's discovery of the circulation of the blood, published as *De Motu Cordis* in 1628, which first visualized the heart as a pump, stimulated mechanical thinking. It was probably no coincidence that Harvey had studied for some time in Padua during the period of Galileo's ascendancy. Robert Hooke in London performed a famous experiment, in which he opened the thorax of a dog and demonstrated that it could be kept alive if air was driven through the lungs with a bellows. (He refused ever to repeat the demonstration because of the suffering that it caused the animal.)

But if one individual can be said to have set the course of modern medicine, it was Herman Boerhaave of Leiden, the most famous doctor of his day.[3] Boerhaave, the son of a Calvinist minister, was born in 1665 in the village of Voorhout, just outside Leiden. The father intended his son for the same calling, but in the course of his studies of philosophy and theology young Herman fell foul of an adversary whom he had humiliated in a debate on the teachings of Spinoza. This man accused Boerhaave publicly of following Spinoza's condemned doctrines and leaning towards atheism. This was happily the end of all plans for a career in the Church, and Boerhaave turned instead to science, and in particular botany and medicine. He graduated in medicine at the university in Harderwijk, some 100 km from Leiden, to which he returned in 1701, never to leave again. Unusually for a scholar of the period, he never travelled to other centres of learning, but Leiden gave him all that he needed. It was not long before he acceded to the professorial chairs of medicine and of botany, and after another four years he acquired a third chair, that of chemistry. Boerhaave made contributions in all these fields, and gained an immense reputation as a teacher.

He is credited with introducing the method of clinical instruction adopted throughout Europe, and the Caecilia Hospital, in which he practised, became the first academic teaching hospital. He himself was revered by his students and patients alike for the warmth of his personality, as much as for his skills. He was famous for his diagnostic acuity and sureness in relating the symptom to the cause. He was called upon to treat many important patients, and gave his name to a disorder discovered at autopsy in the Grand Admiral of the Dutch Fleet and Prefect of the Rhineland, Jan van Wassenaer, who died in great agony after taking an emetic following an over-indulgent meal. The cause of his demise was a ruptured oesophagus (known to this day as Boerhaave's syndrome), induced by violent vomiting.

Boerhaave's fame spread around the world, even as far as China, and he attracted a multitude of students. Peter the Great, who came from Russia to the Netherlands to study ship-building, was one who sat at his feet. Dr Johnson, the lexicographer, was an ardent admirer of Boerhaave, conducted chemical experiments in his lodgings in London, guided by his hero's writings, and after Boerhaave's death wrote his biography. Boerhaave died after many years of intermittent illness, in 1738, lamented by his innumerable disciples and admirers. It is related that a notebook found among his effects bore a legend promising that in its pages would be found every secret of medical practice. But the pages were blank, all except one, on which was written: 'Keep the head cool, the feet warm and the bowels open.'

Boerhaave's views on nutrition built on those of the foregoing mechanists. His fellow-countryman, Antoni van Leeuwenhoek, had discovered with his microscope that blood is a mass of suspended corpuscles – mainly the red blood cells. This appeared to support the view that all matter was composed of minute, perhaps infinitely divisible, granules. The organs (or, as he called them, 'instruments') of the body, Boerhaave conjectured, were constructed from such granules, so arranged as to serve the special function of each 'instrument'. Because of the continuous fluid stresses and the grinding motions required to break down food, the organs were continuously abraded. The lost matter was replaced by the particles of the mechanically processed food. Boerhaave's conception of the body called for no vitalist spirit, no *archeus*, or as he called it, 'spiritual cook', and Galen's teachings were consigned to the intellectual refuse heap of history. The function of bodily matter was governed primarily by Newtonian mechanics, and to some extent by chemistry. Boerhaave held nutrition to be the very key to life and health: if the formula could be found for diets that would maintain the condition of the organs, a long life, untroubled by ill-health, would be the reward.

Boerhaave's picture of the digestive process was perfectly explicit. The food is first chopped, sliced, and ground by the teeth. Thus prepared, it passes into the stomach. There, and in the intestine, it is reduced to the particles of the chyle, which is exuded into the lacteals (the tiny lymph vessels into which fluid drains from the involutions of the intestine). It then undergoes further mechanical and chemical transformations on mixing with venous blood, is worked on by the right ventricle and auricle of the heart, and is driven into the pulmonary artery, where the high fluid pressure to which it is subjected converts it finally into the state of matter needed to replace the worn parts of the body and replenish its fluids. The nutritive particles are carried in the blood to all corners of the body, and undergo, primarily through the action of pressure, the changes that will allow them to act at the specified sites.

Boerhaave's was an excellent hypothesis on which future scholars could build, bearing in mind that the atomic concept of matter had not yet been formulated (leaving aside Democritus and the Greeks). The most distinguished of Boerhaave's immediate successors was a Swiss polymath, Albrecht von Haller. Born in Berne in 1708, the fifth and precocious child of a lawyer, he graduated from the University of Tübingen and headed for Leiden, where he was awarded his doctorate under Boerhaave's tutelage in 1727, at the age of 19. His travels then took him to the great centres of learning in England and France, and finally back to Berne to study mathematics with the contentious Johann (or Jean) Bernoulli, a member of the celebrated Bernoulli clan of mathematicians. Haller's reputation as a savant, and also indeed as a poet, grew until in 1736 he was called to the position of Professor of Anatomy, Botany and Surgery at the new Georg-August University of Göttingen. He taught there for 16 years, in the course of which he was ennobled by the Emperor, having created in Göttingen a major centre of teaching and research. But in 1754 Haller chose to return to his home city of Berne, where he served in mainly minor official capacities, culminating in a position as the superintendent of the local salt refinery, until his death in 1777.

During his career Haller made many notable contributions to physiology. He studied the circulation of the blood and clarified the relation between nerves and muscles (mainly with the aid of rather gruesome animal experiments). His mighty eight-volume treatise on physiology, *Elementa Physiologiae Corporis Humani*, completed in 1766, remained the definitive text for many years. Like Boerhaave, whose works he edited, Haller regarded the body as an engine, and therefore, like man-made devices, subject to wear and tear. Muscles, for instance, expanded and contracted, second after second, year after year, so must surely wear out if not regularly serviced and replenished by the blood.* Different foods

* The modernity of this representation can be gleaned from the following passage, taken from a lecture delivered in 1969 by an eminent physiologist, Professor D.R. Wilkie of University College, London:

> 'Available now. LINEAR MOTOR. Rugged and dependable: design optimized by world-wide field testing over an extended period. All models offer the economy of "fuel-cell" type energy conversion and will run on a wide range of commonly available fuels. Low stand-by power, but can be switched within msecs [milliseconds] to as much as 1 KW [kilowatt] mech/Kg (peak, dry). Modular construction, and wide range of available subunits, permit tailor-made solutions to otherwise intractable mechanical problems.

service different organs. Body fat is produced from fat in the food, as are the red blood cells. The fat particles are carried in the blood, but being viscous, this fluid with its suspended matter cannot be forced into the small vessels (capillaries) and is consequently deposited on the exterior. The mechanical members of the body, by contrast, are maintained by gluten in the food. The fluid components of the body need to be replaced as they are lost in urine, sweat, and the water-laden exhaled breath. Body heat, Haller suggests, can only be generated by friction, as the red blood cells, driven through the arteries, rub against one another and the vessel walls.

Mark Twain owned that he liked science because ‘one gets such wholesome returns of conjectures out of such a trifling investment of facts’. The time was ripe for the entry of quantitative chemistry.

From chemical revolution to French Revolution and the guillotine

Antoine Laurent Lavoisier did for chemistry what Isaac Newton had done for physics. He more than anyone placed it on a rigorous quantitative footing. In truth, therefore, it is Lavoisier, more than Boyle, who can most justly claim parentage of modern chemistry. Neither would he have been prevented by any false modesty from concurring, for he was a man of implacable self-esteem, not given to acknowledging the contributions of others. Lavoisier was born in 1743, son of a prosperous Parisian advocate, and at his father’s behest studied law and qualified to practice it. But his interest all along was in science, and in 1763 he immersed himself in the study of geology. He also wrote papers on agriculture,

> Choice of two control systems:
>
> *Externally triggered mode*. Versatile general-purpose units. Digitally controlled by picojoule [million-millionth of a joule] pulses. Despite low input energy level, very high signal-to-noise ratio. Energy amplification 10^6 approx. Mechanical characteristics: (1cm modules) max. speed: optional between 0.1 and 100 mm/sec. Stress generated: 2 to 5×10^{-5} newtons m^{-2} [the newton is the unit of force and is equal to 1 joule per metre].
>
> *Autonomous mode with internal oscillators*. Especially suitable for pumping applications. Modules available with frequency and mechanical impedance appropriate for:
>
> Solids and slurries (0.01–1.0 Hz) [the Hertz is the unit of frequency: 1 Hz is 1 cycle per second].
>
> Liquids (0.5–5 Hz): lifetime 2.6×10^9 operations (typ.) 3.9×10^9 (max.) – independent of frequency.
>
> Gases (50–1,000 Hz).
>
> Many optional extras e.g. built-in servo (length and velocity) where fine control is required. Direct piping of oxygen. Thermal generation. Etc.
>
> Good to eat.’[4]

devised an improved method for the preparation of saltpetre, the primary constituent of gunpowder, and was awarded a prize for a paper on improving the street-lighting of Paris. On the strength of these achievements he was elected in 1768, at the age of 25, to the august *Académie des Sciences*. That same year he joined the *ferme générale,* the private company that enriched itself by collecting taxes for the state. Three years later Lavoisier took an important step in his life: he married the clever and high-spirited 13-year-old daughter of a rich *fermier*, Marie-Anne Pierrette Paulze. She identified herself with her husband's work from the very outset, became his assistant and collaborator, and even mastered English to help him understand the publications of Joseph Priestley and others from across the Channel. She maintained the laboratory notebooks and acquired the specialized skills of draughtsmanship and engraving demanded for the preparation of illustrations for publication.

Over the next years Lavoisier made a series of monumental advances in the understanding of the material and living worlds. In 1775 he was made responsible for the manufacture of gunpowder, and was given room at the Paris Arsenal to set up a laboratory, with facilities for the building of apparatus far superior to anything seen before. In his Arsenal laboratory he established the principle of conservation of matter ('nothing is lost, and nothing is created'), demolished the reigning theory of combustion (the phlogiston theory), and clarified the chemistry of the common gases. Besides all that, he laid the foundations, as we shall see, for an understanding of respiration and of energy consumption in metabolism. At the same time he and his wife created a salon in their town house, in which the Parisian intelligentsia regularly gathered. On these occasions the Lavoisiers were not above exalting their recent achievements, and on one notorious occasion Mme Lavoisier laid on, as part of the evening's entertainment, a masque in which a figure representing phlogiston was triumphantly slain by oxygen. (Oxygen, a term invented by Lavoisier, was, as already remarked, a misnomer in that it imputed to the new element a presence in all acids.) As recognition came to him, Lavoisier was increasingly called upon to serve the State – on the Treasury Commission, the advisory board for training in the crafts and professions (the genesis of the Collège des Arts et Métiers), and the Academy's Committee on Weights and Measures, on which he did much to bring about the introduction of the metric system.

With the coming of the Revolution a shadow fell over Lavoisier and his ménage, for all that his inclinations were unmistakably liberal, and he was largely at one with the *encyclopédistes*. He had, indeed, argued for the necessity of social reform, for a more equitable system of taxation and for the reform of the prisons and hospitals, and had sat on commissions to consider these matters. In one of his most notable scientific papers, written with his assistant, Séguin, he introduced by way of digression a disquisition on the patriotic services that a savant could render through his discoveries. These might lead, in particular, to the alleviation of disease and prolongation of citizens' lives.

But several years before, Lavoisier had made a powerful enemy. Jean-Paul Marat was a fanatical and violent man, probably unhinged, who nurtured an aspiration to make his mark as a savant. Born in Switzerland, he had journeyed around France and then qualified in medicine at Edinburgh. Returning to Paris, he had earned the gratitude of a wealthy and influential patient, who helped him to set up a successful medical practice. This gave him the time and means to perform experiments on heat and electricity, and, convinced of his own genius, he began in 1778 to inflict himself on the Academy of Sciences. He eventually prevailed on some academicians, and in particular on Benjamin Franklin, then a social lion in the City of Light, to be allowed to present a demonstration. It consisted in projecting a swirling, magnified image of a flame onto a smooth surface, namely the shining bald pate of Franklin himself. This purported to reveal the expulsion from a burning body of igneous fluid (or caloric, in the language of prevailing theory). When no further invitations from the Academy ensued, Marat was incensed and fired off a succession of angry, intemperate missives. The outcome was that the Academy adjudged Marat's submitted works unworthy of its approval and therefore of publication. Marat responded by publishing his observations privately in a treatise entitled *Recherches physiques sur le Feu*. The newspaper, the *Journal de Paris*, mistakenly reported that this work had the Academy's approval, an assertion that was swiftly repudiated by Lavoisier on the Academy's behalf. From then on Marat incubated his hatred for the Academy, and for Lavoisier in particular, and plotted revenge.

Nearly four years were to pass before an opportunity finally presented itself. By then Marat was editor of an inflammatory journal, *Le Publiciste*, and as the self-styled *Ami du Peuple*, he preached that the Revolution had been betrayed by the treachery of, among others, Lavoisier, whom he also denounced as a bloodsucking member of the hated *ferme générale*. On 8 May 1794 all the *fermiers* were arrested, brought before a tribunal, and summarily sentenced to death. It has gone down in folklore that when sentencing Lavoisier the presiding judge declared: '*La République n'a pas besoin de savants*', but no record of this statement has ever been discovered. Lavoisier had to witness the execution of his beloved father-in-law before he himself calmly mounted the scaffold. The mathematician Joseph-Louis Lagrange pronounced Lavoisier's epitaph: 'They needed only an instant to cut off that head, and a hundred years may not suffice to produce another like it.' Marat did not live to gloat over his revenge, for he was already dead, assassinated in his bath by Charlotte Corday the previous July.

Lavoisier's biology

Lavoisier began to think about the cycle of life quite early in his career. He had arrived at some of his ideas about combustion from observations on simple substances. He devised methods of determining the carbon and hydrogen contents of plants and their constituents from the weights and volumes of products of combustion (gases and ash, essentially). He analysed wood, olive oil, brandy, and waxes, and came to the conclusion that vegetable matter consisted largely of carbon, hydrogen, oxygen, and occasionally sulphur and phosphorus. It was another eminent French chemist and contemporary of Lavoisier, Claude-Louis Berthellot (1748–1822), who added nitrogen to this list, as a major component of living things. Lavoisier tried to force the processes of life into a chemical scheme, which he first expounded in relation to fermentation, the products of which he analysed with considerable precision.

He saw the cycle of life on the planet in terms of the assimilation of dead, or inorganic, matter – air, water, and matter from soil – by plants. The plants flourish and start a food-chain: animals eat the plants or other animals that have lived off plants. The remaining plants die, rot, and ferment – a process that Lavoisier likened to combustion, whereby the materials derived from 'the inorganic kingdom' are returned to it. The synthetic part of the cycle, in which plants and then animal bodies develop, he saw as the reverse of combustion. This was as far as he went, for by the time he had published his theory in his *Traité élémantaire de Chimie* in 1789 he had little time for his own research.

Lavoisier discovers physiology

By 1773 Lavoisier, then 30 years old, had developed an absorbing interest in the properties of gases. It was Joseph Black in Edinburgh who in 1750 first prepared a gas in the pure state. Carbon dioxide, which he called 'fixed air', was produced when hydrochloric acid ('spirit of salt') acted on limestone (calcium carbonate). But there appeared also to be other ways of coming by fixed air. Lavoisier knew of four: combustion was one, as when a candle burned in air, another was by the heating of metal carbonates (calcination), and then there was the 'air' emitted in the putrefaction and fermentation of vegetable matter, and lastly the respiration of animals. In the course of these studies he showed that oxygen was essential for combustion and explained how it was consumed in chemical oxidations; he clarified the concept of an element; and he saw off the long-established phlogiston theory (according to which all inflammable substances contained a weightless 'combustible principle' that escaped during combustion in a swirling motion, perceived as flame). It was the apparent relation between combustion and respiration that tickled Lavoisier's imagination. In 1777 he produced his first paper on respiration: 'Experiments on animal respiration and the changes that occur when air passes through the lungs'. More publications followed, until in 1790 Lavoisier and his devoted assistant, Armand Séguin, brought out a synthesis of all their results, 'The First Memoir on Animal Respiration'.

One of Lavoisier's most famous experiments was conducted with the young Laplace, who was to emerge as a central figure in the development of physics and a panjandrum of the French science establishment during the Napoleonic era.* They knew, of course, that respiration produced 'fixed air' (carbon dioxide), the amount of which they could measure. They placed a guinea-pig in an ice calorimeter. This simple device, invented by Laplace, was no more than a metal vessel inside an insulated container filled with ice, so that heat generated in the inner vessel would cause some of the surrounding ice to melt. If the outer container was properly insulated so that no heat escaped into the laboratory, the amount of ice melted would be a measure of the heat generated, in this case by the guinea-pig. After 10 hours they found that 13 g of ice had melted. In a separate experiment they burned a known weight of charcoal and measured the volume of 'fixed air' that it produced. Next they burned a known weight of charcoal in the ice calorimeter and so were able to relate the heat emitted to the amount of fixed air that was formed. It was then easy to work out how much fixed air was produced by the guinea-pig for every gram of ice melted. The answer was 10½ g. Therefore most of the body heat (10½ thirteenths of it) came from the respiratory process, and it confirmed Lavoisier in his opinion that respiration was no more than slow combustion. And certainly the function of the blood, which carries oxygen to the tissues, was not, as the physiologists of earlier times had taught, to cool the body. The combustion that went on in the body must, Lavoisier wrote in his First Memoir, cause the substance of the body to disappear if it is not replaced by nutriments.

The First Memoir went further: it was blood that carried the fuel for the combustion. The oxygen deficit in the exhaled air represents the amount that is used in turning carbonaceous and hydrogen-containing material into carbon dioxide and water (since the inspired air ('vital air') contains neither carbon nor hydrogen). In return, 'caloric', that is heat, is supplied to the body, and so 'animal heat' derives from the chemical processes of respiration. The nitrogen in the air, which remains behind when oxygen is consumed during combustion *or* respiration, Lavoisier concluded, plays no part in either process. He proved this by another experiment: an animal in an atmosphere consisting of oxygen and hydrogen (instead of oxygen and nitrogen) remained alive.

Experiments continued, adhering to Lavoisier's principle that everything must be measured in a kind of double-entry book-keeping approach to physiological research. He and Séguin determined the volume of oxygen consumed by a guinea-pig, and then they decided to extend their studies to man, with Séguin as volunteer. He was totally enclosed in a rubber-coated taffeta suit, sealed at the top, so that no gas or water could escape. Air was supplied through a tube sealed

* Pierre-Simon, marquis de Laplace, as he later became, was indeed a favourite of Napoleon's. He was a determinist, who enunciated the tenet that if the position, speed, and direction of motion of all the particles in the Universe were known, he could (in principle, if not of course in practice) calculate their position at any time in the past or the future. When the Emperor asked what part God played in his mechanical scheme, Laplace was supposed to have replied, 'Sire, I have no need of that hypothesis.'

into the suit and cemented to his lips with putty, and the ambient temperature was raised to a tropical level. The amount of oxygen that Séguin consumed in this state, on an empty stomach and while immobile, was measured. This is what is now known as the 'basal metabolism rate'. The experiment was then repeated after the persevering Séguin had taken a meal or exercised, and also when the surroundings were made more temperate. The oxygen consumption rose in all cases – by as much as threefold during strenuous exercise (lifting weights). The subject's heart rate was also recorded, and rose in parallel with increasing oxygen consumption. The essence of Lavoisier's observations can still be found in today's physiology texts.

But what of the water expelled from the body, besides the carbon dioxide? To find out, the stolid Séguin was subjected to further ordeals. Having been weighed in the nude he donned his rubberized suit and was weighed again on Lavoisier's precision balance. At the end of the day's work, Séguin, complete with casing, was weighed. He was then extracted from his suit and weighed again in the nude. The weighings gave the amount of sweat that he had generated and was retained in the suit, and the total loss of body water during the experiment. The outcome was that a large proportion of this water had been exhaled from his lungs as part of the respiratory process, and implied that the heat consumed in its vaporization accounted for a certain part of the total heat generated by respiration.

At this point, in 1791, Lavoisier's official functions began to overwhelm him, and Séguin was left to his own devices. His enthusiasm was beyond question but his analytical powers were limited. He weighed the food that he ingested, and everything that emerged after lying motionless for many hours, or after exercise. He immersed himself (and members of his family) for hours in tubs of water or oil, in the hope of discriminating between sweat and a hypothetical 'invisible perspiration'. This continued for a year, but little came of it.

Lavoisier on nutrition

Lavoisier felt that what had been discovered about respiration was all very well, but the bulk of physiology, which he equated mainly with 'animalization' – the conversion of food into the tissues of the body – was still a closed book. He was beginning to formulate some ideas on how it might be opened. The method by which in the 18th century the *Académie des Sciences* sought to promote research in an area that it deemed ready for an assault was much the same then as now. But where today grants are handed out on a competitive basis for proposals (those judged to have the greatest merit), the *Académie* would announce that a prize was to be awarded for the best research on the given topic completed within a specified time. And so in 1792 Lavoisier asked the academicians to vote a prize of 5000 livres – a considerable sum – for the best finished work on animalization submitted in 1794.

Lavoisier, though, did more: he outlined in some detail the state of knowledge, and defined the questions that should be addressed. It was known, he stated, that successive steps in the digestive process were effected by the saliva, by the acidic stomach juice, by bile, and by pancreatic juice. When acted upon by these agents the ingested food was transformed into chyle (p. 23). A part of this went into replenishing the blood, which was continuously depleted during respiration (in which he was wrong), and the rest was rejected as faecal waste. The balance of what entered and what came out needed to be definitively established. But the central question, Lavoisier felt, was the function of the liver, the seat, he was in no doubt, of many vital processes. This should be studied in its normal state and in diseases. The properties of the bile must be examined, and also the anatomical and chemical nature of the liver itself and of the gall bladder. Differences between the composition of the blood in the portal vein of the liver and venous and arterial blood elsewhere must be sought, and in general the methods of chemical analysis should be applied to the constituents of blood, liver, and bile. Lavoisier showed again that in his thinking he was a century or more ahead of his contemporaries. It was at the chemistry of nutrition that he was clearly now starting to direct his energies, but the sands had run out.

At about the time of Lavoisier's execution Claude-Louis Berthollet independently developed a coherent, but over-simple, metabolic scheme, whereby the carbon content of food is attenuated during digestion, a part being lost as carbon dioxide, with some accompanying loss of hydrogen. Nitrogen from food and from the air is incorporated, in combination with hydrogen, into the tissues. When present in excess, the nitrogen is disposed of as urine (of which the main dissolved constituent is the nitrogenous urea). Excess hydrogen, on the other hand, combines with other elements to form fat, while phosphorus is oxidized to the substance of bones. (Berthollet, it might be mentioned, was a scientist of great versatility, who made many contributions to fundamental chemistry and to technologies, including dyeing and iron smelting; he also invented bleach.)

Berthollet's supposition that nitrogen could be assimilated from the air was shared by another contemporary of his and Lavoisier's, Antoine-François Fourcroy, who was troubled by the observation that the bodies of carnivorous and herbivorous animals were equally rich in nitrogen, while the nitrogen content of vegetation is much lower than that of meat. He later persuaded himself that plant nitrogen combined with carbon and hydrogen from the blood to yield the required nitrogenous material, a view that was believed for some time. Fourcroy (1755–1809) was famous for his studies on adipocere, or grave-wax, exuded by decomposing corpses in damp soil. He and another chemist collected this substance in an overcrowded Paris cemetery, and showed that it was fatty in nature and resembled soap.

Fourcroy was an ardent revolutionary and a member of the Committee of Public Safety during the Terror. He is thought to have played an ignoble part in the condemnation of his friend, Lavoisier, and was accused of cowardice, or worse, opportunism. Neither he nor other close scientific associates, such as Berthollet, came to Lavoisier's defence. A notable irony was that after the excesses of the Revolution had subsided and normalcy returned, it was Fourcroy who spoke the eulogy at a memorial for Lavoisier, during which he sought obliquely to exculpate himself.

Count Rumford and his famous soup

Count Rumford of the Holy Roman Empire, *né* Benjamin Thompson, had many claims to fame, for his work in physics, in nutrition, and in politics, and for his unsurpassed cheek. He also invented thermal underwear, the drip coffee-maker, and many other aids to the good life. Franklin Delano Roosevelt thought that 'Thomas Jefferson, Benjamin Franklin and Count Rumford are the greatest minds that America has produced.' In the world beyond the confines of the physics laboratory Rumford's name is remembered chiefly for his ideas on nutrition for the masses, and in some parts of Europe it is commemorated still by the name of the soup that he invented. His interest in the subject was ignited, relatively late in his remarkable career, when he was in the service of the Elector of Bavaria, charged with reorganizing the Elector's army. Rumford made an exhaustive study of the economics of running such a large organization, and of how to clothe and feed the men satisfactorily with minimum expenditure. He went on to apply the same principles to the poor in the workhouses. He alluded to the importance of '*the investigation of the science of nutrition*; – a subject so curious in itself, and so highly interesting to mankind, that it seems truly astonishing it should have been so long neglected'.

The basis of Rumford's nutritional theory was a false premise. He accepted the principle, first enunciated by van Helmont (p. 49), that water was the primary food of plants, and that its constituent elements, hydrogen and oxygen, were converted into the substance of the plant. He set out his nutritional scheme in a long essay, entitled *Of Food and Particularly of Feeding the Poor*. Water, he asserts, 'serves not merely as the *vehicle* of nourishment, but constitutes at least one part, and probably an essential part, of the *Food* of plants'. The plant decomposes the water, while 'manures serve rather to prepare the water for decomposition, than to form of themselves – substantially and directly – the nourishment of the vegetables'. Why, he asks, should the same not be true of animals, with solid foods playing the part of manure? The logical inference is that soups should make the most nutritious of meals. Rumford accordingly tested such a diet on his soldiers, and concluded '... what surprised me not a little, was the discovery of the very small quantity of *solid Food*, which, when properly prepared, will suffice to satisfy hunger, and support life and health', and he noted 'the very trifling expence at which the stoutest, and most laborious man may, in any country, be fed'.[5] He did not, at the same time, neglect to consider how this cheap food should taste. 'The enjoyments which fall to the lot of the bulk of mankind are not so numerous as to render an attempt to increase them superfluous.' After five years of experimentation, he wrote, he had arrived at a formula, according to which 'the *cheapest*, most *savoury*, and most *nourishing* Food that could be provided, was a soup composed of *pearl barley, pease, potatoes, cutting of fine wheaten bread*, vinegar – salt and water in certain proportions'.[5] He goes on to describe in minute detail how the soup should be prepared, in what manner of vessel, and how and in what quantity served. For full satisfaction it was important to eat slowly, and the purpose of the bread, which was baked until hard and added last to the soup, was to encourage mastication, which had the additional virtue of promoting digestion. Some scraps of meat were also in practice often added. The soup was evidently a huge success, and its inventor received the gratitude of the poor who consumed it.

Rumford also believed in moderation, not only for reasons of thrift, but also of health. Even the rich could learn a thing or two, so 'If a glutton can be made to gormandize two hours upon two ounces of meat, it is certainly much better for him, than to give himself an indigestion by eating two pounds in the same time.'[5] Then the article leads into a breakdown, to the last farthing, of the cost of provisioning the army and the poor. For 1200 people the cost of a portion of the soup turns out to be 'a mere trifle more than *one third of a penny*, or exactly 422/1200 of a penny'. Rumford then expatiates on other nutritious dishes, including several kinds of 'dumplins' and breads, as well as variants on his soup and 'brown soup', made with rye bread. He finds it absurd that wholesome, nutritious, and economical foods so often fail to cross national borders, except as luxuries. He complains:

> I never see *maccaroni* in England, or *polenta* in Germany, upon the tables of the rich, without lamenting that cheap and wholesome luxuries should be monopolized by those who stand least in need of them; while the Poor, who, one would think, ought to be considered as having almost an *exclusive* right to them, (as they were both invented by the Poor of a neighbouring nation) are kept in perfect ignorance of them.[5]

Some part of these observations arguably holds true today.

Rumford launches into a long dissertation on the value of maize, or 'Indian corn', which he holds to be more nutritious and more palatable than rice or wheat, as well as being an undemanding crop of high yield. Where both rice and maize are on offer, he tells his readers, 'the negroes ... have invariably preferred the latter.' For 'they say that "*Rice turns to water in their bellies,* and runs off;" – but "Indian Corn *stays with them, and makes them strong to work.*"'[5] There follows a long series of recipes, especially for 'hasty-pudding', which is merely polenta paste, improved with molasses and other additions, and also for pasta, based on macaroni. To conclude, Rumford turns his attention to potatoes, which in his opinion should be boiled, and might then serve as the basis for a variety of good dishes. Potatoes were viewed with the greatest suspicion by the European peasantry (p. 46), and it was probably because Rumford surreptitiously introduced them into his soup that they became a staple in Germany and elsewhere.

The tone of Rumford's essays is genial and humane, and displays none of the abrasive persona for which he was notorious. He seems to have a genuine concern for the welfare of the poor. His dietary formulae were thought out with his habitual thoroughness, but whether they impinged, for good or ill, on the vitamin deficiencies that afflicted the European peasant and urban poor in times of dearth, is by no means clear. Scholarly opinion is divided, and it has certainly been argued that they may have aggravated vitamin malnutrition for a century after their promulgation. 'Rumford soup', at all events, is still to be found in some cookbooks, and the cooking stoves that he devised were the progenitors of the modern oven.

Rogue, cad, and savant[6]

Rumford's tempestuous career, and his impact on the course of science, are both so extraordinary that he cannot be allowed to vanish, unremarked, from the narrative. Here in brief is the story of his life. Benjamin Thompson was born on a farm in Woburn, Massachusetts in 1753. He showed, when barely into his teens, a remarkable curiosity about the physical world, and addressed to his schoolmaster written questions such as this: 'Please to Give the Direction of the Rays of Light from a Luminous Body, to an Opake, and the Reflection frome the Opake Body to Another Equally Dence & Opake vizt the Direction of the rays of the Luminous Body to that of ye opake, and the Direction to the other opake Body.'[7] The boy served as an unruly apprentice to several tradesmen, and was finally dismissed for performing a noisy experiment with gunpowder. He thereupon escaped from the shopkeepers into an apprenticeship with a doctor in Boston. There he and a friend formed themselves into a scientific society of two, read voluminously, experimented fearlessly, and attended some lectures at Harvard University. Benjamin seems to have been sent packing after one experiment too many (on a pig), and next surfaced in Concord, New Hampshire as a self-taught and unqualified schoolmaster. Only a few months after his arrival, at the age of 19, he married the recently widowed Sarah Rolfe. She was 11 years his senior, and her late husband had been the richest land-owner in the region. When accused later of shameless knavery, he replied with characteristic insouciance, 'She married me, not I her.' Benjamin Thompson was now well provided for, and quickly showed a remarkable talent for ingratiating himself with influential people. As a member of the local *haut monde*, he became a favourite of the Governor of New Hampshire. He seems to have accomplished this by manifesting an interest in the Governor's passion for new methods of farming. Young Thompson was soon rewarded for his attentions with a commission in the rank of major in the New Hampshire militia. The Governor was a royalist, and with rebellion now in the air he wanted Thompson to spy on the disaffected locals and report on their intentions. Thompson's responsibilities for his own and the Governor's estates brought him into contact with many of the district's tenant farmers and artisans. But his activities aroused suspicion and, having surmounted a number of crises, he got wind of a plan by a band of revolutionaries to march on his house, tar and feather him, and drive him out of the town. He quickly mounted his horse and rode into the night, abandoning his wife, baby, and father-in-law, never to return.

Thompson reached Boston and offered his services to the British army commander there. Whatever was agreed, he made his way back to his home town of Woburn and resumed the life of a country gentleman, while continuing his activities as a spy, now in the service of the British. The first shots in the War of Independence were fired in nearby Lexington, and Thompson sent despatches with his observations of local activity to the British command by innocent-seeming letters, on which he wrote with invisible ink. This was an extract of nut gall, which could be developed by soaking in a solution of iron (ferrous) sulphate. Thompson was never found out, and although he came under suspicion of treachery nothing could be proved against him. Soon, however, he defected openly to the British in Boston, and when they were driven out by Washington's militias he sailed with the remnant of the force to Halifax, and thence to London. There he presented himself to the Colonial Office, and with customary adroitness found a new patron in the Secretary of State at that office, Lord George Germain. In time he attained an official position in the Government as an Under Secretary of State. His duties were not so arduous as to prevent him from turning his mind again towards science, and he embarked on experiments designed to test the force generated by exploding gunpowder. He measured the recoil of rifles and cannon, and his publications on the subject secured him election to the Royal Society. They also served to bring him to the notice of the Royal Navy, and he was invited on a patrol, during which he made observations on the distances that cannonballs could carry. He also devised innovations to the signalling methods of the fleet, and to shipbuilding.

Thompson left England abruptly when, after the capture of a French spy, he came under suspicion of complicity in the affair, although, yet again, nothing was ever proved against him. He returned to America, charged with raising a new regiment for the British, and acquired an evil reputation for brutality against their adversaries. The War at an end, he surfaced once more in London, but tarried only briefly before setting out for the Continent of Europe as a military mercenary. By charm and persuasion Thompson secured an introduction to the Elector of Bavaria, who was delighted to accept his services as military adviser. Not content with this modest office, Thompson returned briefly to London to persuade the King that it would be an embarrassment to the Crown if his exalted position at the Bavarian court were not reflected by an English title. Back in Munich as Colonel Sir Benjamin Thompson, he performed various tasks for the Elector, while spying again for the British. He reorganized the Bavarian army and did experiments on heat insulation by different kinds of uniform fabric with the aid of his newly designed thermometer, which he also used for the study of heat transfer in liquids. This led him to the important discovery of convection currents (local flow provoked by temperature disturbances).

Next Thompson emerged as a social reformer, reorganizing arrangements for the welfare of the poor, and setting up workhouses and schools. The multitudinous beggars of the city were rounded up from the streets and given employment in the workhouse, making military uniforms. Thompson laid out for the Elector's pleasure, but also for the citizenry of Munich, a gigantic 'English garden', which contained a farm and agricultural institutions and became one of the wonders of Europe. At the same time he developed an interest in efficient lighting, improved domestic and institutional lamps, and devised an instrument for the measurement of light intensities, the Rumford photometer, forms of which are still in use. When Thompson's social and technological innovations came into effect the Elector began to recognize the man's prodigious qualities, and honours showered down on him. By 1791 he was a Major-General and Minister for War, Minister of Police, and Court Chamberlain. As Thompson's powers widened and his arrogance and contempt for lesser men swelled, the jealousy of the many enemies he had made intensified. But the Elector protected him, and on becoming interim Regent of the Holy Roman Empire, which survived more as a concept than as a political power, conferred on him the highest honour: Benjamin Thompson was reincarnated as Count Rumford (the original name of Concord in New England), Count of the Holy Roman Empire. It was soon thereafter that Thompson–Rumford, in his capacity as Inspector-General of Artillery in the Bavarian army, began his most famous work. He had decided that the caloric theory of heat was inadequate. 'Caloric', the supposed essence of heat, was a weightless fluid contained in greater or lesser amount in all materials, and emitted when the material was burned. Rumford had noticed that when cannon barrels were bored (or when the cannon were repeatedly fired) they became hotter and hotter and went on producing heat, rather than eventually running out of caloric as the theory posited. By a long sequence of brilliant experimentation and reasoning Rumford not only disposed of caloric, but also laid the foundations of the modern theory of heat.

In 1794 Rumford visited London and was hailed as a celebrity and hero. He gave lectures and demonstrations to the natural philosophers of Britain, and invented the 'Rumford stove', an improved version of the English fireplace, the inefficiency of which had outraged him. Then, in 1796, he received an urgent summons from the Elector to return to Munich, for war had broken out between the French and the Austrians, and their two armies were converging on Bavaria, each apparently intent on seizing Munich. Rumford was appointed commander of the small and under-prepared army there, and the Elector and Government decamped to safer parts. Rumford's enemies waited for him to drain the poisoned chalice, but they had underestimated their man. He withdrew his army behind the city walls and negotiated alternately with the French and Austrian commanders, until in due course the French turned away to deal with a military setback in the north, and the Austrians withdrew. Rumford was now a war hero, even though there had been no engagement. It remained for him to perform one more service for the army: he invented a portable field oven, which he also adapted as a soup kitchen to feed the war-weary population of the city. But the dazzling splendour of his achievement and his overweening arrogance were making Rumford's presence insupportable to the governing class, and the Elector felt it prudent to get him out of the country. Thus it was that Rumford was appointed Minister Plenipotentiary to the Court of St James – much to the displeasure of George III, who saw him merely as a spy for the enemy and a traitor.

Rumford evidently had not considered that there might be a certain animus against him in official circles in London, but it was enough to make his position at court intolerable. He began to look around for other outlets for his unquenched ambitions. He offered his services to the country of his birth: he would establish for the United States a military academy. The suggestion was transmitted to the Secretary for War, who endorsed and forwarded it to President Adams. The President reacted with enthusiasm and Rumford duly received his letter of appointment. But then someone evidently remembered Benjamin Thompson's activities as a spy for the British, and he was advised that it might be prudent to decline. This, nevertheless, was the origin of the West Point Academy. Rumford seems not to have repined, but busied himself with a new project – the foundation of an institution in London for the dissemination and furtherance of scientific knowledge through lectures, demonstrations, and research. His reputation outside government was still high, and he received support and funds for the purpose. A building was procured in Albemarle Street off Piccadilly and the Royal Institution was born. It thrives there still and its public lectures are famous. The first Professor of Natural Philosophy and Chemistry was appointed, but his series of lectures was not adjudged a success. He was dismissed and replaced by the 22-year-old Humphry Davy from Bristol (p. 26), whose greatest work was performed in Albemarle Street. (He was eventually succeeded there by his assistant, the great Michael Faraday. Faraday's laboratory is kept as a shrine and museum at the Royal Institution, and can be visited.)

But Rumford's overbearing and offensive manner had again alienated too many people, and it was time once more to move on. He headed as before for Munich at the invitation of the new Elector, but made a leisurely stop in Paris, where he charmed Marie-Anne, the widow of Lavoisier. She joined him on a European tour, and even on what was in effect a honeymoon, Rumford, presumably with her assistance, conducted experiments in the Alps on the freezing and density of water. After a stay in Munich, he returned to Paris and (his first wife having meanwhile died) married Marie-Anne, henceforth Mme Lavoisier de Rumford. With her money he bought and converted a mansion near the Champs Elysées to house a laboratory. One of his inventions there was an improved table lamp, which not only involved him in an acrimonious lawsuit but also cast so bright a light on the countenance of his spouse that, as he recorded in a letter, the effects of age on her once-handsome features became all too apparent. This, added to other sources of friction – for the strength of Marie-Anne's personality matched that of her husband – led to tempestuous altercations and before long a separation. Rumford withdrew to a new house in Auteuil and began a further series of scientific and technological investigations. Among the most famous was a steam-heated radiator and the drip coffee-pot, much as still used today. Rumford died unexpectedly in 1814 at the age of 61, and was buried in Auteuil. He left the bulk of his estate to Harvard College, which retains responsibility for tending his grave. His eulogy was pronounced by Baron Georges Cuvier, the great biologist and secretary of the *Académie des Sciences*, who referred with some candour to the defects of Rumford's personality, but also of course to the splendour of his scientific accomplishments and his work for the betterment of mankind. 'It was without loving or esteeming his fellow-creatures' Cuvier concluded, 'that he has done them all these services.' Rumford assuredly, at all events, left his mark on the social history of Europe, as well as on the course of scientific progress.

5

The Savants' Disputes

The stomach laid bare

It was probably the illustrious savant with the mellifluous name of René-Antoine Ferchault de Réaumur who made the first careful attempt to see, rather than merely imagine, what goes on in a stomach. Réaumur's activities were numerous. He was born in 1683 in La Rochelle, and studied law before his interest in science prevailed. His first notable work concerned the manufacture of steel. He showed that it contains carbon, and developed a procedure for manufacture of steel from cast iron, thereby initiating the steel industry in France. He interested himself in other practical crafts, including production of glass and porcelain. He devised the alcohol thermometer and invented the temperature scale, long favoured in France, which bears his name. (The freezing-point of water was 0° and its boiling-point 90°R.) But it was for his biological work that Réaumur is most celebrated. He published, in 12 volumes, a natural history of insects, and he gave much thought to the genetics of hybrid plants, in some measure anticipating Mendel's epoch-making researches on sweet peas.

To study digestion, Réaumur chose birds as his experimental subjects. He pushed a metal tube containing fragments of food down the bird's gullet and into its stomach, and at intervals examined the contents. The food progressively dissolved with no sign of putrefaction. This proved that the mechanism of digestion was not mechanical, as the mechanists had insisted (Chapter 4), for the food, isolated by the metal from the stomach walls could not have been processed by 'trituration', or grinding. Thus the transformation must have been engendered by chemical means. Réaumur then mixed pieces of meat with some of the bird's gastric juice, which he retrieved by squeezing out a sponge lowered into its stomach, but could not demonstrate any digestive changes. From this failed experiment he (incorrectly) inferred that some special quality of the stomach was also indispensable.

The contradiction was resolved by another of the cast of remarkable characters who populate this history, the Abbé Lazzaro Spallanzani (1729–99), cleric and professor of natural history at the University of Pavia. He appears as a malign magician in the work of the German writer and poet E. T. A. Hoffmann, but he was in reality an affable man of great personal charm. Spallanzani was born in Modena and attended the University of Bologna, where he fell under the influence of his cousin Laura Bassi, one of several remarkable women scholars of the period, professor of natural philosophy and mathematics. It was she who deflected him from the study of literature to that of science. His precocious accomplishments brought him to the attention of the Austrian Empress, Maria Theresa, who pressed him to accept a position at the University of Pavia, a city that was then a part of her realm.

Spallanzani's most enduring achievement was to put paid to the theory of spontaneous generation of life, a matter at the time of both scientific and theological debate. He had been opposed in this by two eminent naturalists, the great Georges-Louis Leclerc, comte de Buffon and the English cleric, John Needham, who – unusually, for both were formidable intellectual bruisers – received him courteously and framed their reactions to his critique without acrimony.* Spallanzani also interested himself in physiology, and especially in the processing of food by the body. He investigated the digestive power of the gastric juices of many animals, and in particular repeated Réaumur's experiments with great care and under more controlled conditions. He chose in addition to conduct studies on his own falcons, on the grounds that birds of prey, which cough up pellets of rejected, unassimilable parts of their meals such as bones, feathers, or fur, also regurgitate partly digested food to nurture their young. This material he was able to examine. He sought, like Réaumur, to find evidence of digestion in the extracted gastric juice of animal stomachs. To match the temperature at which the stomach operates Spallanzani kept tubes containing minced meat, bathed in gastric juice, in his armpits for periods of up to three days, and, unlike Réaumur, his experiments were successful: digestion proceeded in the tubes just as in the stomach. Why Réaumur's efforts in the same direction failed was never explained, although the lower temperature at which he made his observations would certainly have slowed the digestive process. The outcome of Spallanzani's investigations brought him into conflict with the combative anatomist and surgeon, John Hunter, in London. Hunter disagreed passionately with the view that digestion was a chemical and not a mechanical process. He had a weak heart, and became convinced that his life hung by a thread. 'My life', he declaimed, 'is in the hands of any rascal who chooses to annoy or tease me.' But the choleric Hunter was clearly won over by the evidence and, mollified by Spallanzani's good nature, became his ardent supporter. He made an important contribution of his own, moreover, when he showed that mysterious stomach lesions, discovered in post mortem examinations, arose from the degradation of the decaying stomach tissues by the gastric juice.

* Despite all the evidence, the theory would not lie down, and the argument kept re-emerging at intervals, even into the 20th century.

Spallanzani's final contribution to the field was his work on human digestion. He enclosed pieces of meat in a small perforated tube and swallowed it. When he retrieved it from his stools he found the meat largely digested. His attempts to follow the time-course of the digestive process foundered, for he could not bring himself to induce vomiting at regular intervals. It was nevertheless a major achievement to have established beyond all doubt the chemical nature of digestion and to have laid to rest the teachings of the mechanists. Not long after this, England's leading surgeon, Sir Astley Cooper, a pupil of John Hunter, added a postscript when he recovered partly digested meat from the stomachs of dogs and found that the rates of digestion varied from one kind of meat to another.

The fist-sized fistula

On 6 June 1822, in a fur-trading post on Mackinac Island in Lake Huron, a French-Canadian 'voyageur', Alexis Saint-Martin, received a volley of bird-shot from an accidentally discharged fowling-piece full in the midriff. A voyageur was a transporter of furs, who conveyed his cargo by canoe from the trappers in the far north to the depot of the trading company. Saint-Martin was then a sturdy young man of 28. The shot blew a hole in his stomach and diaphragm, smashed two ribs, and tore the surface of his left lung. His shirt caught fire, and wadding, powder, and lead shot entered his stomach. By good fortune a doctor was on hand, for the island housed a US army post, complete with a small and very primitive hospital. The incumbent was Dr William Beaumont. He had acquired his training as an apprentice to two practitioners in Vermont (the usual route at that time to qualifying as a physician or surgeon). In 1812 he was granted a licence to practise as both, and that same year enlisted in the army as a regimental surgeon's mate. He tended the wounded in the war against the British, and was reputed to be a skilled and resourceful doctor. When peace returned Beaumont left the army and set up in private practice, but in 1819 he re-enlisted and was rewarded with the bleak posting of Fort Mackinac.

Beaumont arrived at the scene of the shooting within a few minutes. He cleaned the wound, which he described as being the size of a man's palm, removing from it wadding, lead shot, and shreds of clothing. Parts of the burnt and lacerated lung and stomach protruded from the cavity, and 'breakfast food was escaping' from the hole in the stomach. It appeared, Beaumont later wrote, 'an appalling and hopeless case'. But to general astonishment Saint-Martin survived, tended by Beaumont. For 17 days all that was given to him to eat emerged from the wound, and he was nourished with enemas. After another week had passed he was retaining and apparently digesting his food and was well on the road to recovery. The wound was healing, and the edges of the hole in the stomach were attaching themselves to the surrounding tissue, thereby isolating the hole from the abdominal cavity. This was fortunate, for, had the stomach contents leaked into the abdomen, Saint-Martin would have died of peritonitis. The hole was now 'about the size of a shilling piece', and to prevent the stomach contents from leaking out Beaumont inserted a kind of stopper, secured by a bandage. The hole – a gastric fistula – remained in Saint-Martin's side for the rest of his long life, but it was many months before healing was complete and Beaumont was no longer having to remove slivers of bone and fragments of cloth or wadding from the abscesses that kept forming.

Saint-Martin was still too weak to work, and the local county authorities had no wish to take responsibility for him, or perhaps the means, and so ordered that he should be returned to his home, 1500 miles to the north. Beaumont, who was evidently maintaining a proprietorial interest in his patient, took him into his own home and eventually engaged him as a servant. He treated Saint-Martin generously, but whether his concern for his patient was totally disinterested is not altogether clear. At all events, some three years after the accident, it seems to have occurred to Beaumont that the gastric fistula afforded a remarkable opportunity to advance medical science. And indeed Beaumont was the first man to be offered the privilege of observing human digestion in what is now called real-time. He followed the events that went on in Saint-Martin's stomach after meals, and then he tied pieces of food to a silk thread and inserted them into the fistula. These comprised beef and pork, fat and lean, raw and cooked, and even 'high seasoned alamode beef', not to mention bread and uncooked cabbage. At intervals Beaumont would haul out the thread and examine the eroded morsels. These ordeals caused Saint-Martin severe indigestion, at which point Beaumont would end the experiment.

Beaumont undertook a series of further investigations. He measured the stomach temperature, examined the state of the gastric juice after a prolonged fast, removed the juice, and soaked pieces of meat in it at stomach temperature. He confirmed the results of the early experiments with gastric juice of animals, showing that human gastric juice had the capacity to dissolve food. After a year of this kind of treatment, Saint-Martin mutinied. Eventually he absconded, returned to Canada, and got married. Beaumont, who by this time had acquired considerable fame on the medical scene through his publications on the gastric fistula, was beside himself. He tried to trace his errant patient, who had returned to his occupation in the fur trade, and after four years he succeeded. Saint-Martin was persuaded to return, accompanied by his family, to Beaumont's service, this time at an army post in Wisconsin. Beaumont took what had become his walking stomach on lecture demonstrations, and conducted many more experiments, which enabled him to compile data on the digestibility of a wide range of food-stuffs. But, nagged by his homesick wife, Saint-Martin once more levanted, and was again coaxed back, and this time Beaumont was able to secure for his mobile orifice a sinecure, as a sergeant in the US army. Beaumont next applied to Congress for a research grant to conduct further studies, and planned a programme of demonstrations of the digestive dramas that daily unfolded in Saint-Martin's stomach. It was too late: Saint-Martin could take no more, and made his third and terminal exit to Canada. With his book, *Experiments and Observations on the Gastric Juice and the Physiology of Digestion*, Beaumont made sure of his place in the medical pantheon.

Saint-Martin, by then an ageing alcoholic, resisted all attempts to lure him back, and also refused what would have been a lucrative trip to London to display himself to the Medical Society there. He fell briefly into the clutches of an infamous quack, one T. G. Bunting, who showed him off (or in the vernacular of the day, 'Barnumized' him) as a medical freak, and later a few more experiments were done on his gastric juice in Philadelphia. Alexis Saint-Martin died destitute in Canada in 1880 at the age of 86. His family allowed the corpse to decompose for some days before the interment, and also had the grave dug to a depth of eight feet to ensure that the departed would not be the subject of any further, posthumous experimentation. William Beaumont for his part, having left the army in 1839, set himself up in medical practice in St Louis, and died after an accident in 1853 in his 68th year.

Some 60 years later, a physiologist at the University of Chicago, Anton Julius Carlson, chanced upon 'a second Alexis Saint-Martin'. He was 'fortunate to have in his service', he wrote, Fred Vlacek, a 27-year-old immigrant from the Austro-Hungarian province of Bohemia, possessed of a gastric fistula 'large enough to permit direct inspection of the stomach, and the introduction of balloons, rubber tubes, and small electric lights for various investigations'. For some years, starting in 1912, Carlson paid his human guinea-pig a retainer and performed a series of experiments, mainly, it seems, observing the motion of the stomach. Carlson was evidently a compassionate man, who would not subject Vlacek to the unpleasantness (to which he himself submitted) of several days' starvation, leading to stomach contractions. These could have been viewed through the fistula in the electrically illuminated cavity beyond. It is not clear that results of lasting value accrued from Carlson's investigations.

The march of science

The first rigorous experimental inquiry into the types of substances inseparable from life was probably that of the French physiologist François Magendie (1783–1855). His achievements were formidable and varied. They extended over many branches of animal physiology, notably the study of the nervous system, and his name is venerated also among psychiatrists. Magendie's nutritional experiments were motivated by the food shortages that threatened much of French society towards the end of the Napoleonic Wars. He was appointed in 1815 to chair the Gelatin Commission, made up of distinguished savants charged with finding out whether gelatin – cheaply and abundantly available from bones and skin of slaughtered animals – could serve as a food. The premise, enunciated as an axiom in its report, was that 'everyone who is familiar with broth knows that its nutritious properties are due primarily, if not entirely, to gelatin'. But Magendie soon found otherwise. He fed dogs on a diet devoid of proteins – on sugar, olive oil, and water. The dogs died of what was evidently starvation. When gelatin was added to the diet they fared little better, whereas other proteins (the so-called 'albuminous' substances, defined during the previous century by Fourcroy (p. 61) in terms of their physical characteristics) allowed the animals to survive. This proved that proteins were needed to sustain life, although not all proteins were equal in this regard. Proteins, as was already clear, differ from the other constituents of food recognized at the time – the carbohydrates and fats – in containing nitrogen. Magendie also found, however, that proteins alone did not suffice. Egg whites his dogs rejected, even when almost mad with hunger, but they ate, reluctantly, blood fibrin, which also failed to keep them alive. The dogs did remain healthy on gelatin supplemented with bones, and the report of a second commission (the Magendie Commission of 1841) concluded that blood and marrow fat possessed elements essential for life. (The basis for this will become apparent later.)

As the 19th century progressed, more and more chemists discovered an interest in physiology, and physiologists recognized the need for chemistry. Out of the encounter was born the new science of biochemistry, or as it was then called, physiological chemistry. In Germany, especially, a certain tension arose between the practitioners of physiological and of pure organic chemistry (the chemistry of carbon compounds). To the German organic chemists, who led the world in their discipline, physiological chemistry was *Schmierchemie*, a derisive expression, implying the study of impure and insalubrious mushes.

The leading German physiologist of his time foreshadowed the shift towards chemistry. Johannes Peter Müller came of humble stock. He was born in Coblenz in 1801, the son of a cobbler, studied medicine, and ascended in 1830 to the chair of physiology at the University of Bonn, rising three years later to the pinnacle of his profession, as Professor of Anatomy and Physiology in Berlin. It was there that he did his best work, and wrote the textbook that was to dominate the field in the decades to come, his *Handbuch der Physiologie des Menschens*. There were few aspects of human biology with which he did not engage, but his study of nerve function brought him the greatest fame. He was consequently also held in high esteem by the early psychiatrists. Müller, it has to be said, was a vitalist, who believed that living organisms were governed by an *élan vital*, or *Lebenskraft*, that was not derived from material, animal, or plant sources.* Müller rejected the

* Vitalism was the belief that reactions in the living body required the participation of a 'life force' or *élan vital*, on the lines of Galen's *pneuma*, and could never be reproduced in the laboratory. This theory should have been buried once and for all by an experiment performed in 1828 by Friedrich Wöhler, professor of chemistry in Göttingen. Wöhler heated an inorganic compound (so classified, even though it contained carbon), ammonium cyanate, and found that it was converted (without loss or gain of weight, and thus of any other atoms) to urea, the primary product of excretion in man and other mammals. Wöhler wrote to his former patron, Berzelius: 'I must tell you that I can prepare urea without requiring the kidney of an animal, either man or dog'. Vitalism did not altogether fade away. Some biologists, like Müller, cleaved to it, and even much later it reappeared from time to time as a somewhat mystical doctrine.

mechanical model of bodily processes and especially digestion, which, Réaumur and Spallanzani notwithstanding, still exerted an unaccountable influence on some physiologists. Nutrition, he declared in his Handbook, was not a subject for microscopic observation: it was, he implied, chemical in nature. Müller, who was a highly skilled anatomist, did use a microscope to search for holes in the blood vessels through which nutritious material from the blood might pass, and finding none, concluded that food must, by the time it enters the blood, have been converted into a soluble state. This, he conjectured, would allow the products to pass through submicroscopic pores to where they were needed. Soluble elements in the digest were particular to the various tissues, which they continuously replenished. Clearly this all occurred by way of chemical transformations, but these Müller could see no way to define.

It was Müller's venerated compatriot, Justus von Liebig, who first attempted to grapple with the chemistry of nutrition. Liebig was a giant, often wrong, but seldom in doubt, whose school of chemistry in the small town of Giessen sent into the world a whole generation of famous chemists, proud of their schooling as *Liebigschüler*. Liebig was the first of the academic generals, who commanded an army of research students and visiting workers. The aspirants were expected to work hard, and Liebig had discovered the ideal location to encourage that; 'We worked from daylight until nightfall', he wrote; 'distractions or diversions were not available in Giessen'. The results of his myrmidons' labours appeared mainly in the journal that Liebig himself had founded, the *Annalen der Chemie*, which became the leading outlet for chemical research, and still exists.

Liebig was a complex personality, vain and petulant, often embroiled in epic disputes over priority, yet thoughtful and often generous to his associates, if with a touch of cynicism. He made it a custom – unusual for the time – to allow his students sometimes to publish their work, carried out under his supervision, under their own names alone. But, he told a friend, 'if it is something good, a part of the credit is allocated to me, and I do not need to defend the mistakes. You understand?' For all that, it was Liebig's personal charm that his disciples remembered, and they remained for the most part deeply loyal to him.

He began life as plain Justus Liebig in Darmstadt in 1803. He was not born to the academic purple. His father was a dealer in paints, varnishes, polishes, and household chemicals, and prepared paints and other materials in his own workshop. Justus did not do well at school, and was written off as 'hopelessly useless' by his teacher, who called him a *Schafskopf*, or sheep-head. This apparently stuck to him as a nickname. But he was fascinated from his earliest years by the processes that he could watch in his father's workshop, and those that went on in the local dyestuff and soap factories. He was particularly excited by the chemical tricks that travelling entertainers performed in the city markets, most of all (or so he later wrote) by one who prepared and detonated the highly explosive mercury fulminate for the entertainment of passers-by. The starting materials were probably available in Justus's father's workshop, and the boy tried, if his own account can be relied on, to entertain his school friends with a similar performance, though evidently on a larger scale. The demonstration supposedly worked all too well and Justus was expelled. Whatever the truth of such stories, Justus certainly left school at 14 without gaining his *Abitur*, the school leaving certificate.

The boy was said also to have experimented with explosives at home, to injurious effect. His father evidently gave up all hope for his son's future schooling and sent him to the nearby town of Heppenheim as apprentice to an apothecary, but this arrangement, too, was abruptly terminated after a year. Another explosion may or may not have been the cause, and Liebig did later produce firecrackers, sold in his father's shop. Fulminates continued in fact to engage Liebig's interest, even at university. But very possibly the reason for the termination of his apprenticeship was that the indenture fees overtaxed Liebig *père*'s limited resources. At any rate, Justus spent the next two years working in his father's business, while devouring science books, which he was allowed to borrow from the court library of the Grand Duke of Hessen. He evidently made a sufficiently good impression on the local authorities, despite his meagre scholastic record, to qualify for a stipend that allowed him to enter the University of Bonn. It was 1820 and Liebig had just turned 17. Even before he graduated he was allowed to embark on research towards a doctorate, which he completed in two years, having moved with his research supervisor to the University of Erlangen. At university Liebig found time for political activities, joined a *Burschenschaft*, or student fraternity, and at one point served three days in jail after a confrontation with the police.

In these early years of the 19th century France was the centre of much of the most enterprising work in chemistry, and it was to Paris that Liebig, having secured further financial support from the Grand Duke, directed his steps. His principal patron there was the great chemist Joseph Louis Gay-Lussac, with whom he carried out further studies on fulminates. While in Paris, he also attracted the interest of the influential scientist and traveller, Alexander von Humboldt, who was immensely impressed by the young man's abilities. Humboldt recommended him to the Grand Duke of Hessen, and urged that an academic appointment should be found for him in the Duke's realm. The Duke complied: Liebig, still only 21, was appointed Extraordinary (that is assistant) Professor at the University of Giessen. (Today it is the Liebig-Universität Giessen.) Only a year later, in 1825, he was a full professor (Ordinarius). His impact on a staid institution was that of a whirlwind. He converted a disused barracks into a laboratory, devised a syllabus that was grounded not in the intellectual abstractions that had characterized the chemistry course in previous years but in a severely practical approach which emphasized the application of analytical methods, and allotted research projects to the students. Liebig's laboratory became famous and drew a stream of aspiring chemists from around the world, many of whom later founded their own schools, based on Liebig's methods. Their students in turn carried on the tradition, and (in the sense that in Germany a student's doctoral supervisor is his 'thesis father') Liebig stands now as the founder of perhaps the most effulgent scientific dynasty of all time.

Liebig's wars

Liebig and his swarm of acolytes made many important advances in his chosen fields of inquiry. He was widely regarded as the ultimate authority on matters relating to chemistry, but as time went on he became increasingly intolerant of dissenting opinions and would attack their proponents with intimidating acerbity. This led him into a series of prolonged, often undignified, vendettas. The fact, moreover, is that he was not infrequently wrong, and his refusal to contemplate such a possibility, allied to the respect that his utterances commanded amongst third parties, ensured that progress was sometimes held back. This was especially true of Liebig's work on animal nutrition and agriculture – practical matters to which he was always much drawn. In 1840 he presented his theories concerning the chemical cycles of growth and decay of living matter in a book with the title (in translation) *Organic Chemistry in its Application to Agriculture and Physiology*. Two years later a second opus appeared, this time called *Animal Chemistry*, the first attempt at an enveloping view of metabolic chemistry, which did away with the vitalism of earlier years. Liebig's scheme, as elaborated in the two books, was based on the rigorous methods of elementary analysis that he had developed in his laboratory, especially for the estimation of the main constituents of natural molecules, carbon, hydrogen, oxygen, and nitrogen. Liebig and his pupils had been analysing fats, which contain predominantly the first three of these elements, but the greater inspiration came from the work of a Dutch chemist, Gerrit Jan Mulder (1802–80).

Mulder had qualified and practised as a doctor, but as time went on devoted himself increasingly to chemistry, and especially to the analysis of foods. At the universities of Rotterdam and later Utrecht he taught tirelessly (at Utrecht it was botany, chemistry, mathematics, and pharmacy) and carried out meticulous elementary analyses on many natural products, such as teas and coffees. In time, his interest focused on what we now recognize as proteins, a name that, indeed, he coined from the Greek for 'foremost' or 'going before'. Fourcroy (p. 61) had in the previous century identified and prepared several similar nitrogen-rich substances, including albumin from blood and from egg white, fibrin from blood, and gluten from plants. Mulder analysed preparations of the same sort for the known elements in foods – once again carbon, oxygen, nitrogen, and hydrogen – and also for the additional elements, sulphur and phosphorus. The relevance of the last two was urged on him by the doyen of the era's chemists, the illustrious Swede, Jöns Jacob Berzelius, with whom Mulder conducted an animated correspondence. The most striking outcome of Mulder's labours was that all his proteins appeared to have the same composition (though of course within the rather generous limits of his experimental precision), and, based on his analyses, he allotted them the formula, $C_{40}H_{62}N_{100}O_{120}S_{1(\text{or } 2)}P_1$ (C, H, N, O, S, and P being the modern symbols for carbon, hydrogen, nitrogen, oxygen, sulphur, and phosphorus – which indeed complete the list of elements present in proteins). The atoms of sulphur and phosphorus, Mulder thought, were somehow attached to the bulky *Grundstoff*, or core substance, of the protein, and could be extracted from it. His faith in his analyses was considerable, for he might well have been content to settle for the formula, say $C_2H_3N_5O_6$, which would have preserved closely enough the ratios of the atoms he had determined (with the assumption that the sulphur and phosphorus were merely impurities). But Mulder would have none of this, and his rejection of such a simple interpretation led to his first remarkable conclusion: that the proteins must be very large molecules; for the number of atoms in the formula adds up to 323 or 324 – many more than were in the molecules with which chemists were accustomed to deal. (The atomic formula in fact gives the minimum possible size of the molecule, since the numbers of atoms of each element could be doubled or trebled or more without changing the ratios. Nearly all proteins are in actuality much larger than Mulder's formula, on the face of it, implies.) This, though, was not Mulder's concern; what impressed him was that all the proteins had (within the limits of his accuracy) the same formula. All the 'albuminous' molecules, whether from plant or animal, were, he concluded, one and the same. From that he inferred, reasonably but nevertheless wrongly, that animals received their nitrogenous substance ready-made from plants. His work received the ultimate accolade – praise from his teacher, Berzelius, in Stockholm.

Liebig also conducted a long correspondence with Berzelius, to whom he dedicated his *Animal Chemistry* treatise. He had improved on Mulder's analytical techniques, but did not dissent from the conclusion that animals derived proteins, chemically unchanged, from plants. They were, however, in some sense plastic and could be remodelled into the substances of blood or flesh. Sugars, which also came direct from plants, and fats played no part in the formation of the animal tissues, and served only as sources of energy through oxidation. Their carbon was exhaled as carbon dioxide, and some was stored in the liver as the constituent of bile, to be mobilized at times of high energy demand. The Inuit, who needed to generate much heat, lived mainly from fats, while the gauchos of the pampas, who were lean and muscular, required primarily the nitrogenous substance of plants or animals to replenish their muscles. Beef was adjudged the sovereign source of muscle protein. The material of the muscles, consumed by wear and tear, ended up as the nitrogenous waste material, urea. Thus the amount of urea in the urine reflected the amount of work that the muscles had performed. In 1840 Liebig approached his patron, the Grand Duke, for permission to experiment on the soldiers of his regiment of bodyguards. For a month Liebig recorded the intake of carbon by the 855 stalwarts and the amount that they excreted. He equated the difference with the weight of carbon, as carbon dioxide, exhaled, but this, unfortunately, he did not have the means to measure.

His assertions about energy balance lacked experimental support, and were hostages to fortune, but Liebig never lacked self-belief. Berzelius, the dedicatee, did not care for the book and its numerology at all and said as much in a letter to another of his correspondents, the German chemist Friedrich Wöhler. Wöhler was (and remained throughout his life) a close friend of Liebig's, and let slip, though in guarded terms, that Berzelius had not altogether approved of Liebig's work. This was enough to turn Liebig against Berzelius, whom he could never forgive for the criticism. He savagely denounced Berzelius and all his works in his house journal, the *Annalen*. By contrast, Liebig's epigones in England (chief of whom was an influential but scarcely distinguished chemist by the name of Lyon Playfair) would countenance no criticism of the master or of his book, which in a matter of weeks had been translated into English.

Liebig now turned also on Mulder, whose analyses he had at first warmly commended, and whose conclusion concerning the ingestion of the body's proteins in preformed state he had enthusiastically embraced. Liebig's students had performed elementary analyses on protein samples like Mulder's, almost certainly with greater accuracy, and they did not agree with his conclusion that all had the selfsame formula. Nor were they able to eliminate from the preparations the small amount of sulphur, which, Liebig concluded (like Mulder), must be an integral part of the molecule. He attacked Mulder with some venom in the *Annalen*. Mulder – himself an obstinate and abrasive man – was shocked and responded with something approaching hysteria. He demanded a retraction and apology. His attacker, he declared, was acting as prosecutor, witness, and judge. Liebig collected Mulder's letters and printed his objurgations in the *Annalen*, with his own rejoinders, under the heading, *Zur Charakteristik des Herrn Professor Mulder in Utrecht*. Such brawling did not sit well with many members of the academic community, and Liebig was perceived as arrogant and overbearing, as he undoubtedly was. Nor had it been forgotten that he had used immoderate language in attacking other scientists, notably the distinguished biologist, Theodor Schwann, pioneer of the cell theory of living organisms, which Liebig denounced as absurd.*

Mulder found an ally in another Dutch doctor, who, while in practice in Utrecht, was spending every moment of his spare time in Mulder's laboratory,

* Liebig and Wöhler had convinced themselves that fermentations were purely chemical processes. So when Schwann and a French engineer, Charles Cagniard de la Tour (chiefly famous as inventor of the siren) simultaneously announced that they had seen in the microscope yeast cells dividing as fermentation proceeded, their scorn knew no bounds. Liebig allowed an anonymous lampoon to appear in the *Annalen*, now known to have been penned by Wöhler, which purported to describe the formation of wine by animalculi; they took in the solution and the products of digestion were 'most definitely recognized by the subsequent expulsion of excrement. In short, these infusoria eat sugar, void wine alcohol from the intestinal canal, and carbonic acid from the bladder. In the full state the bladder takes the form of a champagne bottle.' Schwann was so mortified by this misplaced ridicule that he abandoned his studies on cells and did not return to them for several years.[1]

conducting analyses. This was Jacob Moleschott (1822–93). He had studied medicine in Heidelberg, and arrived at a materialist view of biological function, at odds with that of the German school. He respected Liebig, but was at one with Mulder on the identity between plant and animal proteins. Moleschott voiced his criticism of Liebig's theory-driven outlook courteously in a prize essay, when he was still only 22 years old, and sent a copy to Liebig, who uncharacteristically did not take it amiss and replied in amiable terms. Moleschott now offered to translate Mulder's polemic against Liebig (which bore the inflammatory title of 'Liebig's Question to Mulder Tested by Morality and Science') into German, but having read it he changed his mind; he simply could not stomach the tone. Mulder never spoke to him again.*

The stage was now set for Leibig's next and most celebrated quarrel. The adversary was the ranking French chemist of his time, Jean-Baptiste Dumas,[3] born in 1800, just three years before Liebig, in the small town of Alès in the Gard department. Like many a chemist of the period he had been apprenticed to a pharmacist, but it was not long before he headed for one of the founts of contemporary learning, Geneva. There he threw himself into the study of chemistry, botany, and physiology, and assisted the discoverer of iodine, Charles Candolle, to prepare the new element and its compounds for treatment of goitre. In 1823 Alexander von Humboldt, who the next year gave the young Liebig a helping hand, persuaded Dumas to take the path to Paris. There he was found employment as assistant to Louis-Jacques Thénard, Professor of Chemistry at the *École Polytechnique*, while also working in a dyestuff factory to supplement his income. In 1826 Dumas married into a distinguished scientific family: his bride was the sister of his friend the botanist Adolphe-Théodore Brogniart, and daughter of Alexandre Brogniart. Dumas's father-in-law was one of the founders of mineralogy and exceedingly well-connected. Son of a famous architect, designer of the Paris Bourse, he was now professor at the Museum of Natural History, director of the Sèvres porcelain factory, and a friend of many of the foremost figures in the French scientific and political establishment, such as Baron Georges Cuvier. The family connection secured Dumas an entrée into the most socially and professionally influential circles.

Dumas made his mark on several branches of science, but especially on chem-

* Moleschott was an altogether more emollient and good-natured man than Mulder, though strong in his socialist opinions, which led to his precipitate departure from Heidelberg, where he had received a lecturing appointment. From Heidelberg he moved to a professorial chair in Zurich, and thence in due course to Italy, for he embraced the principles of the *risorgimento*. He held chairs in Turin and Rome, took Italian citizenship, and became a senator. He did notable physiological work, but his concern was always with the nutritional needs of the population. He wrote three particularly influential books, in one of which he forthrightly tackled Liebig's dogma that all foods fall into one of two classes – the 'plastic' or nutritive, and the respiratory or 'combustible' – noting, for instance, that even in starvation an animal will continue producing new blood cells. And proteins (the 'plastic' food) are surely 'combustible', for do not muscles, which are made of protein, waste away during starvation?[2]

istry. He developed the theory of group functions in organic compounds, that is to say the characteristics imposed on compounds by ubiquitous groups of atoms; he determined the formulae of some of the commonest and most important compounds, he developed improved analytical methods, including (like Liebig) techniques for the measurement of nitrogen contents of organic compounds, and he corrected the grossly erroneous determinations of atomic weights by earlier workers. He also devoted himself to the analysis of biological materials – of blood, milk, oils, and fats. He became politically influential. In 1848 he was appointed Minister of Education, and the following year was elected Member of the National Assembly; soon thereafter he was President of the City Council of Paris and director of the mint. His apotheosis came in 1879, five years before his death, when he was elected one of the Immortals of the *Académie Française*.

The clash with Liebig followed Dumas's first publication on the chemistry of life, which summarized the outcome of several years of research. Much of this was a collaborative effort by Dumas and another well-known chemist, Jean-Baptiste Joseph Dieudonné Boussingault (1801–87),[4] a scholar with equally wide interests. Boussingault had received his training at the School of Mines, and had spent 12 years in South America as a mining engineer, some of the time attached to the military staff of the Liberator, Simon Bolívar. On his return to France, Boussingault was appointed to the Chair of Chemistry at the University of Lyon. Later he succeeded Dumas at the Sorbonne and ended his career at the *Conservatoire Nationale des Arts et Métiers* in Paris. Like Dumas, he was politically active, as Deputy in the National Assembly and a Counsellor of State. His fame rested on his studies of nitrogen balance in agriculture and nutrition, carried out for the most part on his own farm. He recognized that nitrogen was essential for the growth of plants, though he wrongly supposed that they obtained it directly from the air. This may have misled Liebig, who was of course also concerned with the chemistry of agriculture and set out to develop an artificial fertilizer. Liebig was persuaded that the virtues of manure came from its content of minerals, and that nitrogen, being in plentiful supply from the air, was not needed. His mineral fertilizer, which he patented in England, was ineffective and drew down on him the scorn of the very agriculturists whom he had derided as ignoramuses. Liebig had based his formula on substances of very low solubility, wrongly assuming that soluble mineral salts would be washed away by rain. When it was later revealed that the soil had the capacity to absorb and retain soluble minerals, Liebig brought out a new prescription, which had some success. Boussingault also modified his conclusions: leguminous plants did indeed derive their nitrogen from the air through the action of nitrogen-fixing bacteria, but wheat and other cereals did not. This convinced Liebig, who amended his conclusions in a new edition of his *Agricultural Chemistry*. Boussingault did show later that there was no reason to suppose that farm animals derived any of their nitrogen from the atmosphere: he kept animals on a diet that just maintained their weight, determined the nitrogen content of what they consumed and what they excreted, and found that the former was much the greater, and sufficient therefore to meet the animal's needs.

The conflict between Liebig and Dumas bubbled up out of broad agreement, for Boussingault and Dumas had concluded, as had Liebig, that animals do not create complex organic substances, only destroy them. According to this scheme, they receive the substances of their tissues ready-made and intact from plants, and it is green plants that represent 'the great laboratory of organic chemistry'. The synthetic processes require only the simplest of starting materials and energy supplied by the sun. It was, Dumas thought, the chemistry of nitrogenous compounds in the life of plants that demanded the attention of physiologists. The scholarly work, entitled *Essai de statique chimiques des êtres organisés*, was met in France with rapture and immediately translated into several languages. Dumas entrusted the German translation to his pupil, a noted chemist from Alsace, Charles Gerhardt, who was much taken with the work and expressed his pride at having been selected to spread the wisdom abroad. The problem was only that Liebig's publications on the subject did not feature in the bibliography, and Liebig, as was his wont, saw red. He wrote to Gerhardt, complimenting him on the hard work he had put in on the translation, but then assuring him that 'like Boussingault, Dumas and all the other Parisians understand nothing about agricultural chemistry or physiology'. Two months later a French journal, the *Revue Scientifique et Industrielle*, carried an article by Liebig, which began:

> Last year there appeared simultaneously in Paris and in Brunswick my book, *Organic Chemistry Applied to Plant Physiology and Agriculture*, wherein the functions of plant nutrition are set out according to the current state of science. I was happy to see my opinions adopted by one of the most famous French chemists, who made a particular point of it, professedly for inclusion in his course. The material, compiled with every care, was printed serially in its entirety in the *Journal des Débats*, and then published in part in a very elegant booklet.
>
> In this booklet he makes no mention of the researches carried out by me and others on the subject at issue, and this is natural, since the citations are excluded from the course notes. But I am at risk through this of losing my title to the opinions enunciated by me.
>
> It is in any case very singular to see this chemist declare never, over 17–18 years, to have occupied himself with plant nutrition, then discourse on this subject for the first time, and by chance six months after the appearance of my book. This can certainly not help to disseminate the truths that I have established, for they have, in the lessons to which I refer been very adroitly admixed with errors.
>
> It is stated in the French edition of my book that I was engaged in similar work on animal physiology. On this subject I gave a course in the winter of 1839–40: the contents of these lectures were widely disseminated by the listeners. And I am thus compelled, so as not to plagiarise myself, and instead of compiling my material in the necessary state of tranquillity, to communicate a part to this journal in abbreviated form, without giving citations or analyses.[5]

This was a devastating accusation of plagiarism, and downright intellectual theft, which Dumas could scarcely ignore. Dumas hit back as best he could, concentrating on the honour of French science, though cleverly avoiding an excess of chauvinism:

> 1. To whom is due the discovery of ammonia in vegetation? It is Schattenmann, whom Liebig should have cited in his course and whose observations were known to him in 1835.
> 2. To whom should be attributed the discovery of the decomposition of carbonic acid and of water by plants? It is Bonnet, Boussingault and Sennebier.
> 3. To whom should be attributed the discovery that nitrogenous organic substances formed by plants pass through the digestion, unchanged [sic], of animals? It is Prévost and Leroyer, in 1824.
> 4. To whom should be attributed the new theory of respiration of animals? It is Mitscherlich [a student of Liebig's with whom the master had fallen out, believing him to have engendered the hostile attitude of Berzelius], Magnus, Mazingue and Alibran.
> 5. To whom is due the true theory of animal heat? It is Laplace, Lavoisier, Dulong and Despretz.
>
> All this, then, precedes the lectures given by M. Liebig in 1840 and published in 1842.[5]

The scientific world did not have to wait long for Liebig's answering broadside. He did not deign to address himself to Dumas's points, but –

> In the same winter term of 1840–1, among my listeners was an able young chemist, M. Marignac of Geneva, now Professor of Chemistry there, but then still a pupil at the École des Mines in Paris, who had been recommended to me by M. Dumas. M. Marignac returned to Paris at the end of February 1841, and I have therefore every reason to believe that my views were no longer unknown to the French chemists, notably M. Dumas. [This was a shrewd, and probably a valid charge.]

Then Liebig put on the knuckle-dusters. Dumas had not even himself done the analytical experiments that he had claimed, for

> M. Dumas is not in the habit of doing his own experiments; he employs compliant assistants, who adorn his table with dishes which they prepare, and are satisfied with the crumbs that fall from it. Who was it who in actuality examined for him before 1841 the compositions of nitrogenous principles of plants? *Now* he seizes on the use that he can make of the experiments and publications of M. Boussingault, who achieved without him results of high significance; *now* he conjures up the instrument and wants to convince us that only he, M. Dumas, knows how to play it. While he was expending his own resources in arid theoretical speculations, we in Germany dug to new depths the techniques of which I myself reported in my *Annalen*, vol. 38, p. 195. It is only after that that M. Dumas changed direction. But does this change allow him to wrest from us our priority? Why did M. Dumas not resolve the matter at a time at which there would have been merit in

> announcing his opinions of today, when, I say, one can see him, while not upsetting his colleagues in the Academy, prostituting himself in the eyes of the world? But this can be explained, because at that time he knew nothing of the principles in plants that lead to sanguinification [generation of blood]. Only later did he recognise the truth.[5]

Such strong language seldom featured in the academic discourse of even those less circumspect times. Stung, Dumas devoted much of a lecture course that year to exculpating himself. Thrashing around for rebuttals, he demanded:

> Prove to us, M. Liebig, that these are your ideas; prove to us that one has appropriated your work, that one has plundered your publications, and we will bow to your genius, and we will sacrifice to you all our savants, back to the memory of Lavoisier. Shake, then, the dust from your old books, open the works that date back to Lavoisier, and you will be persuaded, M. Liebig, that you have said nothing new, that it was Lavoisier who originated this theory that you have annexed, that it was Frenchmen who made the discoveries on which your ideas rest.[5]

This undignified brawl had few parallels in history, at least since the 17th century, when accusations of plagiarism and theft flew and the bile flowed free. Liebig had the better of the argument, but neither he nor Dumas were fastidious in giving credit to others, and Dumas is undoubtedly right in claiming that Liebig had assimilated earlier theories into his physiological scheme. Dumas's co-author, Boussingault, maintained a dignified silence, and came off best.*

Fats: a self-inflicted wound

Less attention had been given to the third principle in the nutritional canon, the fats. Their chemical nature had been brought to light as early as 1811 by the great French chemist Michel Eugène Chevreul, one of the most remarkable scientific figures of his time. He was born in 1786 and became a protégé of a leader of the profession, Louis-Nicolas Vauquelin, Professor of Chemistry at the Natural History Museum in Paris. Chevreul did much work on plant and animal substances. He found that fats were esters (p. 253) of fatty acids – organic acids with long chains of carbon atoms – bonded to glycerine (known to chemists as

* The animosity between Liebing and Dumas did not soon abate, and the quarrel rumbled on for the next two decades. But in his old age Liebig showed a greater inclination to forgive his enemies, and a reconciliation came about during the great Paris Exhibition of 1867. There was a glittering banquet for the leading chemists of the day, at which Liebig's famous meat extract was served. In an after-dinner speech Liebig paid charming tribute to the French chemists, and especially to Gay-Lussac from whom he had learned so much 45 years before. 'When we had finished a successful analysis', he related, '(you know without my telling you that the method and the apparatus described in our joint memoir were entirely his), he would say to me, "Now you must dance with me just as Thénard and I always danced together when we had discovered something." And then we would dance.' The mellow mood of the occasion brought about a reconciliation between the two, now aged, adversaries.[6]

glycerol). The fatty acids, when chemically split off from the glycerol, Chevreul recognized as soaps. Each glycerol molecule had three fatty acid chains attached. He separated the total fatty acid content of the fats into two fractions, a solid which he called margarin (which later gave its name to margarine (p. 106)), and a liquid, which he called olein. We now know that these fractions can be subdivided into several fatty acid molecules, differing in the lengths of their carbon chains, and also that the liquid fraction consists of 'unsaturated', and the solid of 'saturated' chains.* Another celebrated French chemist, Marcellin Berthelot, showed in 1854 (in his doctoral thesis) that the glycerol could be recombined with the fatty acids to regenerate fat, and moreover that the process of hydrolysis (heating with alkali to split the glycerol from its fatty acids, as Chevreul had done) could be encompassed by pancreatic juice.

Chevreul went on to do more work on fatty substances. He extracted shorter, and volatile fatty acids from butter and vegetable oils by a similar hydrolysis process, and he also isolated from human bile an abundant component, which could not be hydrolysed and was therefore chemically quite different. This same material, he found, was the principal constituent of gall-stones. He called it cholesterine, and it was of course cholesterol.†

The fats, then, were viewed by Dumas as an essentially French topic, and very suitable ground on which to chasten the insufferable Liebig. For Liebig had stated in his *Animal Chemistry* that fats are synthesized from carbohydrates by elimination of oxygen. Dumas and Boussingault would prove beyond further argument, using the fats as their exemplar, the error of their central tenet that

* By 'saturated' is understood a chain of carbon atoms bearing their maximum complement of hydrogen atoms (two for each carbon in the chain), whereas in 'unsaturated' fats or fatty acids, there is a double chemical bond between one or more adjacent carbon atoms. The latter can then only associate with one hydrogen atom apiece. This will appear clearer in the formulae given in the Appendix.

† Chevreul then turned to other interests, for he had been appointed director of the Gobelin tapestry works. There he did notable work on dyestuffs and the chemistry of dyeing, and developed the theory of primary colours. His work captivated the painters Seurat and Signac, and gave rise to the *pointilliste* school. Chevreul retired from his directorial position at the Gobelin factory at the age of 97, but retained his professorial chair, and continued to contribute to the advance of chemistry. His centenary was an occasion for celebrations in France, and a gold medal was struck in his honour. He received marks of esteem from the academies of the world and from monarchs, including Queen Victoria. Chevreul died in 1889 at the age of 102, and is buried in the Panthéon. As for Berthelot, he passed most of his scientific career at the Collège de France, where he occupied a chair of organic chemistry specially created for him. He did important work in every branch of chemistry, and rose to a position of dignity and power as President of the Scientific Defence Committee during the siege of Paris in 1870–71 (in which capacity he appears, thinly disguised, in Émile Zola's novel *Paris*) and head of the national explosives committee, Perpetual Secretary of the Academy of Sciences, Senator, Minister for Public Instruction, and Foreign Minister. His huge influence over science in France was not always exercised benignly, and in particular he rejected, indeed vehemently opposed, the atomic theory, which was not taught in France until some years after it had been absorbed into the scientific canon in Germany and Britain.

animals received their body substance in preformed state from plants – that they were capable of nothing more than oxidative breakdown of the ingested nutrients. The experiment that Dumas conceived was an ingenious one. He and an assistant started with a closed hive, occupied by about 2000 bees. First they took 100 of the insects at random, killed them, extracted their body fats with ether, and determined its quantity. Then they gave the bees a supply of honey and allowed them 44 days to lay down their honeycombs. At the end of this time they weighed the wax comb, and again killed a sample of 100 bees and assayed their fat. They were disconcerted to find that the bees had lost no fat to speak of. The wax, then, must have been made by the bees from something other than fat – a something that could only have been the sugar in the honey.

Boussingault was meanwhile trying in his own way on his estate. He fed his pigs on a diet of potatoes together with whey and other essentially fat-free waste matter from the farm. At the start of the trial he selected an average pig, slaughtered it, and analysed its tissues. Then, after 90 days on the fat-free regime, another pig, which had been matched at the outset with the first for weight, was killed and the carcass analysed for body fat. Again the outcome was the wrong one for the French, for the second group contained considerably more fat than the first. Fat therefore was perforce synthesized in the body from the carbohydrate in the food. Dumas especially was mortified: he and Boussingault had set out to establish the primacy of their loudly proclaimed theory of nutrition, and by elegant experimentation they had actually buried it, and handed their German adversary another rod with which to belabour them – though, of course, the theory that they had destroyed was also his. Both turned away from nutrition, Dumas to pure chemistry, but more to politics as Mayor of Paris, and Boussingault to practical husbandry.

The elder statesman at bay

In the intervening years Liebig had become very rich, for his later work had taken a practical turn and led to industrial innovations. The change of direction in his research may have stemmed partly from the mounting evidence against his theory of animal nutrition. One of its central tenets was that the amount of nitrogen in the urine (in the form of the nitrogenous compound urea) was a measure of the erosion of tissues, and the energy generated by the muscles derived from degradation of the ingested protein that served to maintain them. The first doubts came from the observation that fasting animals excreted less urea than those on a diet rich in protein (and therefore of nitrogen). The author of this work was a physiologist, Friedrich Bidder at the University of Dorpat (now Tartu in Estonia). But Bidder recognized that his result was not altogether conclusive, since other reactions involving nitrogen might be occurring in the body. Then a former student of Liebig's, Moritz Traube, intervened. He was a wine-merchant whose heart was still in research, a passion to which he dedicated all his spare moments (and which indeed was in the family, for his elder brother, Ludwig, was a distinguished physician and researcher). Moritz Traube is now remembered mainly for his work on the chemistry of fermentation, but in 1861 he had a remarkable brainwave, which dealt a blow to Liebig's scheme. It occurred to Traube that draught animals, horses and oxen, which put out huge amounts of energy, were herbivores. Traube calculated how much nitrogen such animals consumed in their fodder, and discovered that it was far too low to account for the muscular energy output (which he could calculate from the work that they performed).

Worse was to come. In 1852 Liebig had left Giessen for the richer pastures of Munich, where he had the means to promote physiology, and to recruit a bright young doctor, Carl Voit, who had also studied chemistry. Initially, Voit and T. L. W. Bischoff, the Professor of Physiology, had convinced themselves that output of urea in the urine was indeed a quantitative reflection of the metabolic activity going on in the body, and they also endorsed Liebig's dictum that there were two kinds of food, 'plastic' and 'respiratory'. But then Voit did a critical experiment: he measured the urea output of a dog running on a treadmill and found that it was no greater than that of the resting animal. Groping for an explanation that would preserve Liebig's design, he imagined that muscles were degraded continuously, regardless of the level of physical activity, and that the resulting chemical energy was stored in electrical form. This reservoir of energy could then be called upon as required. It was a less than compelling argument. More evidence against Liebig's model began to accumulate. A London doctor and social reformer, Edward Smith,[7] had been contemplating the nutritional problems of deprived families and of prisoners. Throughout much of the 19th century prisoners had been forced to drive a treadmill for some days each week, with alternating spells of 15 minutes' work and 15 minutes' rest. Smith set out to determine the energy consumed in this arduous exercise by performing it himself. He measured the amount of carbon dioxide that he produced, by exhaling through a face-mask into a solution of a calcium hydroxide ('lime water', which fixed it as insoluble calcium carbonate), and also his excreted nitrogen (in urea). He did this both on the treadmill regime and when resting. The outcome of his observations, and of additional studies on the felons, revealed that the exhaled carbon dioxide reflected the amount of work that the body performed, while the excreted nitrogen corresponded only to the food intake.

It may have been Smith's results that led to the final eclipse of Liebig's theory, but Liebig caved in only after the outcome became known of a carefully planned exercise in self-experimentation by a chemist and a physiologist from Zurich. Johannes Wislicenus (1835–1903) was a professor of chemistry at the Polytechnikum, which later became the famous ETH, the federal technical university. He had an unusual career. Born in Germany, the eldest of eight children of a Polish Lutheran pastor who had left his country to escape religious persecution, his studies were interrupted when his father was again driven out. The family migrated to Liverpool, and thence in 1854 to America. The young Johannes worked at Harvard as assistant to an analytical chemist, Eben Horsford (whom we shall meet again), and then found a position in the Mechanics Institute in New York. After two years the family returned to Europe and settled in Zurich. There Wislicenus resumed his studies, and then sought to broaden his experience by further study in Germany, where, however, his political activities incurred the wrath of the authorities and resulted in his return once more to the more liberal climate of Zurich. His talents were recognized and he rose quickly to a professorial chair, first at the university, then at the ETH, where he established a flourishing school, notable especially for its work in the new field of stereochemistry (the shapes of molecules) and in organic synthesis. The experiment, which took place in 1866, was planned with his physiological colleague, Adolf Eugen Fick (now best remembered for the equations of diffusion that bear his name), and Fick's brother-in-law, the noted English chemist, Edward Frankland.[8] The plan was to undertake the strenuous climb to the top of the Faulhorn in the Bernese Oberland, and compare the work involved, as calculated from the gain in height, to the energy produced by breaking down protein. (The heat equivalent of the work done in climbing to a known height could be easily calculated, for the principles of thermodynamics were by then understood, and the Manchester physicist James Prescott Joule had explicitly determined the quantitative relation between work and heat – the mechanical equivalent of heat, as it is called.) The destruction of protein would be revealed by the output of nitrogen in the climbers' urine. It was stormy in the Alps and to Frankland, looking out of the window of his hotel in Geneva, the prospects of a climb appeared hopeless. So he telegraphed to his brother-in-law that he would be returning to Zurich. But the weather cleared, and Frankland was chagrined to learn a few days later that Wislicenus and Fick had done the climb. The calculation of energy balance demanded many approximations and assumptions, but the result was clear: far more work was expended than could possibly have come from the nitrogenous ('plastic') food. Frankland recorded his conclusions: when the muscles are activated by an instruction from the brain, they proceed to convert their stored energy into movement and heat. He likened the muscle to the piston and cylinder of a steam engine, which generates motion from heat. Both require fuel, but neither derives its mechanical power from the oxidation of its own substance. Or, as Fick and Wislicenus put it, the opposite view (Liebig's hypothesis) would imply an absurdity: a locomotive 'consists essentially of iron, steel, brass, &c.; it contains but little coal; therefore its action must consist of burning iron and steel, not of the burning of coal'. Frankland later improved the calculations by measurements of the heat that would have been given out in forming the urea that Fick and Wislicenus had put out in their urine. For this he used what was then an advanced instrument, a bomb calorimeter, to find how much heat was generated in the combustion of a representative protein, albumen, as well as some related substances.* (Frankland went on to devise a method of achieving complete combustion of foods and measuring their energy yields. He tabulated the calorific values of a series of foods, and could be said to be the father of calorie counting.)

Liebig had a high respect for Wislicenus and Frankland, and conceded defeat, at least on the issue of whether urea could measure energy output. Their work, he wrote to Wöhler, 'sent to the grave my former theory'. It would have been too much to expect Liebig to abandon his scheme totally, however, and for a while he and Voit cleaved to Voit's lame concept of electrical energy – or rather, in Liebig's case, to the even more nebulous notion of some form of storage mechanism that could gradually yield up its energy like a 'spring in a clock'. Voit soon had second thoughts, and turned his back on Liebig's ruminations. Liebig reacted with a characteristic volley of obloquy, but Voit did not flinch and merely quoted Liebig's own riposte to Berzelius's strictures a quarter of a century earlier – that they came from an old man stranded in the past. Critics in Germany and Britain picked at the results of the Faulhorn experiment, searching for weaknesses. One criticism appeared in the form of a doggerel verse, which primly concluded:

> I think that these Doctors, though plainly no fools,
> Have ignored both Digestion's and Logic's first rules:
> And, comparing to urine the use of their tools,
> They entirely forgot to examine their st_._.
>
> So while thus Wislicenus and Fick make a bustle,
> And in victory's plumes Frankland's trying to rustle,
> Many think that they haven't the best of the tussle,
> But that Liebig and Playfair are right as to muscle.[9]

(Lyon Playfair, Professor of Chemistry in Edinburgh, was still Liebig's most devoted and vocal British disciple.) It was assuredly not the end of protein-centred nutritional thinking in Germany, and even Voit remained convinced that a large protein intake would boost nervous energy. The right daily intake of protein for a man engaged in physical work, Voit asserted, was 120 g, more if his exertions were particularly strenuous. Other highly placed experts agreed. It was left to an American researcher to prove otherwise. Russell Henry Chittenden (1856–1943) had read chemistry at Yale University, and after a formative period of study in Willy Kühne's laboratory in Heidelberg, where digestive enzymes were being isolated, he returned to Yale, there to remain for 40 years. Chittenden gathered data on protein consumption and nitrogen excretion by athletes and soldiers on taxing training regimes, on his faculty colleagues, and, most extensively, on himself. He discovered that a low-protein diet over several months had no adverse effects on physique or performance. The muscles did not waste

* To make allowance for the fact that the end-product was nitrogen and not urea he had to subtract the heat generated by combustion of urea to nitrogen.

away, as Voit and others had predicted, and Chittenden felt fit and healthy. The upshot, he reported, was that a protein intake of 64 g per day was more than sufficient to maintain health. In the United States the dictum emerged: 'Look after the calories and the protein will look after itself.' This dangerous doctrine was supplanted only with the arrival of calorimeters large enough to contain an animal, and indeed even a man on a bicycle (p. 111). (Chittenden became more and more extreme and dogmatic in his pronouncements and, as will emerge (Chapter 10), was linked with some disreputable nutritional movements.)

Liebig's extractum carnis

Liebig began to interest himself in the composition and quality of meat at a quite early stage of his career. As early, in fact, as 1841 he had written to his friend Wöhler that he was tired of laboratory work, and it was only applications of science that now held any appeal for him. Over a period, he identified several chemical substances in meat, both raw and cooked. When meat is boiled, the water often sets to a jelly, and this Liebig equated with gelatin, which is extracted from skin and bone. As we have seen (p. 73), François Magendie in Paris had shown that, although gelatin was rich in nitrogen – more so indeed than blood fibrin and other proteins – and had all the characteristics of a protein, it could not sustain the life of dogs. Gelatin therefore had no capacity to replenish the tissues, but, the conjecture went, it was probably required to form bones and tendons. Liebig was persuaded, at all events, that both the soluble and insoluble parts of meat played a role in nutrition. Meat exposed to high temperature before roasting, or immersed in boiling water, Liebig thought, would retain more of its soluble matter and would therefore be more nutritious, because the coagulation of the protein ('albumen') in the outermost layer would create a seal. This entirely erroneous notion of 'sealing' a steak, for instance, persists to this day, but was regarded at the time as an incisive *apperçu*.

Liebig's faith in meat extracts grew perhaps out of an incident that occurred in 1853. An English friend, James Muspratt, a purveyor of chemicals, had sent his three sons to Giessen for their education, and he now despatched his 17-year-old daughter, Emma, to stay with the Liebig family, by then in Munich, to learn German. Liebig and his wife and two daughters entertained Emma with all the amusements that the city had to offer, but in the midst of this she fell ill with scarlet fever, and for a time the doctors despaired of her life. Liebig, beside himself with anxiety, tried to find a source of nourishment for the enfeebled girl. Beef tea was an option, but if the soluble fraction of meat was so nutritious, might one not do better to extract it in the cold and thereby minimize degradation of the nutritious principles? Liebig procured some chicken meat, minced it, and softened it in a little acidulated water. The pressed juice was administered lukewarm to Emma at half-hourly intervals, and lo! she was quickly on the road to recovery. This extract was afterwards used in the Munich hospitals. Nevertheless beef tea, extracted from meat by boiling and concentrating down, was undoubtedly more palatable and remained a favoured food for the sick and feeble, and no doubt for many who simply liked it.

Some eight years later Liebig's interest in meat took a new turn. A German acquaintance, an enterprising engineer by the name of Georg Giebert, working in South America, wrote to Liebig telling him of a glut of beef cattle in Uruguay and Brazil. In Uruguay the cattle were being slaughtered for leather and most of the meat was thrown away. Why not then salvage the meat and use it to prepare a nourishing and flavoursome extract? A company was set up, registered in London, while at Fray Bentos in Uruguay, where Giebert bought land for the purpose, a plant was constructed. The meat was crushed between rollers, and the pulp was steam-heated for one hour and strained. The fat was separated and the fluid was concentrated by heating for some hours. The dark brown, viscous liquid was filtered and sealed in this sterile state in tins. This *extractum carnis Liebig* took Europe by storm, initially as an additive to give savour to the dishes of the rich. In England, where it was launched with a great fanfare in 1865, the very concept of concentrated beef had a special appeal, for it was widely held that copious consumption of beef was what had made the English great. In Germany Voit again annoyed Liebig by analysing the extract and showing that it had little nutritional value.

Liebig's beef extract exercised a mesmeric effect on the healing professions, and many believed it to be something of a panacea for all manner of ills. Liebig himself had already stoked the fire by claiming in an anonymous article in a respected German newspaper that beef tea had cured patients of typhus. William Beneke, a physician at the German Hospital in the London suburb of Dalston, declared in *The Lancet* under the title 'On extractum carnis': 'I scarcely trust in any medicine more than in beef tea in this disease [typhus].' He urged that it should be included in the Pharmacopoeia, and, he continued with mounting enthusiasm, 'it is not to be replaced by any other food or remedy. I have administered it in scrofula and phthisis [tuberculosis], especially in those cases in which derangements of the digestive organs were present, such as ulcerations, dyspepsia, tuberculosis of the intestinal glands, etc.'[10] The medical establishment concurred. An editorial in the *British Medical Journal* proclaimed that the entire medical profession owed Liebig 'a deep debt of gratitude', that probably no other food was 'as effective in restoring the tissues of the sick.' At St Thomas's Hospital in London 12,000 large jars of the extract were, at the height of its reputation, bought in each year. The military was also keenly interested in the implications, for if *extractum carnis* was indeed the very essence of nutritional virtue, then perhaps it could serve as a convenient and economical means of providing for soldiers in the field or sailors on cruises. A study reported that one pound of extract would feed 128 men in the field.

It was not long, however, before murmurs of dissent were heard. One of course emanated from Carl Voit, while in England two analytical chemists,

Thomas Vosper and Arthur Hill Hassall (who was also (see Chapter 7) a qualified physician), independently reported the results of analyses of Liebig's extract. It had, they both agreed, no nutritive value at all. Edward Smith, too, denounced the stuff as useless. Vosper set out the evidence in *The Lancet*, and Liebig responded in typically combative fashion. But Vosper had an unanswerable case, and exposed the evasions of 'the learned Baron' in cutting terms. As for supplying the nutritional needs of soldiers and workers, he observed, 'Heaven help the poor men taken into the field, either to fight the battles of their country or till its soil, fed upon the homeopathic or infinitesimal diet prescribed by Liebig.' He might also, he concluded, 'remark upon the want of courtesy which has characterized [Liebig's rejoinders], and the very unusual proceeding of charging those who differ from him with incompetence, without at the same time producing a shred of evidence in support of such a charge'.[11] The Baron backed down and rather disingenuously changed his position: he had never, he now stated, claimed great medicinal values for his extract; rather, its use was to provide flavour and thus stimulate failing appetite.

Medical opinion now divided into two camps, for and against Liebig. A distinguished scientist, Edwin Ray Lankester, zoologist and disciple of Darwin, who was to occupy professorial chairs in London and Oxford, and was eventually, as Sir Ray Lankester, to be appointed Director of the British Natural History Museum, also intervened. He, too, contributed an analysis of Liebig's extract, 'an article of food [that] has lately found its way into every grocer's and chemist's shop in the country', and noted the absence of protein and fat. Yet the action of this 'all-potent juice' on 'exhausted nervous systems and debilitated frames is no delusion'. Whence, then, its potent curative properties? They could reside, Lankester asserted, only in its high content of minerals. 'It is to these, then, that we must look for an explanation of the marvellous powers which the extract of flesh exerts on the human frame.' But by then Eduard Kemmerich in Germany had already published the result of an experiment showing that dogs fed on the extract quickly died (though he misinterpreted the outcome, concluding that the elixir contained a toxin). By 1870 the sceptics had the upper hand. Dr J. Milner Fothergill in his *Manual of Dietetics* frothed at the mouth: 'All the bloodshed caused by the warlike ambition of Napoleon is as nothing compared to the myriad of persons who have sunk into their graves from a misplaced confidence in beef tea.'

But Liebig's extract was by then so firmly established, both in Europe and America, that there was no stopping it, and, assisted by slick marketing, it sailed serenely on. Rival products sprang up, to be sure, on occasion with Liebig's august name attached, and writs flew. Sometimes Fray Bentos won and sometimes it lost. At all events, extracts were still promoted up to the Second World War with the aid of testimonials by long-dead doctors. The Fray Bentos company later diversified, marketing corned beef, as well as the ubiquitous Oxo cubes from beef extract. These were cheaper to produce, and made initially from beef of the poorest quality. Fray Bentos thrived for a century, but ended calamitously when an outbreak of *Salmonella* poisoning in Aberdeen in 1964 was traced to tins of its corned beef.

Liebig – baker and dairyman

The meat extract was by no means the only profitable culinary enterprise that Liebig undertook. He started to think about bread in about 1851, but it was only in 1867, when the harvest failed in East Prussia and famine threatened, that he gave it his serious attention. The baking methods used in Europe were, he thought, archaic, and in 1868 he published a newspaper article asserting that fermentation with yeast resulted in a serious deterioration of nutritional quality. To circumvent this he tried leavening a mixture of whole, unrefined rye flour, complete with its bran, and some white wheat flour, salt, and sodium carbonate, by adding a dilute hydrochloric acid solution. This caused an effervescence of carbon dioxide in the dough (just like yeast fermentation). Liebig reported that this bread, baked in his house, was preferred by the entire family to baker's bread. Later he replaced the hydrochloric by phosphoric acid. He demonstrated the preparation of his bread at one of the series of public lectures that he gave in Munich.

This process, though, was scarcely feasible for the bakers, and it was Liebig's former American pupil, Eben Horsford (the employer of Liebig's nemesis Wislicenus during his time in the United States), who gave the world baking powder. This mixture of sodium bicarbonate with acidic calcium and magnesium hydrogen phosphates could be substituted for yeast as a source of carbon dioxide to make the dough rise. Horsford, when young, had obtained a stipend to work in Liebig's laboratory in Giessen, and held his master in high affection. They corresponded regularly, and Horsford communicated to Liebig every advance in the development of his baking powder, which Liebig in return fully endorsed. He regarded it, he stated in his journal, the *Annalen*, as 'one of the most important and beneficial discoveries that has been made in the last decade'. Horsford set up a manufacturing plant, the Rumford Chemical Works, and marketed the product as Rumford Baking Powder, a name that still survives. So also does George Borwick's baking powder and MacDougall Brothers' self-raising flour in England, both from the same period. Curiously enough, and to Liebig's chagrin, the German industrial chemists were less adept at producing a satisfactory baking powder, and it was some years before a pharmacist by the name of Oetker succeeded. 'Doktor Oetker' is now a large concern manufacturing a wide range of culinary products. As for Horsford, he succeeded to the Rumford Chair of Chemistry at Harvard University, the endowment of the egregious Count (Chapter 4), and died rich.

Liebig's views on bread were in fact sound. Black bread had greater nutritive value than white because of losses of minerals (and, as later emerged, other essential substances) when the flour was refined, but was unpleasing to the bulk of the population. An exception, he noted, was in Westphalia, where a strong black wholemeal bread, *Pumpernickel*, was a staple part of the diet, and the digestive apparatus of the citizenry was (he told the world) in far better order than elsewhere. Liebig suggested that calcium, magnesium, and potassium phosphates should be added to white flour, and potassium bicarbonate be preferred in baking powder. (The recommendation was widely adopted.) He was particularly shocked by the custom of many British bakers of adulterating the flour with alum (potassium aluminium sulphate). John Snow, the London doctor credited with discovering that cholera is transmitted in drinking water by tracing the source of a devastating epidemic in London to the Broad Street pump in Soho, wrote an article denouncing this abuse in *The Lancet* in 1857. In it he quoted Liebig's words: 'Since phosphoric acid forms with alumina a compound hardly decomposed by alkalis or acids, this may perhaps explain the indigestibility of the London bakers' bread, which strikes all foreigners.' Snow conjectured that the sequestration of phosphorus in unassimilable form was the cause of the ubiquity of rickets throughout the country (see p. 3). Liebig, he added, had not mentioned rickets in this connection – although he believed, for inscrutable reasons, that scurvy might originate from phosphorus deficiency. Liebig's suggested additive was calcium hydroxide solution (lime water), which would curb acidity and supply a healthy admixture of calcium.

Next, Liebig went after coffee. He tried to prepare an 'instant' coffee by evaporation, but was left with only an oxidized mass, which had an unpleasant taste and poor solubility. It was an idea before its time, and coffee drinkers had to wait a century before a chemist and coffee addict, Tadeusz Reichstein, succeeded in producing a moderately palatable concentrate in his laboratory in Zurich. (Freeze-drying was the innovation that led to the instant coffee we know today.) Frustrated, Liebig did at least examine the various methods then, and now, used to brew fresh coffee, and compared their merits and demerits. He advocated boiling up the ground coffee beans from cold to ensure efficient extraction, and then adding a further third of the weight of coffee to the boiling liquid for the aroma.

The power of Liebig's name was such that his advice and endorsement of food products was widely sought, and delivered only for a fee. In England he was called on by two of the largest brewers to scotch the rumour that the bitter taste of their ales was being enhanced by the addition of strychnine. The founder of the Australian wine industry sent samples to Liebig, who was willing to declare that they could stand comparison with the best European vintages. But his most profitable venture, which came late in his life, was into milk substitutes for infants, a concept of his own that started an entire industry, and added a twist of irony in that it caused a rise in the incidence of rickets (p. 95).

With the improvements, devised by Liebig and others, in the analysis of foods, a renewed interest sprang up in the comparative compositions and qualities of the milk of different animals. At the same time Liebig's principles, regarding the types of food required to satisfy nutritional needs, raised in the minds of several physiologists and chemists visions of an artificial baby food that could replace, and perhaps even improve on, mother's milk. One of the first to concern himself with this project was Edward Frankland, whose wife was unable to breast-feed Percy, their firstborn. Frankland devised a formula, consisting of cow's milk with several additives, which evidently did Percy no harm, for he grew up to become – like his father – a distinguished chemist. 'Frankland's Milk' was marketed, and remained in British shops for the next 40 or so years. It never achieved the popularity of Liebig's 'Perfect Infant Food' ('Liebig's Malt' in England or *Liebig's neue Suppe für Kinder* in Germany), which hit the market in 1867. Liebig's interest seems to have been prompted, like Frankland's, by family concerns, for his daughter also experienced difficulty in breast-feeding her babies. The 'perfect food' was compounded of wheat and malt flour, (malt for extra sugar) and cow's milk, heated with potassium bicarbonate to reduce acidity and supplement the potassium intake, in accordance with Liebig's precepts. Initially it was marketed as a liquid, later as a dried-down powder. Its commercial success stimulated a long line of similar (or worse) competing products, generally deficient in vitamins, as will be seen (Chapter 6).

Freiherr (or Baron) Justus von Liebig died in Munich in 1873 of pneumonia. He was one month short of his 70th birthday, immensely rich, laden with honours, active and pugnacious to the end. He appears to us now a flawed genius, as disastrously wrong in many of his pronouncements as timely and penetrating in others. He excited loyalty and dislike in equal measure. George Eliot and her companion, the scientifically minded scholar George Henry Lewes, visited him in his laboratory in 1858, and described his appearance, 'with his velvet cap on, holding little phials in his hand and talking of kreatin and kreatinine [creatine and creatinine being metabolically important compounds] in the same way that well-bred ladies talk of scandal'.

What is digestion?

By the middle of the 19th century the theory of Liebig and Dumas that proteins were incorporated unchanged into the body had had its day. It was still generally believed that nature had evolved only four proteins, although these could be somewhat modified in different tissues. (We know now that there are between 10,000 and 30,000 in man, not counting modified forms.) It was known from the early animal experiments (Chapter 4) that solid proteinaceous (and other) foods were rendered soluble during digestion, and their form seemed to be changed when treated with gastric fluid. But this was widely supposed to result mainly, if not entirely, from the action of the stomach acid. Then in 1843 the great French physiologist Claude Bernard, at the beginning of an illustrious career, performed a critical experiment. He showed that the protein egg albumin, injected into a dog's vein, was excreted, apparently unchanged, in the urine, whereas if it had first been exposed to the action of gastric juice no trace of it could be found. (Dumas, too, had suggested that proteins undergo a far-reaching transformation in digestion, before changing his mind and converging on Liebig's view.) It was soon recognized that the digestive process converts the protein into smaller molecules, which diffuse freely through membranes. A German physiologist, Carl Lehmann, called these protein fragments peptones. Yet the principle that at least some protein had to pass into the blood in its intact state if the tissues were to be kept in good repair died hard. It was found that proteins could pass through the intestinal wall, though extremely slowly compared to the peptones, and Bernard showed that injection of proteins into the blood of animals caused kidney damage, and so their transport in undegraded form was probably not a physiological process.

Several German physiologists had managed to purify, at least partially, a substance from gastric juice with high capacity to digest proteins. This was called pepsin, and was described as a 'soluble ferment'. *Ferment* is still the German word for enzyme, a protein that expedites biochemical reactions. The discovery attracted little attention for some years, but Bernard showed that proteins were also digested by pancreatic juice, and later Willy Kühne in Heidelberg purified a proteolytic (that is to say, protein-digesting) enzyme from this source, and called it trypsin. Soon more such enzymes were discovered. When it transpired that no trace of proteins or peptones were to be found in the blood emerging from the liver into the portal vein after a proteinaceous meal, the principle gained gradual acceptance that the body must synthesize its own proteins from scratch: they are neither absorbed whole nor reassembled from peptones.

Invalids nourished at both ends

The discovery of digestive breakdown of proteins into small fragments caused a light to switch on in the brains of many clinicians. It stood to reason, they thought, that if nutritious proteinaceous foods were to be treated with an extract of stomach or pancreas, the resulting predigested protein would spare an enfeebled body the rigours of digestion. Surely, then, it would be good for invalids. By the latter half of the 19th century most doctors had reluctantly conceded that beef tea and Liebig's concentrated meat extract had negligible nutritive value, even if it was still regarded as a valuable stimulant to weak appetites. There was therefore now a demand for new invalid foods.

Pancreatic juice, which, unlike the gastric fluid, was not acidic, was deemed preferable. It was marketed in various forms for treating food in the home, and 'peptonized' diets were also widely touted. 'Peptonized milk' was considered highly wholesome for invalids, though cream had generally to be added to mask the bitter taste of the fragments of broken-down milk proteins, and other foods, based on cornstarch for instance, were similarly treated. For babies there was

Volmer's Mother's Milk (also with added cream), and other similar products. The Pharmacopoeia included *Liquor Pancreaticus* as a useful digestive aid. It was added to the food, which was then briefly incubated at blood temperature. Later Fairchild's Pancreatizing Powder appeared on the market and attained wide popularity. Moreover, there had always been a school of thought in medical circles which held that food given in enema was kinder to the fragile digestion, and some concoctions, such as raw eggs beaten up in brandy, were often administered by this route. Some doctors accordingly maintained that enemas of peptonized foods would afford the same advantage. The grounds for supposing that food could be digested during an upward, as well as a downward, journey were never made clear.

Claude Bernard

Claude Bernard was the pre-eminent physiologist of the 19th century, perhaps of all time.[12] He was born in 1813 in a village near Villefranche in the Rhône Valley. His father was a wine-grower, but his business failed and he died leaving the family destitute. The village *curé* saw to the boy's education and enrolled him in a local Jesuit college. From there he went to Lyon as apprentice to a pharmacist, but his ambitions were literary, and he tried his hand as a playwright. His first effort, a comedy called 'La Rose du Rhône' was staged with some success. Thus encouraged, he wrote a drama, 'Arthur de Bretagne'. He took the manuscript to Paris and showed it to a literary critic, who found some merit in the work but explained to its author the hazards of his intended career, and urged him to secure his future by qualifying in medicine. Bernard took the advice and abandoned his play (which was published only after his death), but study did not at first come easily to him, and he was hampered by a shy and retiring personality. His marriage to the daughter of a prominent doctor enabled him to survive as a humble research assistant. He did, however, develop unusual skill in dissection, and this brought him to the attention of the foremost physiologist in Paris, François Magendie, who eventually, in 1847, appointed him his *préparateur*, or assistant demonstrator. Thereafter there was no holding Bernard, who suddenly revealed extraordinary talents. By the time he inherited Magendie's chair at the *Collège de France* in 1854 a series of groundbreaking discoveries had flowed from his laboratory bench.

Yet Claude Bernard's life was not an easy one. He loved his work and declared that scientific discovery afforded more delight to the spirit than any other human experience. 'Physiology! I feel it in me', he wrote to a friend. But his work, like that of Magendie, was based on experimentation on living animals before anaesthetics came into general use, and in France, as in England and elsewhere, a strong antivivisection movement had sprung up. Its adherents pursued Bernard relentlessly, and worse, his wife and daughter joined their ranks. By way of expiation of Bernard's sins, Mme and Mlle Bernard founded a sanctuary for stray dogs and cats. Bernard's domestic life became intolerable, not helped by his practice of bringing home mutilated dogs from the laboratory for observation. It was only in late middle age that, having separated from his wife, he found solace in a warm relationship with a sympathetic older woman.

Whereas Magendie took the Cartesian view that animals were insensate machines, incapable of pain or fear, and could therefore be used as one pleased, Bernard seemed to think otherwise and to regard the suffering of his experimental animals as an inevitable price of human aspiration. 'A physiologist', he declared, 'is no ordinary man. He is a *savant*, a man possessed and absorbed by a scientific idea. He does not hear the animal's cry of pain. He is blind to the flowing blood. He sees nothing but his idea.' Moreover, 'science allows us to do to animals what morality forbids us to do to our own kind'. He could not help but be aware of the revulsion provoked by the means through which physiologists made so many remarkable discoveries. The science of life, he observed by way of justification, was 'like a superb salon, resplendent with light, which could be reached only by way of a long and ghastly kitchen'. His accomplishments, with all their implications for the treatment of human diseases, were recognized nevertheless, and the Emperor Napoleon III personally arranged for him to be set up in a fine new laboratory at the Museum of Natural History. Bernard's final service to physiology was his great treatise, written when his health was beginning to fail, *An Introduction to the Study of Experimental Medicine*, which remained a standard work for several decades. Claude Bernard died in Paris in 1878, and was the first French man of science to be accorded a state funeral.

There are few areas of physiology on which Claude Bernard did not leave his mark, but, as to nutrition, his most important, indeed brilliant, contribution concerned carbohydrate metabolism. He regarded the preoccupation of the German school with material and energy balance as footling; such 'mere statistics' could shed no light on what went on inside the machine. It was 'like trying to tell what happens inside a house by watching what goes in by the door and what goes out by the chimney'. Bernard's first milestone was the demonstration that sucrose (cane or beet sugar) gives rise to glucose, also known as grape sugar. (The glucose molecule comprises one simple sugar unit, based on a ring of six carbon atoms; it is linked in sucrose to another very similar sugar unit of fructose, or fruit sugar.) When sucrose was injected into a dog's vein it was recovered unchanged from the urine, while glucose, similarly injected, simply vanished from view. Glucose then was evidently utilized in the body. The next discovery was that glucose emerged in the blood after a meal of starch. It had previously been supposed that the appearance of sugar in the blood was a diagnostic mark of diabetes. For 10 years Bernard continued to study what happened to carbohydrates in the body. He was astonished to find glucose in the blood of a dog that had been starved for a day. Whence, then, this glucose? Bernard thereupon fed a dog only meat for some days and then examined its tissues for the presence of glucose. He found large amounts in the portal vein, which carries nutrients derived from food to the liver. The livers of animals, he then discovered, were rich in glucose, regardless almost of diet, and by tying off the portal vein he established that the glucose had indeed diffused back from the liver, and had not come from food products entering the vein from above. Next Bernard found, quite accidentally, that glucose continued to be formed in the excised liver. To track down the cause of this totally unexpected phenomenon, Bernard began to experiment on isolated livers, a highly innovative procedure that demanded all his, by now legendary, dexterity. The liver, washed free of blood, continued to produce the sugar, and in time Bernard isolated its source – a starchy substance, now known as glycogen, the storage form of sugars, waiting to be utilized for energy. These momentous discoveries were the last that Claude Bernard made in the field of nutrition. Thereafter he devoted his energies to the study of glands, nerves, and oxygen transport by blood. In the words of one of his contemporaries, he was not merely a physiologist, he was 'physiology itself'.

6

The Poor, the Rich, the Healthy, and the Sick

The staff of life and other hazards

The progress of science and the concerns of the Victorian social reformers stimulated an increasingly close scrutiny of the methods by which staple foods were prepared and the purity of the products that went into them. The quality of bread was an enduring preoccupation. In England the weights of loaves were standardized and their price regulated as early as 1266 by a body called the Assize of Bread. Bread at this time was most often baked with rye or wheat wholemeal flour, produced by grinding up the entire grain. By the 17th century a preference had developed for a more or less white bread, made with wheat flour from which a large part of the husks, or bran, had been separated. In Britain and in some other countries there arose a curious and persistent aversion to dark bread. White bread was equated by the poor with luxury, and viewed by the rich as more fitting to their station than the coarse, plebeian fodder that was dark bread. But in truth dark bread, and in particular the rye bread favoured in most of central Europe, was a great deal more nutritious. In the course of his nutritional studies, carried out when food was scarce in France during the Napoleonic Wars, François Magendie (Chapter 5) had found that dogs on a diet of white bread, baked with high-quality wheat flour, died of malnutrition in a matter of weeks, while others fed black bread or hard-tack biscuits survived.

The production of white bread was refined by a series of mid-century inventions. First, sifting, or bolting, to remove bran (husks) from the pulverized wheat grains was made more efficient by the introduction of silk mesh as a sieve (still used today when not replaced by nylon). Then machinery was devised to facilitate the separation of husks from grains; this allowed flour to be recovered almost devoid of bran. And finally grinding between stone mill-wheels was largely supplanted by roller-milling, an innovation first tried in Budapest and quickly disseminated throughout Europe. Whereas stone-grinding reduced the entire seed to small fragments from which neither the remains of the bran nor the wheat germ could be easily separated, the rollers, driven by steam engines, flattened and sheared the seeds. The broken, but largely integral husks could then be eliminated from the flour, as also could the germ, which was reduced to flakes, readily filtered out by a mesh. Although analysis revealed a diminution in protein and mineral content of the flour prepared in this way, no-one could then have guessed, of course, that trace nutrients, in particular the B vitamins and the vitamin A precursor, carotene, were also lost. Any remaining nutrients were finally destroyed when bleaching was introduced to make the flour even whiter. Among the bleaching agents were sulphur dioxide from burning sulphur, blasts of warm air, and later nitrogen dioxide and chlorine. Mere storage of the flour also caused the yellow colour due to the carotene to fade.

An attempt to resuscitate wholemeal bread as a staple for the British was initiated in 1911 by an English eccentric, Sir Oswald Moseley. He was the fourth baronet of his line and the father of another Sir Oswald, the ill-famed founder of the British Union of Fascists. Moseley believed that it was 'white Viennese flour' that had caused the deterioration in the health of the nation, brought to light in the Army recruiting centres. The *Daily Mail* took up the cry, and launched its campaign for a legally defined 'standard bread', which should be purveyed to Buckingham Palace so that the Royal Family might set an example to the hoi polloi. The endeavour failed. For one thing the millers objected on the grounds that producing wholemeal flour would reduce their profits, for there would then be no wheatgerm, a marketable by-product used as cattle feed. The poor continued to regard coarse bread as a symbol of penury. Only the long-established Hovis company, which had its own flour mills and was therefore independent of the large millers, continued to fill a niche market with its product (originally 'Smith's Old Patent Germ Bread'), which was made from white flour enriched with extraneous wheatgerm and other additives.

The use of yeast as leavening agent was largely supplanted at about this time by baking powders (developed, as related in Chapter 5, by Eben Horsford in America and Liebig in Germany), and then in England a new technique was invented: the dough was inflated with carbon dioxide, blown into it in a pressure-chamber. This was the method patented in Britain by the Aerated Bread Company, the ABC, which continued manufacture of this spongy white bread, and served it, together with other delicacies, in its chain of tea-houses until 1951.

Milk and other perils

Milk for town-dwellers presented a considerable health hazard, nor was its nutritive quality satisfactory, for the cows, especially in the winter, were kept indoors, commonly in cellars, and were poorly fed. In Britain this was especially the case during and after the Industrial Revolution. The milk, often in any case watered, contained pus from inflamed udders and hooves, and blood. (These are still constituents of much of today's milk from the industrial dairy producers.) When the railways came, the practice of keeping cows in the city declined, and the quality of the milk improved. At around the same time suppliers hit on the idea of adding preservatives to the milk, and in some countries formalin (formaldehyde) was favoured. This is toxic and a severe irritant of the mucous surfaces, and in England it was eventually (in 1901) banned. Elsewhere sodium borate (borax) and boric acid were used, and in the United States several patent preserved milk products were marketed under such names as Aseptine and Freezine. All the same, diseases caused by infected milk did not finally disappear until the advent of pasteurization.

Louis Pasteur had shown in about 1860 that spoilage of food was caused by micro-organisms, and that most of these could be killed by heating to about 60°C. Wine and beer, Pasteur demonstrated, could be kept for long periods after such treatment, but he seems not himself to have tried it on milk. The method appears first to have been used for milk in Germany in 1880, though with the sole aim of extending shelf-life and thereby increasing the profits of the suppliers. The benefit to health was stressed only later, first apparently by a German chemist, Franz von Soxhlet, and, most influentially, by Abraham Jacobi, a German-born paediatrician who had emigrated to New York. He advocated breast-feeding, while insisting that where this was impossible cow's milk would serve but it must be pasteurized. A debate developed, however, between doctors who urged pasteurization, others who believed that the milk should be sterilized by prolonged boiling (which destroyed much of its nutritive value, as well as its taste), and others again who declared that only 'natural', or raw, milk was a healthy food. This was also the loudly proclaimed view of the dairy industry. It took many more years for pasteurization to establish itself around the world, and only well into the 20th century did it become mandatory in most countries of the West. Much later (in the 1930s) irradiation with ultraviolet light was developed as a practical alternative to heat treatment, and offered the additional advantage of elevating the vitamin D content of the milk, thereby making it more effective in combating rickets.

The provision of free school milk early in the 20th century was hailed as an enlightened and revolutionary measure. Three generations of children have been brought up on it in most developed countries, and its virtues have never been questioned until quite recently. It now appears that milk is a common cause of hidden gastrointestinal bleeding, sufficient to cause anaemia in many children. The proportion of the population with impaired capacity to digest lactose (milk sugar), common in Asia and among Jews and some other groups, seems also to have been underestimated in the West. It causes discomfort and diarrhoea and is a frequent cause of persistent complaint by babies. It has been argued that a child, once weaned, is not equipped by evolution for a milk-rich diet. There have been recurring assertions in the medical literature that milk may increase the incidence of a host of ills, from acne to cancer. The debate, in which the dairy industry has always engaged with furious indignation and with scientific testimony of its own, shows little signs of a resolution, even in the 21st century.

Poor and rich

Throughout the 19th century bread was the principal food with which the poor fended off hunger. Jams of dubious quality were also cheap, but meat, even the most inferior, such as very fat bacon, could only seldom be afforded. Malnutrition was endemic. The families of the poorest agricultural labourers often had to make do with a supper of dumplings made from flour and water. William Cobbett in *Rural Rides*, his description, published in 1830, of life in the English countryside, painted a dismal picture of the plight of the rural population, most of whom barely subsisted and, according to one observer, not so much lived as just not died. The urban population in Britain during the first decades of the century fared better, and the fall in wheat prices following the repeal of the protectionist Corn Laws in 1846, allied to the arrival of cheap imports from the United States and later Canada, brought the price of bread down sharply. But with mechanization such trades as hand-weaving became unprofitable and destitution was the common consequence. Many, perhaps a quarter, of the urban population ended their days in the workhouse. Infant mortality in the towns ran at a level of some 20%.

The rich in the early years of the 19th century lived primarily on meat, especially beef; they generally despised dairy products and anything green seldom appeared on their tables. Root vegetables were viewed as peasant fodder. It was widely supposed that the gross consumption of meat was the cause of gout (though see p. 115), the most prevalent affliction of the prosperous classes. In the cities meat was expensive and commonly putrescent. Trichinosis and other diseases from undercooked meat, especially pork, were rampant. Fish was little better, but quality improved enormously with the arrival of the steam trawlers, and especially when cold-store facilities were constructed on ships and on land. Oysters remained cheap until the late Victorian period, by when stocks had been largely exhausted. In England a 'Society for the Propagation of Horse Flesh as an Article of Food' was founded, with the aim of making good the shortage of the customary types of meat, and no doubt because horse-meat had been a prized item of diet across the Channel ever since the time of famine towards the end of the 17th century. By way of publicizing its virtues a great banquet was organized in the Langham Hotel in Langham Place, London in 1868, by which time French-inspired gastronomy had changed the attitudes of the rich. Here is how the menu read:[1]

POTAGES
Consommé de cheval. Purée de destrier
Amontillado

POISSONS
Saumons à la sauce arabe. Filets de sole à l'huile hippophagique
Vin du Rhin

HORS-D'ŒUVRES
Terrines de foie maigre chevaline. Saucissons de cheval aux pistaches syriaques
Xérès

RELEVÉS
Filet de Pégase rôti aux pommes de terre à la crème. Dinde aux châtaignes.
Aloyau de cheval farci à la centaure et aux choux de Bruxelles. Culotte de cheval braisé aux chevaux-de-frise
Champagne sec

ENTRÉES
Petits pâtés à la moëlle Bucéphale. Kromeskys à la Gladiateur. Poulets garnis à l'hippogriffe. Langues de cheval à la Troyenne
Château Pérayne

RÔTIES
Canards sauvages. Pluviers
Volnay
Mayonnaise de homard à l'huile de Rossinante. Petits pois à la française.
Choux-fleurs au parmesan

ENTREMETS
Gelée de pieds de cheval au marasquin. Zéphirs sautés à l'huile chevaleresque.
Gâteau vétérinaire à la Ducroix. Feuillantines aux pommes des Hespérides
Saint-Péray

GLACES
Crème aux truffes. Sorbets contre-préjugés
Liqueurs

DESSERT
Vins fins de Bordeaux. Madère. Café

BUFFET
Collared horse-head. Baron of horse. Boiled withers

In France and Italy horse-meat had developed into something of a luxury food, and was preferred by some to beef, but whether the banquet at the Langham Hotel inspired a lust after horse oil or horse liver amongst the rich or poor is doubtful.

Preservation: first steps

A vastly more important aid to nutrition was the emergence of preserved foods. The origins of dried meats is lost in the fog of history. The practice was certainly common in medieval times. Air-drying or slow drying in wood smoke were the traditional methods in much of Europe, and other procedures were introduced in the 18th and 19th centuries. In North America, and in particular in Canada, pemmican was produced in bulk by the indigenous Indians, especially the Metis people. It was made from bison meat, at least until the bison were hunted practically to extinction, and the trading organizations, such as the Hudson Bay Company, relied on it through most of the 19th century. Bison meat, cut into strips and dried in air (*viande seche*) was popular, but pemmican was more palatable and more durable. The strips of meat were dried over a fire and when they had become brittle were pounded to a powder, mixed with bison fat, and sometimes with berries, and formed into bars. These could be kept for many years without deteriorating.

Before the invention of canning, meat was often salted or pickled, not always effectively, and the cheap pickled pork, imported into Europe from Australia and elsewhere, had a bad reputation. Sulphur dioxide fumes and even chlorine were widely used during the 19th century to assist preservation. Oxygen in the air was held (until the effects of bacteria were recognized) to be the agent of putrefaction; its exclusion, it was supposed, would ensure perfect preservation. An English physiologist, Arthur Gamgee, thought that if animals were purged of oxygen before slaughter the resulting meat would resist corruption, and he accordingly devised a system for asphyxiating the beasts with carbon dioxide. No beneficial effect could have resulted. Another, more rational device was to perfuse the animals immediately after slaughter with preservatives, such as salt solutions, but again it is unlikely that a sufficient concentration in the muscle tissue could have been attained to do any good.

Refrigeration, for those who could afford it, was a more effective measure. Ice (which had been used for preserving food since ancient times) became the basis of a modest industry in European countries early in the 18th century. It was cut from frozen lakes and rivers in large blocks and stored in icehouses. In about 1820 physicists devised practical means to bring about cooling, and throughout the first half of the century industrial refrigeration units were developed. By mid-century refrigerated railway wagons were conveying dairy products around the United States, and a decade or two later the meat producers recognized the virtues of this scheme. Domestic refrigerators did not come into common use until well into the 20th century.

The first synthetic food

In 1869 a patent was granted to a French food chemist, Hippolyte Mège-Mouriès, for the production of '*certains corps gras d'origine animale*', collectively called margarine. His work had been prompted by the demand of the French Navy for a butter substitute that would be cheap and calorific and would not go rancid on a long voyage. Mège-Mouriès was a chemist of some note, who had worked on several aspects of food science. He had developed a method with greatly improved yield of producing bread for the Army, for which service the Emperor had bestowed on him the Légion d'Honneur and a gold medal. Thereafter he turned his attention to fats. Working on the *Ferme Imperiale de la*

Faisanderie, the Emperor's private farm, Mège-Mouriès observed that the fat content of cow's milk appeared to be little affected in under-nourished animals. This led him to suppose (wrongly) that the cow converted reserves of body fat into milk fat in its stomach or udder. Acting on this theory, Mège-Mouriès warmed macerated beef suet with water and chopped sheep's stomach, adding a little potassium carbonate to suppress acidity. After a few hours the enzymes in the sheep's stomach had digested any fibrous tissue in the suet, and the fat floated to the surface. Mège-Mouriès thought at this point that he had produced butter in the laboratory, but the fat that set to a solid when cooled looked and tasted nothing like butter. Next he tried a further treatment with chopped cow's udder, together with milk and sodium bicarbonate. This produced something a little more palatable, that perhaps could serve as a butter substitute. It was white, however, and unappetising in appearance. Either because he took it for the fatty acid margaric acid (at that time an ill-defined mixture), or because he was struck by its pearly sheen, Mège-Mouriès called in *oleomargarine* ('margarine' from the Greek for a pearl). There was at the time an acute dearth of butter in France. War with Prussia was feared to be imminent (as of course it was), and the Army needed to be fed. Napoleon III recognized the value of Mège-Mouriès's invention, and rewarded him with a generous prize. A factory was built for the production of the new food, but the plans were overtaken by the start of the Franco-Prussian War.

The difficulty with margarine at the outset was its pale and uninviting appearance, which resembled that of lard, and so a fat-soluble yellow dye had to be added. In 1871 Mège-Mouriès sold the rights of his invention to a Dutch company, Jurgens. A year later a factory began operation in Germany, and manufacture also started in the United States, and in 1873 margarine was licensed for human use in France. Soon it was being consumed all over the Western world, mainly imported from the United States and Holland. Efforts were made to improve its appeal by incorporating an admixture of plant oils, and also by adding cultures of the bacteria that fermented cream in the hope of supplying a buttery flavour. Then in 1902 a German chemist, Wilhelm Normann, developed a method for the hydrogenation (reaction with hydrogen to turn 'unsaturated' into 'saturated' fat (see Chapter 11)) of vegetable oils. This transformation causes 'hardening' – the liquid oil becomes a solid at room temperature – and this affords an improved alternative to beef suet fat. Margarine has been made from vegetable, commonly soybean, oil ever since.

In the United States the arrival of margarine did not please the dairy farmers, and their clamour led to the passing of the Margarine Act in 1886 to limit its sale and forbid the addition of a yellow dye. In those states in which the dairy industry was most powerful the sale of margarine was prohibited for many years, in Wisconsin until 1967.

The coming of the can

In 1795 the Revolutionary regime in France offered a prize of 12,000 francs for an invention that would lead to the preservation of food. Napoleon, who famously declared that an army 'marches on its stomach', was concerned at the toll that scurvy, and the ravages of malnutrition generally, were taking of his soldiers. The supplies of food that the Army carried regularly went bad, and the soldiers, especially in the Iberian peninsula, were reduced to pillaging and foraging in the devastated land, deliberately laid waste by the peasants. The challenge was taken up by Nicholas Appert, a Parisian confectioner and chef, and the endeavour occupied him for nearly 15 years. Appert held to the prevailing doctrine, that putrefaction was caused by the action of air on the food, and that therefore the expulsion of air would have the desired effect. His method consisted in heating the food in a closed, sealed vessel – essentially a pressure cooker – almost to boiling, then decanting it into glass jars. These in turn were heated with steam and sealed while hot with wax and a cork. He applied this method to many types of food – meats, vegetables, gravy, and more. In 1806 he announced his success and a trial was conducted by the French Navy on a cruise of four months. Appert reported that when jars containing 18 types of food were opened, all had retained their freshness and none had undergone any discernible deterioration during their time at sea. Napoleon, who was by then contemplating his Russian campaign, handed Appert the prize. The process became known as appertization.

Word of Appert's invention spread quickly and excited particular interest in England, where the military had the same concerns as Napoleon. A pair of entrepreneurs, Bryan Donkin and John Hall, set up a business in Bermondsey in London. They were inspired by the ideas of another Englishman by the name of Peter Durand, who had taken out a patent relating to the preservation of food in 'vessels of glass, pottery, tin or other metals or fit materials'. He had in mind, in fact, to use tin-plated iron, but in the event it was Donkin and Hall who followed up the obvious proposition that metal containers – which Appert had tried, with little success – could be more securely sealed and would certainly be less prone to accidents than glass. Donkin had learned his trade at the Deptford Iron Works, which made artefacts for ships, and knew how to shape metal. After a year's experimentation the factory began to produce tinned foods in 1812, although the process was highly labour-intensive and demanded skilled tinsmiths to solder and seal the containers. Nevertheless, within a year Donkin and Hall were victualling the Army and the Royal Navy, which was soon consuming 24,000 tins a year, each containing nearly two pounds of meat or other food. The tins at this time were not only very heavy, but could be opened only with hammer and chisel. With the coming of thin tin-plate matters improved, and by mid-century a machine process was in use for the manufacture of tins on a greatly increased scale. The first tin-opener, a clumsy, bayonet-like knife, was patented only in 1858.

In France the first sardine cannery opened in Nantes in 1820, and within about 15 years was producing 100,000 tins a year. In America canning started a little later, brought over initially by Durand, but was hugely stimulated by the exigencies of the mid-century gold-rush, and again by the Civil War. In 1855 Gail Borden, a dairyman and entrepreneur in New York, was awarded a patent for the preparation of dried milk (produced by spraying milk into a heated chamber), and a year later he took out a patent for 'evaporated' milk. This was merely concentrated milk, sold in tins. Later this evaporated milk, supplemented with 40% by weight of sugar, was marketed as 'condensed' milk. All these products have endured, largely unchanged, into our time. When they first appeared, they were greeted with much enthusiasm. They were, alas, deficient in vitamins A and D, which the processing methods had largely destroyed, and were responsible for an increase in the incidence of rickets.

Among the German states Lower Saxony led the way, when a venturesome nobleman, the Baron Wilhelm Eberhard Anton von Campen, visited France and saw Appert's process in action. In 1830 he set up his own operation for the conservation of slaughtered game. Later the port of Lübeck evolved as the centre of canning and exporting food. In Italy, it was tomato juice that became the favourite canned product, used for soups and sauces. In England, in the latter part of the 19th century, tins of meat for military and civilian consumption began to come in from Australia. The quality was not high: the meat contained lumps of gristle and fat, and there were cases of food-poisoning from infected, improperly sterilized samples. The sailors referred to the meat as 'Sweet Fanny Adams', an allusion to the victim of a notorious murder in Manchester, who had been chopped into small pieces by her assailant. Another unforeseen hazard is thought to have accounted for the men of the Franklin expedition to the Arctic. Two ships under Sir John Franklin sailed from England in May of 1845, charged with seeking out a Northwest Passage connecting the Atlantic and Pacific Oceans. The expedition was lavishly provisioned with every comfort (including a library of 3000 books) and tinned food of the best quality, sufficient to feed the crew for at least five years. Franklin, his ships, and his men vanished into the ice and it took several expeditions and 12 years before the remains of some of them were found. Mounds of tins had apparently been discarded unopened, and there was evidence of cannibalism. Analysis revealed that the contents of the tins, which had been constructed with lead solder, were heavily contaminated with lead. It was inferred that the crew had died of lead poisoning – a condition that could have been mistaken for scurvy. Another theory is that botulism, caused by contamination with *Botulinum* bacteria,* was responsible. On the other hand, contents of some of the tins, recovered from the Arctic, were cautiously tried by historians

* *Botulinum* toxin is the most dangerous of all food contaminants. It was known in ancient times, and the name derives from the Latin *botulus*, a sausage (or little bag), since botulism has been especially associated with rotten sausages.

in recent years, with no ill-effects, so the matter remains open. The quality and reputation of tinned meat, at all events, improved when P. J. Armour set up his vast canning plants in Chicago and Cincinnati, which started production in 1868.

Nineteenth-century diets

Medical opinion of the period adhered in general to Liebig's doctrine that the dietary principles required for health were five in number – protein, fat, carbohydrate, water, and salts (minerals). These last must include sodium, chloride, as well as iron to maintain the blood, calcium and phosphate for the bones and teeth, and iodine for the thyroid gland. In the most astute quarters the suspicion grew that other factors might also be needed, as indeed animal experiments were clearly indicating. The urban poor, who had only limited access to fresh food, and certainly enjoyed no balanced diet, were commonly malnourished, while amongst the rich gluttony prevailed. Disraeli's famous dictum, that the English rich and poor were 'two nations, between whom there is no intercourse and no sympathy', comprehended the food that they consumed. The poor in fact often did better in the workhouses and prisons than in the home. After the Poor Law Amendment Act of 1834, the diets in the workhouses were improved to include meat, bread, potatoes, vegetable soup, cheese and milk, and usually beer. Such perceived luxury caused anger in some political circles. It was even feared that so rich a diet would excite unwholesome urges. A debate developed about whether meat should be withdrawn from the undeserving poor. *The Lancet* was outraged, both at the callousness of the politicians and the unconcern of the Royal Colleges of Physicians and of Surgeons, and denounced them all in the fine style of the day: 'If that sated Vampire flap its wings, the effort is never to arouse or invigorate.'

In prisons matters had been far worse. After 1822 an outburst of pharisaical moralizing in Parliament led to the reduction of the diets to starvation level. Bread and a thin gruel were the staples, with meat or cheese perhaps once a week. Potatoes could be given, especially if scurvy made its appearance. Dysentery, scurvy, and tuberculosis were endemic. In 1835 new legislation allowed some improvement, and in particular more potatoes were supplied to contain scurvy. Thirty years later there was a small further improvement. The energy output demanded of a prisoner on a treadmill could, following the calorimetric data drawn up by Frankland and others, be compared to what the diet supplied. Edward Smith (p. 87) made known the results of his observations, which revealed a large dietary energy deficit, but his work received little attention.

In Germany nutritional science at the time was taken more seriously. Poor nutrition was recognized as the primary cause of ill-health and it was mainly two physiologists who determined what should be done about it. One was Carl Voit (Chapter 5), the other Max Pettenkofer, both of whom had begun their careers in Liebig's laboratory. Pettenkofer, one of eight children of a peasant farmer, was born in a small town in Bavaria in 1818. Destined for a labouring life, he was rescued by an uncle who needed an assistant in his pharmacy. In time nephew and uncle fell out, and after many vicissitudes, which included a spell as an actor, Pettenkofer enrolled in the University of Munich and qualified in medicine. But it was to scientific research that he was more strongly drawn, and he eventually managed to bring himself to the notice of Liebig. He joined the laboratory in Giessen, but lasted only a year before departing, for obscure reasons, to a position in the Bavarian mint in Munich. Liebig had nevertheless recognized Pettenkofer's qualities, and secured for the young man a position as Extraordinary (Assistant) Professor of Medical Chemistry in the same city. In due time he was promoted to Ordinary (full) Professor and Rector of the university. Pettenkofer's name is now most often associated with improvements in public hygiene, especially purity of the domestic water supply, and the prevention of infectious disease.* His work with Voit on human nutrition was nevertheless of the first importance. Their plan was to determine what factors were involved in nutrition, and to this end they had constructed for them (with money obtained from the King by Liebig, then already in Munich) a closed chamber large enough to hold a man. Air was admitted at a fixed rate, and the air leaving the chamber was analysed for carbon dioxide exhaled by the experimental subject within. The volunteer in the chamber took no food, but was allowed water, and not only the carbon dioxide, but also the nitrogen in urine and faeces was measured. The upshot was that the ratio of carbon to nitrogen lost by all routes was too high to be explained by consumption of protein, as we have already seen (Chapter 5), and that the fasting subject processed fat, as well as protein. Pettenkofer and Voit tried to work out the laws of metabolic balance, and both saw it as their duty to relate this work to the nutritional needs of people. It was Voit who published tables of dietary norms for men and women, adults and children, manual and sedentary workers. A typical male adult, for instance, weighing 70 kg and engaged in moderate physical work, should receive per day in his food 105 g of protein, 56 g of fat, and 500 g of carbohydrate. This would yield about 3000 calories. A sedentary man would require only 2400 calories, and so on. The 'Voit

* Yet Pettenkofer could never bring himself to accept that such diseases were caused by germs, as the work of Louis Pasteur and Robert Koch had shown. Pettenkofer cleaved obstinately to his conviction that epidemics generally, and of cholera in particular, were caused by some kind of exhalation from the soil, contingent on a rise in the groundwater level. To demonstrate that Koch's cultures of cholera bacilli were innocuous he drank down a substantial draught, which he had procured from Koch's laboratory, and suffered nothing more than a mild attack of diarrhoea. He was joined in this suicidal libation by two of his assistants and three doctors, none of whom apparently came to any lasting harm. It was generally supposed that Koch's assistant, who provided the brew, surmised what Pettenkofer was about, and made it weak. The episode gave Koch some embarrassment and Pettenkofer much gratification, but failed, all the same, to win him wide support. Not long afterwards he was thrown into a depression by the death of his wife and several children, retired from the university, and in 1901, at the age of 82, died by his own hand.

Standard', as it was called, based for the first time on the energy intake (measured in calories) that was needed to sustain life and work, amounting to an average of 110, 56, and 500 g of protein, fat, and carbohydrate, respectively, per day, was adopted in many countries.

In the United States a calorimeter was constructed at a small liberal arts college in New England, Wesleyan University, by Wilbur Olin Atwater (1844–1907), its first Professor of Chemistry, with considerable help from his colleague the Professor of Physics. Atwater had studied agricultural chemistry at Yale, and had visited Voit's laboratory and seen Pettenkofer's apparatus. Throughout the 1890s Atwater and his students measured the energy output of an athlete on a static bicycle that the calorimeter was built to accommodate. The cyclist, pedalling flat-out for several hours, three days at a stretch, was given in alternation high-fat and high-carbohydrate diets, and put out the same energy on both. This was an important result, for it established that it mattered not where the calories came from. On the basis of his measurements on his cyclist and on calorie values of different foods,* Atwater laid down standards to be adopted in the USA for men engaged in physical work. By general agreement, the human body was now considered, as articulated so clearly by Frankland, a machine. Caloric intake had to be matched to the machine's efficiency, which was recognized to be high – in the range 15–30%, compared to about 10% for steam engines.

In Germany an economic crisis in 1870, aggravated by poor harvests, concentrated the thoughts of politicians, and socially aware scientists like Voit and Pettenkofer, on the need to ensure that the masses – and most of all the poor, the unemployed, and the inmates of prisons and barracks – were properly fed. Food riots took place, and from 1879 there was a keen awareness of the social convulsions going on in France. It was Carl Voit's pupil, Max Rubner (1854–1932), who became the dominant figure in the social movement. He distinguished himself in meticulous calorimetric estimates of food intake and energy output, measured on dogs. He eventually succeeded the great Robert Koch in the Chair of Hygiene at the University of Berlin. Rubner laid down a new standard, somewhat higher than Voit's, for human nutritional requirements, and essentially founded a new academic discipline. This was called *Arbeitsphysiologie*, or physiology of work.

The 20th century – hunger and plenty

Arbeitsphysiologie received its official imprimatur as an independent discipline when an entire institute was dedicated to its furtherance. The *Kaiser-Wilhelm-*

* One of these was wine, and it was Atwater who can be credited with the admirable discovery that alcohol is a food. The American liquor trade was not slow to capitalize on the good news, and it brought down much opprobrium on Atwater's head from the Methodist Church, to which he himself adhered, and from his University, a Methodist foundation.

Gesellschaft (the Kaiser Wilhelm Society, transmuted after the Second World War into the Max-Planck Society) was an expression of the German regime's enlightened belief in the value of science. The Society's institutes, established in many university towns, were designed to embrace all the sciences, and so the *Institut für Arbeitsphysiologie*, under its director, Max Rubner, was established in Berlin in 1912.

The Institute laid down nutritional standards – not forgetting that the diet should be palatable and satisfy hunger, and should also include a modicum of roughage – that would allow workers to maintain a good level of productivity in their work. When war came, and with it the 'Hunger Blockade', it fell to Rubner to work out how to make the best of dwindling food supplies, which eventually fell to starvation level (Chapter 1). He and his collaborator Nathan Zuntz (later famous for his studies on high-altitude physiology) worked out exactly how far working men and women could be starved before becoming incapable of any normal level of patriotically useful activity. In the event, they fought a losing battle, for even by the end of 1914 the Government was urging a 'potato diet' on its citizens; bread was rationed to a half pound per day, and the growing black market soon ensured that meat and dairy products (which in principle were also rationed) were beyond the means of all but the prosperous. The *Kartoffelkrawalle*, or potato riots, followed, and when even the supply of potatoes began to run short the Government ordered that the country's pigs should be slaughtered (the *Schweinemord*) to save the potatoes used in their feed. This produced a glut of pork products, followed predictably by a dearth. The next year saw the first deaths from malnutrition. The winter of 1915–16 was the 'turnip winter', when the roots were made into bread, cakes, jam, coffee, and even, when fermented, into a dismal beer. Other *Ersatz* foods were bread made from dried, ground straw, coffee from acorns, and 'meat' produced from ground animal hides and the woody parts of vegetables.

The British physiologists made their own measurements of energy balance. An important innovation, that came too late for the First World War, was the introduction of the Douglas–Haldane breathing apparatus* for analysing the composition of exhaled breath, which, it was discovered, changed during exercise. The measurements gave the result that sedentary adults consume some 2400 calories per day, while in heavy labour this rises to anything up to 6000 calories. The Royal Society committee, set up at the Government's urging during the War, came up with a table, laying down the energy requirements of workers, male and female, in different trades. For men this extended from tailors, who

* The Haldane in question was not J. B. S. (p. 12), but his father, John Scott Haldane, Professor of Physiology in Oxford. J. B. S. used the apparatus in his dangerous underwater experiments for the Admiralty during the Second World War. Ronald Clark, in his entertaining biography,[2] describes J. B. S.'s rage: when suffering from hallucinations brought on by the conditions of the experiment, he called his father's apparatus 'a purple bitch' and had to be restrained from destroying it.

required 2750 calories per day, and bookbinders (3100 calories) at the lower end to stonemasons (4850 calories) and woodcutters (5500) at the top. Women, of course, had other occupations: a 'head seamstress' expended only 2000 calories per day, and a typist 2100, whereas the greatest exertion was demanded of a 'charwoman', who put out, and required, 2900–3600 calories daily.

Babies, infants, and their woes

The feeding of the babies of the poor was almost uniformly disgusting. The Industrial Revolution had driven mothers into work. They would have had less opportunity to breast-feed their babies, and the combination of debilitating factory labour and an inadequate diet would have impaired the quality and volume of their milk. The diet of watered and often contaminated cow's milk did not conduce to healthy growth, and even less so the tinned condensed milk, made from skimmed milk. The demand for this substitute increased as cows disappeared from their stalls in the cities, banished by the nuisance inspectors, and after about 1870 it became so cheap as to be generally accessible to the poor. It afforded little but protein and sugar, and contained no vitamins to speak of or fats. The incidence of rickets accordingly rose, and it was estimated that at that time about one-third of the children of the poor in London and other British cities were affected. Yet the cause was not identified; calcium deficiency was the most prevalent theory, but the disease was also widely held to have a heritable component. Syphilitic parents were also blamed, and so were insanitary living conditions. Moreover, when conscientious parents, made aware of the dangers of infection, sterilized their babies' food, vitamin C was destroyed and infant scurvy might follow. But as if all this were not enough, the luckless babies had other perils to contend with. Feeding-bottle nipples (at least until rubber teats, invented in the United States, came into use in mid-century) were made of chamois leather, parchment, or sponge, were seldom changed, and were in any case difficult to clean and so became infected. Another source of infection was pre-masticated food, given to the already enfeebled babies, who were often further weakened by laudanum and other opiates. These drugs were administered to the baby so that the exhausted parents, whose working day might be anything up to 18 hours, could enjoy a little sleep. Syrups containing narcotics were cheap, and available from all chemists' shops. They had names like Mother's Helper, Soothing Syrup, Atkinson's Royal Infants Preservative, Dalby's Carminative (suppressor of wind), and the most famous of all, Godfrey's Cordial. This last was a dispersion of opium in treacle; it was made up by the pharmacist or his apprentice, and jars of the sinister fluid stood on shop counters. According to one investigation, 10 gallons, enough for 12,000 doses, were sold weekly in the city of Coventry, and administered to 3000 infants under the age of two. In the United States the practice was equally prevalent, and here, too, a host of preparations could be bought over the counter. They included Ayer's Cherry Pectoral, the candidly styled McMunn's Elixir of Opium, and the popular Mrs Winslow's Soothing Syrup, which contained morphine and was preferred during the teething period. Some parents, especially the Irish, did not approve of these poisons and preferred a dash of alcohol, or even methylated spirits, added to the milk to induce sleep. Another dire practice was to bubble coal-gas through the milk, for then the carbon monoxide would cause anaemia and an ensuing torpor. Similar measures were used in other countries. It was little wonder that infant mortality throughout the 19th century could reach 80% in some places. Legislation to exclude opiates from infant foods was not brought in until well into the 20th century. As late as 1920, a preparation called Gripe Cordial could proudly advertise itself as 'infants' preservative without laudanum'. In the United States even declaring the contents on the label was resisted by Congress as late as 1911, and it required resolute action by the Bureau of Chemistry of the Department of Agriculture (the precursor of the respected Food and Drug Administration, the FDA) to ensure that labelling regulations became more stringent and that the ubiquitous narcotics were excluded from infants' and children's foods and medicines. At length, in 1915 even the manufacturers of the profitable Mrs Winslow's Soothing Syrup were brought to book for mislabelling, and not only morphine itself but even the designation 'soothing' were barred.

The infants of the rich had other problems to confront. Wet-nurses were still widely employed, with more attention than in the past given to their health and character (p. 95). A French savant enunciated the rule, widely accepted, that the milk of blonde women was inferior to that of the dark-haired. Liebig's 'perfect' infant food (Chapter 5) and its successors brought distress to the babies of the emerging middle classes. In England a company – Liebig's Registered Concentrated Milk Co. Ltd. – was established, but did not prosper. To make the product more attractive, that is to say cheaper and more convenient to use, a dried powdered version was devised. This contained less milk, and some pea flour. It, too, did less well than had been hoped, and attracted some criticism on grounds of poor digestibility. Liebig's concoctions did, though, start a whole new industry.[3] Henri Nestlé of Geneva was first off the mark. He adopted Liebig's inflated style to advertise his dried, milk-based food: it would guard against or cure plastic and respiratory ailments. It was made from 'good Swiss milk and bread, cooked after a method of my invention, mixed in proportion, scientifically correct, so as to form a food which leaves nothing to be desired'. Innumerable, for the most part highly farinaceous and aggressively advertised, formula foods followed. Among the most famous and durable were Horlick's Food for Infants, and later Horlick's Malted Milk, invented by two English brothers, James and William Horlick, who set up their business in Chicago, and Mellin's Food for Infants and Invalids, the brainchild of an English pharmacist, Gustav Mellin, who settled in the United States. Many of the preparations were based on digested protein, or 'peptones' (Chapter 5). They included (in addition to the prototypic Volmer's Mother's Milk and the 'pancreatizing powders', referred to in Chapter 5) Eskay's Albuminized Food, Carrick's Lacto-Preparata, and Wells, Richardson and Company's Lactated Food. Others again, such as Imperial Granum and Robinson's Patent Barley (as well as Eskay's), contained variously processed grains. In America, especially, these were given saturation coverage in women's journals, and the manufacturers also distributed pamphlets on infant nutrition. All flaunted fulsome commendations from doctors, and laid stress on the rigorous scientific basis on which they had been conceived. The emphasis was always on the rate at which babies thus nourished gained weight. The advertising sought to foment guilt in mothers whose babies cried frequently or appeared puny compared to the chubby, dimpled, giggling monsters pictured in the advertisement. Because of the high carbohydrate content of the preparations, especially the sugar in many of them, the babies did indeed tend to grow unnaturally fat. One product was even promoted by a picture of a baby's head on a pig's body. What was more, there were pictures of rotund, smiling babies with bowed legs that showed all the signs of rickets. By the last two decades of the 19th century the patent baby foods were attracting unfavourable attention from paediatricians. Analysts were finding them to be, for the most part, an imposture, made of flour, sugar, salt, dried milk, and little else. It was nevertheless many years before these pernicious concoctions slowly fell out of favour.

Rickets, then, was a disease that still afflicted the rich as well as the poor. Its incidence rose during the latter part of the 19th century because of the inadequacy of the milk and the decline in most countries of the vitamin content of bread, following the introduction of the roller-mill. Bandy legs and bad teeth were the most overt signs. False teeth, or as they were originally known, Patent Masticators, could be afforded only by the wealthy, who until then had relied on implanted teeth recovered from cadavers – called 'Waterloo teeth' in the decade after 1815, for the corpses on the battlefield yielded a rich harvest of sound teeth from strapping young men. Scurvy was still apt to strike the children of the poor and the inmates of prisons and sometimes workhouses.

Gout – 'the disease of kings and the king of diseases' – and stone were considered the marks of an affluent lifestyle. 'Gout', Lord Chesterfield instructed his son, 'is the distemper of a gentleman; whereas the rheumatism is the distemper of a hackney-coachman or chair-man, who is obliged to be out in all weathers and at all hours'[4] – and could not, he might have added, afford the heroic quantities of port that his betters consumed. A huge intake of protein and commonly of alcohol did indeed dominate the diet. Both were considered inseparable from well-being in the 18th and 19th centuries. (As Oliver Goldsmith put it, 'Possession's beef and ale …'.) But, as now appears (p. 121), it is lead poisoning, mainly from wine, that may well have been the primary cause. Gout and kidney stones both arise from a defect of metabolism that leads to the accretion of uric acid, a normal intermediate in the breakdown of proteins. It is insoluble and so produces solid deposits. In the case of gout, these take the form of needle-sharp crystals or tophi, which accumulate in the joints. Thomas Sydenham, a leading physician of the

17th century – the great Herman Boerhaave was said always to have doffed his hat when Sydenham's name was mentioned – was himself a sufferer from the age of 25, and gave a famous description:

> The pain is like that of a dislocation and yet the parts feel as if cold water had been poured over them … . Now it is a gnawing pain and now it is a pressure and tightening. So exquisite and lively meanwhile is the feeling of the part affected, that it cannot bear the weight of the bedclothes nor the jar of a person walking in the room.[5]

As the 19th century progressed, the incidence of gout diminished. Today, in the Western world, it afflicts 840 people in 100,000 and is treatable.

Taking a grip

As the 20th century dawned, so also, in Britain, did the realization that something had to be done to relieve the desperate conditions in which so many people still lived. The reports of concerned individuals, such as Seebohm Rowntree (Chapter 1) and Charles and William Booth (of the Salvation Army), could no longer be ignored. In 1911 the statistics still told a dismal story: 1.4 million people had an income greater than £700 a year, another 4.1 million earned between £160 and £700, while 39 million subsisted on less than £160, many of them on the poverty line. A number of reforms were introduced in the decade preceding the Great War. The School Meals Act allowed local authorities to increase rates by up to a halfpenny in the pound in order to provide daily meals for children too hungry to attend to their lessons. Such of the poor as had survived to 70 became entitled to a modest pension.

The rich meanwhile basked in Edwardian opulence. A sizeable proportion of those in work were 'in service', and saw to the demands of the burgeoning middle class, as well as the very rich. The many books on cooking and household management that appeared around this time tell their own story. The Edwardian period was a high point in gastronomy in Britain and in some other European countries – the era of celebrated chefs, of a club-land awash with gourmet food and fine wines, of grand hotels and superior restaurants. The following modest menu[5] was on offer at a London restaurant for 2s 6d (today about £8 in averaged purchasing power, although it might just, in a respectable metropolitan establishment, buy you the consommé):

Hors d'oeuvres variés
Consommé Caroline Crème à la Reine
Sole Colbert
Filet Mignon Chasseur
Lasagne al Sugo
Bécassine Rôtie
Salade de Saison
Glace au Chocolat
Dessert

The Great War, once the submarine blockade had been defeated, brought relief to the poor. Not only did rationing, when it came (Chapter 1), ensure some sort of equity, but it was also a period of essentially full employment, bringing with it a surge in agricultural production. After the War food imports increased in quality and variety, especially in regard to fruit. The food industry expanded, companies such as Unilever and Heinz grew great, and large retail outlets came into being, with the advent of such chains as Lipton and Sainsbury. Convenience foods made their appearance – jellies, pre-cooked porridge, and the new breakfast foods from America, for example. Chains of cheerful tea-houses – the ABC and Lyons Corner Houses – catered for the pleasures of the less affluent. Yet malnutrition did not disappear, and even on the eve of the Second World War the unemployed, and their children especially, still went hungry, as John Boyd Orr (p. 9), to the politicians' great discomfort, revealed. Under the post-war Labour government Sir William Beveridge's plans for a Welfare State found expression, and conditions improved enormously. As we approach our own time, affluence has begun to create its own woes, as we shall presently see.

7

Cheats and Poisoners

Ancient vices

Adulteration began when men started to trade in food. The literature of Greece and Rome contains allusions to the illicit addition of flavouring and colouring matters to wine; Pliny the Elder denounced the bakers who added chalk and cattle-feed to bread. The first legal measures to curb such abuses in Europe date from the Middle Ages. In England concerns were heard about the quality (and selling short) of bread as early as the 13th century: in 1202 King John set up the Assize of Bread to regulate the bakers' trade. The sizes of loaves were fixed by law, and selling short, as well as 'expanding' the flour with ground peas or beans, became a crime. In London, delinquent bakers could be dragged through the streets on a hurdle and put in the pillory, to be pelted with stones and offal. In time laws were promulgated to curb fraud by butchers and brewers as well. In medieval London a group of officials called garbelers (from an Arabic word meaning to select or sift) were entrusted with the responsibility for sniffing out fraud. The *London Letter Book* of 1316 listed the types of malpractice qualifying for punishment – the mixing of wares of disparate kinds, or of dear produce with cheap, and dilution of fresh with stale. And:

> Item. That no man shall moisten any manner of merchandise such as saffron, alum, ginger, cloves, etc., that is to say by steeping the ginger or turning the saffron out of the sack and wetting it whereby any manner of weight may be increased, or deterioration of the merchandise arise.[1]

(In our enlightened times, of course, swelling meat by impregnating it with water is not merely legal but customary.)

In France legislation to regulate the quality of foods was progressively introduced during the 13th and 14th centuries, and laws banning the use of even harmless colouring substances from vegetables to improve the appearance of butter, cheese, and beer were enforced. In some countries transgressions were punishable by death. In Turkey, bakers reportedly paid their apprentices wages commensurate with the occupational hazard of a premature demise, for one of their obligations was to take the blame and discharge the master's penalty, should an offence be discovered. Detection was, of course, always difficult. The Hon. Robert Boyle (Chapter 4) occupied himself for a time with the problem, and devised a hydrometer for measuring the specific gravity of milk in order to determine whether it had been watered. The craft of adulteration reached its peak during the late 18th and the 19th centuries, when the Industrial Revolution was depriving the poor, driven off the land and into the city slums, of the freedom to garner their own food. More victims created more opportunities for skulduggery. As G. K. Chesterton saw it,

> God made the wicked Grocer
> For a mystery and a sign
> That men might shun the awful shops
> And go to inns to dine;
>
> .
> .
> .
>
> He sells us sands of Araby
> As sugar for cash down;
> He sweeps his shop and sells the dust
> The purest salt in town,
> He crams with cans of poisoned meat
> Poor subjects of the King
> And when they die by thousands
> Why, he laughs like anything.
>
> The hell-instructed Grocer
> Has a temple made of tin,
> And the ruin of good innkeepers
> Is loudly urged therein;
> But now the sands are running out
> From sugar of a sort.
> The Grocer trembles, for his time,
> Just like his weight, is short.

Chesterton's verse encapsulates some of the commonest frauds upon the public.

Apart from the generally insanitary standards of food preparation – Eliza Acton, for instance, was disgusted to see the baker's sweat cascading into the dough as he kneaded – pollutants of many kinds were routinely introduced. But it was the advance of chemistry that created new and varied methods of making inferior products appear more inviting (some of which survive even now.) Among the materials often shovelled into bakers' flour (besides pea or potato flour) were chalk, pipeclay, plaster of Paris (gypsum), sawdust, bonemeal, and lime. In England alum (potassium aluminium sulphate) was often added – the practice that so dismayed Justus von Liebig (p. 94). Alum in small quantities is innocuous, but aluminium compounds in bulk are toxic, and their main target is bone. In children they will disturb calcium metabolism and elicit all the symptoms of rickets. In the quantities often found in English bread alum must at times have had still worse effects on adults, as well as children. In those with kidney disease it would have been lethal, and even in healthy people it could, if ingested in sufficient amount, have attacked nerves and brain, with cruel results.

In Germany, flour was a common diluent of more expensive foods. Unscrupulous butchers would make sausages, for example, from meat residues and fat, mixed with the glutinous paste produced by cooking flour with water, adding a dash of fuchsin dye to impart a cheerful red colour.* Meat from the carcasses of sick animals, of course, was often sold. In London in 1862 one-fifth of meat sold by butchers was estimated to have come from diseased animals, or often cows that had died of old age. Dairy products were almost invariably contaminated. The popular Gloucester cheese, for example, would be coloured with red lead. An enterprising Italian merchant even sold grated umbrella handles as Parmesan cheese.

Most common and insidious of all was adulteration of milk, for it struck at the most vulnerable section of the population, the infants. Before pasteurization (Chapter 6), milk was a source of tuberculosis and many other infections. Sick, or improperly fed cows gave milk that contained blood, pus, or mucus. Watering of milk was prevalent, and, when Boyle's milk hydrometer was introduced to measure its specific gravity, it was on occasion discovered that half the volume consisted of added water. An attentive buyer would have noticed a certain translucency, unless measures had been taken to conceal the offence. So flour, sugar, chalk, plaster of Paris, or ground rice, or a mixture of several of these, could be added, and a little later pulped calves' brains were found highly effective, especially to simulate cream. Another repellent stratagem was to throw snails into watered milk to thicken it with their mucus and lend an attractive froth. In the United States during the 19th century adulteration was almost universal. Skimmed milk was sold as the real thing, a nice creamy colour being most often restored by addition of the extremely poisonous yellow pigment, lead chromate. In London in 1877 the Local Government Board discovered that a quarter of all milk sold was heavily adulterated. Small wonder, then, that infant mortality was high. Elsewhere in Europe such practices were equally common, and the children of the poor seldom thrived. In Berlin, for instance, in the period following the

* It is worth recalling, though, that the traditional British sausage never contained more than a modest proportion of meat, much less meat of prime quality. At the Brussels World Fair in 1958 the United Kingdom exhibition presented the British way of life in the form of two 'traditional' pubs, called The Britannia and The Horse and Hounds, both serving traditional English fare, including the cherished 'banger'. But Belgian legislation required sausages to contain no less than 90% meat. The food scientists and chefs of the suppliers, Harrods of Knightsbridge, were enjoined to apply their skills to the urgent task of designing a sausage made with 90% meat, while still tasting as though it contained perhaps 30%. To the credit of British science their efforts bore fruit, and for many years afterwards the meat-rich sausages could still be bought in Knightsbridge.

founding of the German Empire in 1871 some 40% of babies did not survive beyond their first year.

Beer and wine were often a veritable witches' brew of unwholesome chemicals. The practice of watering wine and adding colouring and flavouring matters has its origins in ancient times. Among the materials sometimes employed, according to classical sources, were salt, pitch, and a variety of spices. In Europe dilution and adulteration caused perennial vexation to drinkers. A recently uncovered 15th-century English church mural illustrates the indignation that the watering of beer excited: it depicts devils dragging down to hell a group of ale-wives, guilty presumably of this misdemeanour. Chaucer alludes in 'The Pardoner's Tale' to the common practice of tampering with wine. Winegrowers were always experimenting with methods of enhancing taste, colour, and bouquet. In southern France aloes were sometimes added, and smoke was blown into the barrels to darken the wine and intensify its flavour. In Frankfurt in the early 15th century a body was set up to control what could legitimately be added to wine, and came up with a formidable list that included egg white, clay, milk, mustard, brandy, flour, copper sulphate, loam, eggshells, mussel shells, ginger, juniper, sloes, and alum. The last was particularly valued for sharpening taste and deepening colour, and in the concentrations that would have been used it was probably harmless. Depending on the locality, different additives were allowed or forbidden. In Ulm at the end of the 15th century, for instance, a ban was placed on woad as colouring matter, and on mustard, lime, bacon, vitriol (sulphuric acid), apple or pear juice, and white lead (a form of lead carbonate).

Lead in one form or another found its way into wine all too often, sometimes when lead drinking vessels were used, especially with acidic wines that dissolved the metal, and most alarmingly when lead acetate ('sugar of lead', or in France *sucre de Saturne*) was added as a sweetener and preservative. Lead poisoning, occasioned in this manner, accounted, or so the conjecture goes, for the fall of the Roman Empire, since one of the many unpleasant manifestations of the condition is male sterility. Successive emperors consequently left no heirs, and internecine warfare ensued. It is also now thought that lead may have been a governing factor in the prevalence in Britain of gout, the 'monarch's disease', which so grievously afflicted the prosperous classes throughout the 18th and 19th centuries. Though certainly not helped by over-indulgence, gout can, it seems, be aggravated or elicited by ingestion of lead salts. It was not at the time recognized that the disease often also has a hereditary component, nor was it always accurately diagnosed, for griping pains in the gut, caused by leaded wine, would be referred to as 'visceral gout'. This, as well as true gout, at all events, became the scourge of the rich after the Methuen Treaty of 1703 with Portugal, which caused a torrent of fortified wine, especially port, to engulf the British Isles. Much of this must have been, or become, contaminated with lead. The toxicity of lead compounds was not finally acknowledged until well into the 18th century.

In 18th-century France a fraud was uncovered by the *fermiers* (Chapter 4), who noticed that inexplicable quantities of '*vin gaté*', or soured wine, which was used to make vinegar, were being brought into Paris. On looking into the matter, the *fermiers* discovered that the supposed purveyors of vinegar, who took in the shipments, were selling wine. They were, it transpired, treating the soured wine with litharge, or white lead oxide, which reacts with and neutralizes acids and which thus evidently allowed the now highly toxic vinegar to pass for wine. Lead acetate was also thrown in for additional body.

Some of what passed for wine was totally innocent of grapes, and not necessarily illegal. *The Innkeeper's and Butler's Guide*, published in London in 1805, gave a recipe for 'English claret'. It was prepared by fermenting six gallons of water with two of cider and eight pounds of squashed raisins; the brew was then strained and barberries (a sour wild berry), some juice of raspberries and cherries, and a little mustard-seed were added, and after further fermentation in a warm place the fluid was bottled and allowed to mature. Then 'when it becomes fine and ripe it will be like common claret'. Californian 'sherry', later in the 19th century, was a brew of white wine, grape syrup, an extract of walnuts or hickory nuts for colour, and either quassia, or bitter aloes (both astringent plant extracts) to add intensity. Of course, many additives were, and are still, accepted as legitimate. Pine resin, added to wine in ancient Greece as a preservative, still gives retsina its characteristic astringency, and sodium hydrogen sulphite remains a permissible preservative.

Adulteration with unwholesome substances is not unknown in our time. In the 1970s a scandal erupted in Italy, when analysis of wine marketed by a supposedly reputable shipper made use of the lees recovered from the holds of banana boats and, it was alleged, tincture of iodine as sources of colour and flavour. More enduring damage was done to the reputation of Austrian wines when, in the early 1980s, new intensive cultivation methods had begun to generate a thin, acidic product, unpleasing to the palates of the mainly German consumers. The scandal broke in 1985, when it was discovered that some of the wine had been laced with diethylene glycol. This substance, used as an antifreeze, lent sweetness, body, and an attractively unctuous texture to the wine, but it is toxic and can cause serious liver damage. The adulterated wine was furnished with forged certificates of provenance and loaded into wine-tankers bound mainly for Germany. It was a decade before Austrian wines again found any market abroad.*

* The fraud was discovered when one of the conspirators, whose greed surpassed his intelligence, entered a claim for the cost of diethylene glycol in his income tax declaration. The toxic properties of diethylene glycol were already known by then, for in 1937 a small drug company in the United States had marketed the most widely used bactericidal 'sulfa' drug, sulphanilamide, in the form of Elixir Sulfanilamide. The solvent was a mixture of the sweet-tasting diethylene glycol and water. It caused catastrophic kidney and liver failure, which resulted in more than 100 deaths and many more cases of severe and lasting illness. At first suspicion fell on the drug itself, but it was soon discovered that the diethylene glycol was the culprit. The concentration in the Austrian wine was of course very much lower, and may indeed have been relatively innocuous, at least to moderate drinkers.

A tragic case of adulteration occurred in Spain from the addition of neurotoxic oils to olive oil. The precedent was the common practice during the 19th century of 'expanding' salad oils with machine oil. The physicist, William Wollaston, devised a sensitive hydrometer for identifying different oils by their densities, and for estimating the purity of oils and the strengths of alcoholic drinks; tables were provided to guide the analysts. In the United States cod liver oil, which was given to the children of the better off throughout the same period, was often mixed with train oil.

Beer was a target for swindlers throughout history. Ground marble or oyster shells could absorb some of the substances that gave stale beer its unpleasant after-taste. Bitter roots, herbs, and often the bean from which strychnine is derived might be added, especially in Germany, and sometimes even the bitter strychnine itself. In the mid-19th century a process chemist, Johann Peter Griess, working for Allsopps of Burton on Trent, discovered a new class of dyestuff, the azo dyes. He synthesized the first of these, starting from picric acid, which was itself used as a yellow wool dye. The name derives from the Greek for 'bitter', and Allsopps took to adding picric acid routinely to their beer; it is now known to cause liver damage when ingested in quantity. Bone ash, sheep intestines and other animal matter, even carpenter's glue, and very frequently sulphuric acid, were used to clarify and 'improve' the product. In France powdered dried pimento was used as a colorant, and in England, coperas, that is iron (ferrous) sulphate, also called green vitriol. The highly toxic lead chromate and copper sulphate came into use for the same purpose, or, if the consumer was lucky, turmeric or liquorice.

Tea in the 16th and 17th centuries came from China and only later from India. In England it was much abused. Merchants repackaged and sold used tealeaves, collected from hotels and eating houses, washed and dried down on hot copper plates to impart a lustrous colour. Tealeaves were also often mixed with all manner of extraneous substances, such as blackthorn, elder, ash, or sloe leaves, graphite, and iron filings. Ferric ferrocyanide (or 'Prussian Blue', which in small quantities at least is harmless) would be used to lend colour, and also logwood, which generates a deep red infusion, turmeric, sloes, and many unpleasant chemicals. ('The subtle Chinese', as an 18th-century English writer on the subject called them, had a test for the purity of tea: addition of sulphuric acid, or vitriol, to the leaves produced a characteristic bluish colour, while alien leaves gave rise to other, brighter hues.) Sometimes gunpowder was added, although this may have been chosen for its supposed medicinal virtues. (Francis Galton, eccentric pioneer of genetics and kinsman of Charles Darwin, recommended in his book of advice to travellers a slug of gunpowder suspended in whisky as a restorative for the weary voyager.)

An Act of Parliament in Britain in 1803 forbade the adulteration of coffee, tea, or cocoa with 'burnt, scorched or roasted pease, beans or other vegetable substances' on pain of a fine of £100 — a considerable sum. A variety of alien plants were in fact commonly used for the same infamous purpose. Common contaminants included roasted roots of dandelion, carrot, parsnip, and beet, and baked seeds (acorns, lupins, and others). An English chemist, by the name of Jackson, had reported on such practices in 1758, but the introduction of chemical and microscopic analysis in a systematic way had to wait nearly another century.

Poisoned sweets

During the most creative period in the history of adulteration – the 18th and early 19th centuries – it was confectionery that harboured the most lethal additives, for it was brilliant colours that lured the consumers, especially children. Among the chemicals on the confectioners' shelves were mercuric sulphide (cinnabar, vermilion, or red sulfuret of mercury), which is said to have spread death among the royal families of ancient Egypt who used it as a cosmetic), copper carbonate (verdigris), copper sulphate (blue vitriol), copper arsenite (Scheele's green), red lead (a mixed lead oxide), and the brilliant yellow lead chromate. These and related substances would have caused anaemia, bone disease, and rotting teeth.

Frederick Accum – virtue traduced

One of the true heroes in the long history of food adulteration was Frederick Accum,[2] born Carl Friedrich Marcus in the small town of Bückeburg, near Hanover, in 1769. His father was a Jew, his mother a Huguenot refugee. The father had a modest business, making soap and other household materials, and this may have stimulated his son's precocious interest in chemistry. Carl Friedrich was apprenticed to an apothecary, a member of the well-respected local Brande family, who apparently arranged for him to travel to London in 1793, there to join the establishment of a relative, William Brande, apothecary to the Hanoverian King George III. The premises, fashionably located in Arlington Street off Piccadilly, attracted personages of rank, with some of whom Friedrich Marcus, now Frederick Christian Accum, became acquainted. One who took a fancy to the young man was a chemist and man of culture, William Nicholson. Nicholson was busy as a writer and editor, and had founded a periodical, the *Journal of Natural Philosophy and the Arts*, known to all as 'Nicholson's Journal'. Its success is clear from the remarkable number of distinguished savants who contributed to its pages. Accum translated foreign scientific publications for the intellectually voracious Nicholson and made himself useful in a number of other ways. At about this time he became absorbed by the problem of food and drug adulteration, and he began to ponder how the newly developed methods of analytical chemistry might be put to use in detecting such abuses. In 1798 he began to air these interests in Nicholson's Journal.

It was not long before Accum set up his own analytical laboratory and factory at 10 Old Compton Street in Soho. He made apparatus and chemicals and undertook contract analyses of many products. He had already been appointed 'chemical operator' at the Royal Institution, close to Mr Brande's premises, where he had often attended lectures. In this capacity he assisted the director, Humphry Davy (a frequent fellow-contributor to Nicholson's Journal) in his famous public demonstrations. Accum also found time for yet another activity – teaching chemistry, as Professor at an organization called the Surrey Institution, in Blackfriars Bridge Road by the Thames. The beginning of the 19th century was a period of great intellectual ferment in the country, and scientific discourses were the rage among the *beau-monde* of London, so Accum's lecture demonstrations attracted large audiences. In 1803 his two-volume treatise, *System of Theoretical and Practical Chemistry*, appeared and further consolidated his reputation. Several other books followed, including *Chemical Amusement*, a book of home experiments for amateurs, which also served as a text for his many private students. The business thrived, and because among Accum's students were numbered Benjamin Silliman, first incumbent of the newly created Chair of Chemistry at Yale, William Peck, who was to be similarly elevated at Harvard, and the mineralogist J. F. Dana, he became one of the principal suppliers of apparatus and chemicals to several American seats of learning.

In 1809 Accum was called upon by Frederick Winsor to fight off a challenge to a patent that Winsor had been awarded for a new method of preparing coal gas. Accum appeared as an expert witness before a committee of the House of Commons, followed by another at the House of Lords. He was ill-prepared for the first of these and performed poorly, but prevailed before the second. The result was the passage of a Parliamentary bill that allowed Winsor's Chartered Gas-Light and Coke Company to establish itself as the sole provider of lighting to the streets of London. Accum published a comprehensive book on the subject of street-lighting, and his fame grew. Yet, though admired by his pupils, such as Silliman, who retained the highest regard for his mentor, Accum had developed a certain arrogance and a capacity for making enemies, a trait for which he was to pay a heavy price.

It was in 1820 that he published the work that made him famous. It was called *A Treatise on the Adulterations of Food and Culinary Poisons: exhibiting the fraudulent sophistications of bread, beer, wine, spirits, spirituous liquors, tea, coffee, cream, confectionery, vinegar, mustard, pepper, cheese, olive oil, pickles, and other articles employed in domestic economy and methods of detecting them*, and it pulled no punches. It listed the numerous adulterating agents that Accum and others had detected in foods in England, and it named the names and addresses of the known suppliers of contaminated comestibles. The book was a lurid production: the cover depicted a spider's web, surrounded by a design of writhing snakes, and in the centre was the spider in the act of seizing its prey. Above was a skull and crossbones, surmounting an inscription that read '*There is Death in the Pot*' (a Biblical quotation). The first chapter addressed the quality of water, and there followed a chapter on wine and beer. Beer was clearly a subject of fascination to Accum.

Nearly all beers, he had ascertained, had been tampered with. The porter – considered the prince of beers – that the big London breweries were delivering to their retail outlets contained 7.25% alcohol, but by the time it reached the consumer this had dropped to 4.5%. It had been diluted, therefore, by cheaper brews with lower alcohol content, or with water, and then it received additives, most commonly iron (ferrous) sulphate, alum, and salt. These conveyed a degree of sharpness and helped to generate a frothy ('cauliflower') head. An extract of the poisonous *Cocculus indicus* berry, used in dyeing, served to enrich the colour. These practices were of course illegal, and the dealer would take some care to conceal his activities, smuggling the additives into his premises in the pockets of a loose coat. At the end Accum listed the malefactors who had been caught, detailing the findings of the courts, and he named also the pharmacists who supplied them with illicit materials.

Later chapters of Accum's book considered every class of food in turn. Copper salts were ubiquitous, added either as copper sulphate or from copper vessels in which foods such as acid pickles or fruits were boiled, or from added coins. Confectionery, Accum found, was coloured with green copper compounds, or the bright red lead oxide (red lead or vermilion), the vivid 'red sulfuret of mercury' (mercuric sulphide), white lead (lead carbonate), and perhaps most alarmingly, Scheele's green, or copper arsenite. Grocers did their own chemistry to increase the bulk or appearance of the produce. Red lead was used to impart a more lively colour to faded cayenne pepper (a much-favoured condiment), and sweepings from the floor were added to ground black pepper. In his book and in articles in Nicholson's Journal Accum once more named the miscreants, who were often brought before the magistrates and made to pay hefty fines. This was the practice that earned him a series of powerful enemies, who eventually contrived his downfall. Meanwhile the first printing of a thousand copies of his book sold out within a month. A second edition was in the shops before the end of the year, and it also appeared that same year in the United States and a little later in Germany.

Threats of various kinds having proved ineffective, Accum was charged a matter of months after publication with damaging books in the Royal Institution library, where he had at one time served as librarian. He had, it was alleged, been tearing pages out of the books for years and taking them home. This improbable accusation was taken seriously, and the Royal Institution's librarian, accompanied by police with a search-warrant, entered Accum's house. They identified what was probably waste-paper as pages from the books and Accum was arrested, but discharged by the magistrates. Further accusations followed, however, and Accum, apparently frightened and depressed, failed to appear at the trial, and so forfeited his bail. Soon afterwards he left London to surface in Berlin, where he seems to have been well received, for he taught as professor in two institutions, a trade college and the Building Academy. In 1826 he published his final book, on the subject of building materials. Accum's disgrace in London evidently affected him deeply, for he never published another research paper under his own name. Some that appeared in the proceedings of the Prussian Academy of Sciences were presented either anonymously or written under the transparent pseudonym of Mucca. New editions of his books were printed in London, but with his name expunged from the title pages. Accum's disappearance from the scene allowed his enemies in the food and drink trades to resume their pernicious ways more or less unchallenged for another four decades.

The sleuth at the microscope

Adulteration was so deeply embedded in the culture of the food and drink industry, and pursued with so much ingenuity, that detection demanded much skill, courage, and persistence, the more so after Accum's undoing. His baton passed in the first place to another London-based analyst, John Mitchell, who in 1848 published his *Treatise on the Falsification of Foods and the Chemical Means Used to Detect Them,* and gave evidence before a Parliamentary Select Committee on Adulteration of Food. He had, he testified, analysed 'by chemical means' 200 samples of beer over a period of nine years, and, apart from a few taken at the brewery, none had been free of contaminants. It was no secret by then that watering by publicans was universal practice – understandable perhaps, because profit margins in the public houses were small and the hours cruelly long. An analysis of 150 beer samples had revealed that not a single one was within 20% of the advertised strength – that with which it had left the brewery. The standard procedure was to dilute the beer by one-third and add four pounds of coarse sugar and one pound of salt to each barrel, together with a dash of something that would generate a good head of froth in the glass. Another prevalent practice was to dilute high-quality with cheap beer. The Guinness brewery in Ireland sold their famous stout very selectively and supplied mainly porter, knowing that the stout would anyway be mixed with rather less expensive porter. In the United States adulteration of beer was also customary. Salicylic acid or tannin would be added as a preservative, and the bitter taste enhanced with cheap substitutes for hops – quassia, aloes, and *Cocculus indicus*, as before. To improve the texture, the highly toxic glycol, together with glycolic acid (by-products of rendering of animal fat), were introduced, and often a little of the opioid narcotic valerian was included in the expectation that it would induce an agreeable stupor. A retired brewer in Rochester (New York) told the local newspaper that the brewery office resembled a chemistry laboratory. Wines were treated with no greater respect, and brandy was distilled from soured wines, and suitably 'improved'. One had no greater chance, a retired distiller declared, of coming by pure brandy in New York than being struck by lightning.

In Britain, it was one resolute crusader who eventually turned the tide of fraud. Arthur Hill Hassall[3] is commemorated in the anatomy and pathology textbooks by the microscopic bodies in the thyroid gland that bear his name. He was a highly accomplished microscopist, whose micrographs far surpassed most of the best efforts of his contemporaries. The notion that alien substances in foods might be detectable by microscopy was probably first mooted by John Thomas Queckett, Lecturer in Histology at the Royal College of Surgeons in London (who gave his name to the Queckett Club, the professional British association of microscopists). Hassall took up this challenge. He had already performed an influential study on London water, which had disclosed a dismaying level of alien matter, living and dead. The report was published in 1850 and that same year he began his researches on foods. Hassall was not opposed to chemical methods of searching for adulterants, and indeed made occasional use of them, but they could not in general, as he pointed out, detect foreign plant or animal matter. His first undertaking was to examine samples of coffee, which he bought from several shops in London. To his great satisfaction he found that he could easily distinguish between ground coffee beans and such contaminants as chicory, wheat, rye, beans, and peas; their microscopic structures, it turned out, survived both roasting and pulverizing. Nearly all the coffees, Hassall discovered, were extensively adulterated, and in the worst cases were almost entirely chicory. He reported the outcome of his study to the Botanical Society of London, which made it the basis of a press release. The ensuing publicity caused much indignation, and the Chancellor of the Exchequer, Sir Charles Wood, denied that such infamy could abound in the land. But the evidence presented by the Botanical Society was incontrovertible.

Thereupon Hassall received a communication from Thomas Wakley, the editor, founder, and owner of *The Lancet*. Would Dr Hassall pay him a visit? Wakley, a surgeon and Member of Parliament for Finsbury, had espoused many good causes and ventilated them in his increasingly influential journal. He had been agitating since Accum's time for legislation to curb fraud in the production and sale of food, and his indignation may have been stiffened by a distressing episode in 1850, when a large number of children, housed in the Drouitt Institution for paupers, died of gastrointestinal poisoning. In his capacity as coroner for West Middlesex, Wakley concluded that their deaths had been occasioned by oatmeal 'extended' with ground barley, a notorious laxative irritant. Wakley now invited Hassall to widen his investigations on food adulteration and report the results in *The Lancet*. No revelations, however shocking, he told Hassall, would have any effect unless the malefactors were named. This was a policy that demanded great courage, for a slip by Hassall would lay Wakley and his publishers open to legal action, with probably calamitous consequences. The responsibility also bore heavily on Hassall, but he never wavered. Hassall certainly needed no convincing of the seriousness of the problem. In his memoirs he recounts the story circulating in London of the grocer who would summon his assistant from the cellar with the question, 'Have you watered the treacle and sanded the sugar?' 'Yes.' 'Then come up to prayers.'[4] In 1851 a series of devastating reports started to appear in *The Lancet*, initially anonymous, but later under Hassall's name.

Hassall and his assistant, Henry Miller, an artist engaged to reproduce the projected microscope images, would go out into London at night and call at grocers' shops. Most often one of them would enter and make the purchase, while the other observed the transaction from the doorway, so that he could appear, if needed, as a witness in court. The address of the shop would be inscribed on the package, and the wrapping dated and signed. The first report was again about coffee, and here Hassall extended the scope of his earlier observations. The degree of adulteration varied inversely with cost. Of 34 coffees examined, 2 at the highest price were pure, and the purveyors were given credit. All the rest were adulterated: there was chicory in 31, roasted corn in 12, beans in 1, and potato flour in another. At medium price, Finest Turkey Coffee, for instance, contained chicory, roasted corn, and very little coffee; none were more than half coffee. At the lowest price, coffees with such alluring names as Parisian Coffee, Delicious Coffee, and Superb Coffee contained, at most, only traces of coffee. The street addresses, though not the names, of the traders were stated.

Next Hassall tackled sugar. The main upshot was the discovery in massive profusion of a 'louse-like insect', the sugar mite, in the raw sugar that was then mainly sold. The creatures could be seen in all stages of development, from eggs and embryos to the mature mite. The public revulsion that this revelation occasioned led eventually to the development of sugar refining on a large scale. His microscope technique allowed Hassall to detect minute proportions of contaminants in foods. A mustard manufacturer claimed that his product was pure, but Hassall found a trace of turmeric. When taxed with this, the manufacturer readily admitted that he used turmeric to improve the colour, but the amount was too small to be significant – 1 part in 448.

Milk proved tolerably pure, though sometimes watered and coloured with a little turmeric to impart a more creamy appearance. Sometimes a thickener, such as gum tragacanth was added, and also sodium carbonate to prevent, or conceal, souring. Potted meats and fish were seldom pure. Favoured colouring materials were 'bole Armenian' and Venetian Red, both reddish-brown clays comprising mainly iron (ferric) oxide. At the aristocratic Fortnum and Mason of Piccadilly the rich could buy bloater paste 'of a very deep and most unnatural red colour, arising from the presence of a large quantity of *bole Armenian*; the pot also contained much starch'. Bole Armenian, indeed, appeared in 23 of the 28 samples of potted meats and fish. Boiled starch was also a ubiquitous filler, while in curry powders ground rice made up the bulk, together with salt and often a generous dash of red lead for a more convincing appearance. Samples of cayenne pepper contained abundant amounts of ground rice, brick dust, and a variety of red pigments, some harmless, many noxious. Vinegar was often fortified with sulphuric acid, but pickles told a shocking story. Besides sulphuric acid, most samples contained copper, often in alarming, if not lethal, concentrations. This was equally true of bottled vegetables and fruit, which were heavily adulterated, especially again with copper. Animated by Hassall's report, Mr Albert Bernays, an analytical chemist from Derby, wrote to *The Lancet* to relate his own startling experience:

> I had bought a bottle of preserved gooseberries from one of the most respectable grocers in this town, and had had its contents transferred into a pie. It struck me that the gooseberries looked fearfully green when cooked; and on eating one with a steel fork, its intense bitterness sent me in search of the sugar. After having sweetened and mashed the gooseberries with the same steel fork, I was about to convey some to my mouth when I observed the prongs to be completely coated with a thin film of bright metallic copper.[5]

An editorial note commends Mr Bernays for his observation, but upbraids him for not naming his grocer.

Hassall next turned his fire on jams and jellies, and more disagreeable intelligence emerged. The famous victuallers to the quality, Fortnum and Mason, again fell short, for they were offering greengage jam 'of a much deeper green than natural', containing copper. Samples of bitter orange marmalade contained generous admixtures of turnips or possibly apples, and copper was everywhere, often (in 19 out of 35 samples of jams, jellies, and crystallized fruits) in dangerous amounts. Worse was to come. Artistically fashioned and brilliantly coloured sweetmeats were very popular at the time, coveted especially by children, and, as in Accum's time, they proved to contain enough metal compounds to stock a chemical laboratory. Cake decorations, for example, were often spectacular creations of the confectioner's art. There were ships in full sail, multicoloured lions, Prince of Wales's feathers, and a variety of plants and animals. The yellows were generated by lead chromate ('chrome yellow', available in two forms, pale or deep) or sometimes by gamboge, a resinous gum, the reds by the harmless cochineal or by the poisonous 'red lead', the lead oxide also called minium, or, even worse, by cinnabar or vermilion,* forms of mercuric sulphide. Then there were browns, blues, and greens, among which the persevering Hassall identified van Dyke brown, umber, and sienna, all earths comprising mainly iron (ferric) oxide, indigo (by no means good for the constitution), Prussian blue or ferric ferrocyanide, ultramarine (an aluminium compound), Brunswick green, verditer, and emerald green, also called Scheele's green, all copper compounds, the last (copper arsenite) quite deadly, as we have seen, and various mixtures of these pigments. White lead (lead carbonate) was also a common ingredient of these terrifying confections. It is not surprising that the medical journals reported many cases of life-threatening poisoning, especially of children, and it was surmised that painful deaths from lead and mercury poisoning were far from rare.

The results of his investigations appalled Hassall, Wakley, and the readers of *The Lancet*. The respected *Quarterly Review* declaimed: 'Nay, to such a pitch of

* Vermilion was originally another term for cochineal, derived from the Latin *vermiculus*, a little worm, from the cochineal insect from which the dye was distilled.

refinement has the art of fabrication of alimentary substances reached, that the very articles used to adulterate them are themselves adulterated; and while one tradesman is picking the pockets of his customers, a still more cunning rogue is, unknown to himself, deep in his own'. Parliament was compelled to take notice and in 1860, after tedious debate, the Food and Drugs Act was passed. But this was a largely toothless and ineffectual measure. The reforming Member of Parliament, John Bright, opined that adulteration was inseparable from trade, more, 'the sign of a vigorous and competitive' community. Enforceable regulation of the industry would undermine the economy. The 1860 Act did at least allow for the appointment of public analysts. The first and most distinguished of these was Henry Letheby in the City of London, to whom Hassall often turned for help. At last, in 1875, Parliament enacted more rigorous legislation with the passage of the Sale of Food and Drugs Act, and wider encouragement for local authorities to appoint their own public analysts. The analysts had already created a professional body, the Society of Public Analysts, with its own journal. The Society laid down recommended analytical procedures, but standards of precision and skill remained inconsistent until well into the 20th century, and legal requirements lax. There was nothing, for instance – until the suffragette Sylvia Pankhurst denounced the exploitation of the women employed in the venture – to prevent tiny chips of wood from being fashioned into simulated raspberry pips. These were then added to a nondescript jam, probably made of root vegetables. Sylvia Pankhurst was stimulated to establish a factory for the manufacture of jam from actual fruit, but it did not survive long. This episode is reminiscent of an admirable venture begun in 1881, when a model factory was built in Red Lion Square near Holborn in London to produce wholesome, unadulterated food for adults and infants (in accordance with the prescriptions of the time). It did not thrive, probably because pure food cost more than adulterated, and soon closed.

Milk and meat: the great retch

Milk was for decades the most vexed issue for analysts and consumers alike, for the lactometer, which measured its density, was of little help: a high cream content and watering both reduced the density, and the most likely impurities were the hardest to detect. In the United States milk drinkers did not like it white: a yellowish tinge would have suggested a more creamy quality, so, as in England, the wily dairymen would colour the slightly bluish skimmed milk with a dash of lead chromate. In France watering was all too common, its effect being often disguised by the addition of boiled starch and a decoction of almonds. The milk was watered first by the farmers and then by the delivery boys. Then, in 1881, officials observed railway workers engaged in an unconcealed third watering operation at the fountains outside the station at which the milk arrived for distribution in Paris.

In Britain the debate focused increasingly on bacteria, and the conditions of hygiene in the milking sheds, and on the detection of bovine tuberculosis and other diseases. Milk was thought during the Victorian period to be the most common cause of tuberculosis in the population. After Koch's discovery of the tubercle bacillus in 1882 screening became possible. A. Wynter Blyth's widely used *Manual of Public Health* of 1890 also noted the following:

> The cow passes daily an immense quantity of semiliquid manure and a large quantity of urine; there is also the atmospheric contamination from the lungs, and it is to be remembered that in addition to carbon dioxide, the cow passes from the intestine much marsh gas.

(Marsh gas is methane; it is lighter than air, and the flatus of cows makes, it has been calculated, no small contribution to the greenhouse effect.) All these effluvia often found their way into the milk. An estimate from this period concluded that the milk consumed each day in Berlin contained some 15,000 kg of cow dung. Like so many improvements in public health, the elimination of the hazards that lurked in milk was accomplished more by engineering than by science.

The condition of the slaughterhouses and the quality of the meat in the latter half of the 19th century make for queasy reading. Neither had improved much in the course of the foregoing hundred years (p. 103), nor could much reliance be placed on the suppliers of tinned meats. Here is a report from *The Times* in 1852 of the findings by a board of investigation set up by the Admiralty:

> On Tuesday, 643 of [the 10-pound tins] were opened, out of which number no fewer than 573 were condemned, their contents being messes of putrefaction. On Wednesday, 779 canisters were opened, out of which number, 734 were condemned. On Thursday, 791 canisters were opened, out of which number, 744 were condemned. On Friday (this day), 494 canisters were opened, out of which, 494 were condemned. Thus, out of 2707 canisters of meat opened, only 197 have proved fit for human food, those condemned for the most part containing such substances as pieces of heart, roots of tongues, pieces of palates, pieces of tongues, coagulated blood, pieces of liver, ligaments of the throat, pieces of intestines, – in short, garbage and putridity in a horrible state, the stench arising from which is most sickening, and the sight revolting.[6]

'The few canisters containing meat fit for human beings to eat', the report concludes, 'have been distributed ... to the deserving poor of the neighbourhood', while the rest were taken by boat to Spithead and dumped in the sea. The article conjectures that tinned meat of similar quality might have been supplied to Franklin's expedition (p. 108) and caused the catastrophe with which it met.

The French appear to have been more fastidious about their food and its origins, and there are 19th-century prints, for example, of immaculate slaughterhouses, visited by young women of fashion, who would come to drink the blood of freshly killed beasts. This, it was supposed, would bring roses to their cheeks.

In New York a small group of intrepid women, who had been agitating against various unpleasant practices in the food industry, formed themselves into The Ladies' Health Protective Association. It soon unfortunately became conflated with the temperance movement, but its achievements were for a time considerable. Fifteen women invaded the Manhattan slaughterhouses, then all clustered on Fifth Avenue between 43rd and 47th Streets, and they did not at all care for what they saw. The floors reeked of congealed and dried blood and rotting parts of animals, and onto them were thrown the sides of fresh meat. The campaign soon moved to the stockyards of Chicago, and the lurid descriptions of what went on there inspired a journalist, Upton Sinclair, who became a formidable food-faddist (Chapter 10), to take up residence in the stockyard district and report on what he found. The outcome was his celebrated novel, a sensational *succès de scandale*, *The Jungle*, published in 1906. Sinclair was in truth much more affected by the conditions under which the immigrant stockyard workers – mainly Poles and Lithuanians – laboured than in the loathsome processes by which the meat was produced and handled. But, as he later said, he aimed at the nation's heart and accidentally hit it in the stomach. There were episodes in the book that made a grisly impression on its many readers, such as the moment when a workman slips, falls into the grinding machine, and is tinned, along with the beef. There were calls for investigations, and *The New York Times* was not slow to take up the cudgels. Four months after the appearance of Sinclair's book the paper carried a vivd description of how the products were handled: the meat, it told its quivering readers, was 'shovelled from the filthy floor and piled on tables'. Then it was 'pushed from room to room in rotten box-carts, all the while collecting dirt, splinters, floor filth, and the expectorations of tubercular and other diseased workers'. Scraps of meat, 'dry, leathery, unfit to be eaten' were ground up with pieces of pigskin, rope strands, and other rubbish to make potted ham. Many canned meat samples contained spoiled or parasite-infested products. It was too much. That same year federal legislation came into force, in the form of the Pure Food, Drinks, and Drug Act, and the great Harvey Wiley, Chief of the Bureau of Chemistry of the US Department of Agriculture, and soon to be head of the Food and Drug Administration (FDA), appeared on the scene. His chemists found that most of the food sold throughout the country was chemically adulterated. Sweets, milk, pasta were routinely still being coloured with lead chromate ('egg yellow'), pickles and tinned vegetables with copper sulphate, and so on. In fact most of the abuses identified in England 40 or so years earlier by Hassall and Letheby were still practised. In 1905 *Collier's* weekly magazine had carried an exposé by a journalist, Samuel Hopkins Adams, of the iniquities perpetrated by the food and drug industries, under the title, 'The Great American Fraud'. Further articles followed in 1906, which added fuel to the public's indignation. The 1906 Act condemned the use of mineral colourings in food, and gave a schedule of organic dyes, such as William Perkins's recently discovered mauve, that were deemed acceptable. It took another 50 years or so for most European countries to follow this lead. In Britain it was only in 1957 that legislation was passed specifying which food colourings were permissible. All the same, the 1906 Act demanded only voluntary compliance. Enforcement had to await enactment of further legislation in 1938.

Harvey Washington Wiley, who cleaned up the American food industry, was born in a log cabin in Indiana in 1844. He qualified in medicine at Indiana Medical College, but his interest turned to chemistry, which he taught at Purdue University, recently founded in his home state. He specialized in agricultural chemistry and early in his career began to occupy himself with problems of adulteration. In 1882 Wiley was appointed Chief Chemist at the US Department of Agriculture. He was perturbed by the variety of substances that had come into use as preservatives, and he used his staff (the 'poison squad') as guinea pigs to test the physiological effects of these chemicals when swallowed in large quantities. The publication of the results caused much public concern, and led directly to the 1906 Act and the creation of the FDA. Wiley was a dynamo, a bachelor of unconventional habits, who raised many hackles. His great work, *Foods and their Adulteration*, which appeared in 1907 and in a second edition four years later, alerted the country to the dangers still concealed in its food. Wiley did not remain long at the helm of the FDA. In 1912 he resigned and set himself up in an independent laboratory belonging to *Good Housekeeping Magazine*, an influential organ, which issued the 'Good Housekeeping Seal of Approval', conceived by Wiley. He died in 1930 and is buried in Arlington National Cemetery. The American people owed him and his successors at the FDA an incalculable debt. It is only in recent years that many of its teeth have been drawn through the capitulation of governments to pressures from the giant food, agricultural, and pharmaceutical industries. Wiley would have been astonished and dismayed by the range of chemical additives, and indeed of wholly synthetic foods, that are now legal and commonplace. They will be discussed later (Chapter 11).

Dame Harriette Chick (1875–1977), heroine of Vienna, in her laboratory at the Lister Institute in London.

An early 16th-century painting of the Virgin and Child by Hans Bergmaier (German school). The Christ child shows the characteristic manifestations of rickets.

James Lind (1716–1794), naval surgeon, who was credited with devising, in the course of his studies on scurvy, the first controlled clinical trial.

Sir Gilbert Blane (1749–1834), surgeon and public servant, who persuaded the Admiralty in 1795 that sailors should be dosed daily with citrus juice to avert scurvy.

Andreas Vesalius (1514–1564), the Flemish physician and anatomist, commonly regarded as the founder of modern medicine.

'Keep the feet warm, the head cool and the bowels open': Hermann Boerhaave (1668–1738), Dutch physician and polymath, whose school at Leiden was the world's pre-eminent centre of medical learning.

Advertisement for Rose's Lime Juice Cordial, the first soft drink marketed in Britain, in the wake of the publicity that accompanied the issue of lime juice to the Royal Navy.

LEFT Antoine Laurent Lavoisier (1743–1794), pictured here with his wife and collaborator, Marie-Anne Paulze, in the famous painting by David of 1785. Lavoisier laid the foundations of modern chemistry and of the quantitative approach to physiology. He died on the scaffold at the height of the Terror.

BELOW James Gillray's depiction of a lecture demonstration at the Royal Institution in London in 1802, illustrating the supposed properties of the newly discovered gases ('factitious airs'). Count Rumford, founder of the RI stands on the right, and Humphry Davy, later its director, holds the bellows.

Scientific Researches! — New Discoveries in PNEUMATICKS! — or — an Experimental Lecture o
Powers of Air —
ROYAL INSTITUTION

Frederick Accum (1769–1838), chemical analyst and scourge of the adulterators of food in Britain.

Baron Justus von Liebig (1803–1873), the most powerful figure of his time in the field of organic and physiological chemistry, who used his authority to promote several of his own food products, not always for the public good.

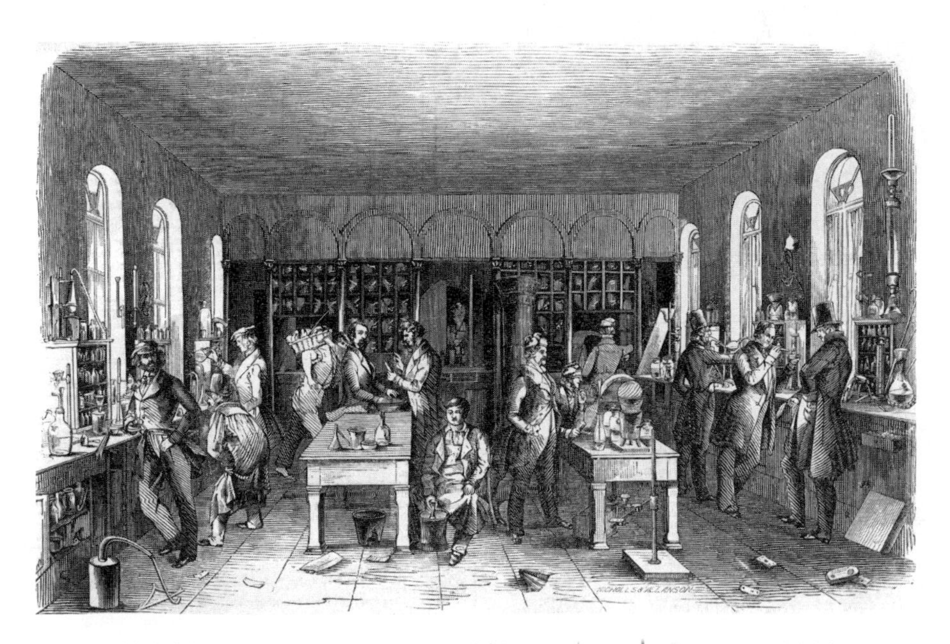

Liebig's laboratory at the University of Giessen (now the Justus von Liebig University). It attracted aspiring chemists from all over Europe and beyond.

A Paris slaughterhouse in 1874, showing young ladies of fashion, who had called in for a glass of warm blood from a freshly killed beast, to nourish their complexions.

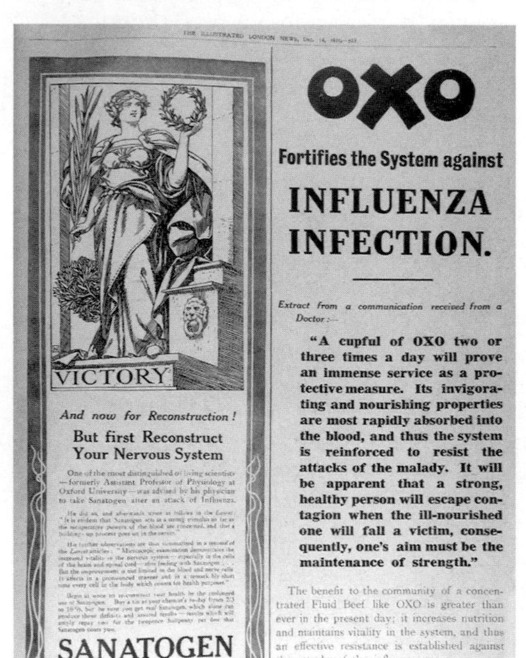

Advertisement for Oxo cubes, a concentrate inspired by the success of Liebig's *extractum carnis*, illustrating the unrestrained claims made for the health-promoting and curative properties (here for the deadly Spanish influenza of 1918) of this and other proprietary products, including also Sanatogen, seen on the left.

Dr John Harvey Kellogg (1852–1943), health crusader and enemy of the dreaded 'autointoxication', who invented the first breakfast cereal at the Battle Creek Sanitarium in Michigan.

A PLATEFUL OF HEALTH
6½d per packet
People don't think nowadays in terms of the hot, heavy breakfasts our grandparents ate. Those were meals that left one loggy and put a heavy tax on digestion. Today we know better. Now it's a plateful of delicious, crunchy Kellogg's Corn Flakes—golden-brown—oven-fresh.
Solid nourishment but in *lighter form*. Good for breakfast, lunch or supper. And always ready—no cooking needed. Serve Kellogg's Corn Flakes direct from the sealed, inside Waxtite bag, with cold milk or cream, adding sugar to taste—or with fruit or berries.
Ask your grocer for the red-and-green packet.
Kellogg's
CORN
FLAKES
OVEN-FRESH
FLAVOUR-PERFECT
Kellogg's CORN FLAKES

Advertisement for Kellogg's most famous product, still going strong in 1934, and even more so now.

A pigeon displaying the effect of a diet of polished rice, and, on the right, the same bird three hours after receiving a dose of a vitamin B preparation (from a paper of 1922 by Casimir Funk).

LEFT Advertisement for Mrs Lydia Pinkham's infamous cure-all. Its contents were never revealed, but alcohol featured prominently.

BELOW Advertisement for Bile Beans. Their contents were unknown, but they were commonly supposed to have consisted of nothing but coloured corn flour.

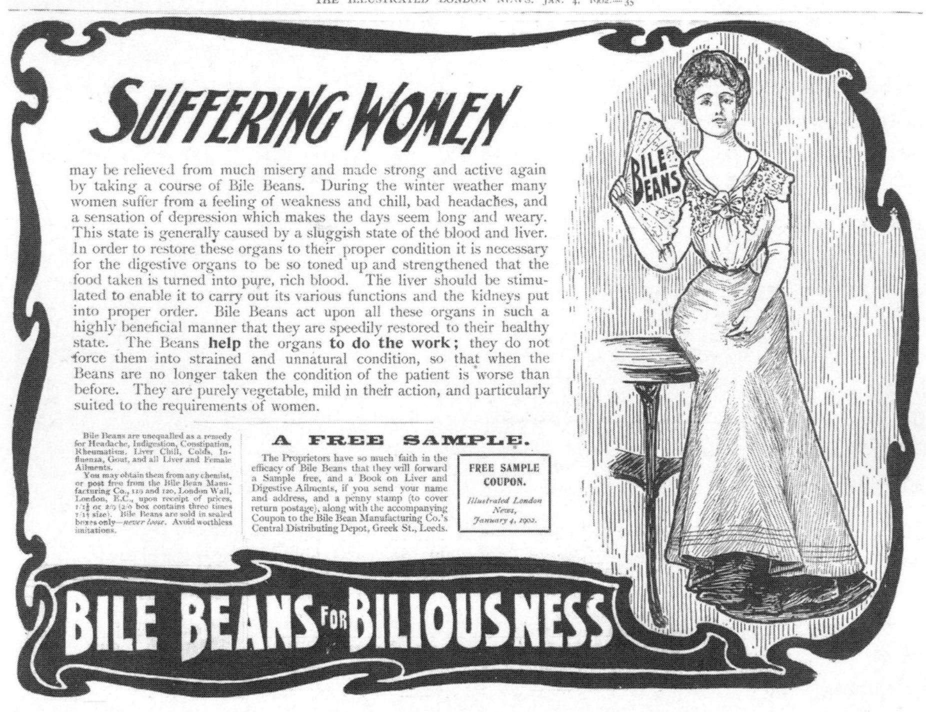

8

Paradigm Postponed: the Tardy Arrival of Vitamins

Intimations of heresy

It may have been Jonathan Pereira who first articulated doubts about the doctrine of Liebig and his successors in Germany that proteins, carbohydrates, and fats – 'the dietetic trinity' – afforded (allowing for minerals) all the nutrition that the human frame demanded. Pereira was born in London in 1804, the scion of a respected Sephardic family. He served an apprenticeship in pharmacy and surgery, acquired a broad knowledge of chemistry, lectured on the subject and on *materia medica* at the London Hospital, and published three influential books while continuing to work as a surgeon. In the last of these, *A Treatise on Food and Diet*, he questioned Liebig's dogma and enunciated his own principles of nutrition, including especially the absolute need for a varied diet, which must include both vegetables and fruit. Pereira might have taken his conclusions further, had he not died prematurely in 1853.

There were indeed many indications before the 19th century drew to a close that something besides Liebig's three principles and an adequate calorie intake was needed to sustain life. There were, for instance, Magendie's experiments on the nutrition of dogs (p. 73), but these were, for the most part, resolutely ignored, as also were Dumas's observations (with similar conclusions) during the Siege of Paris in 1871 (p. 155). This indifference was a consequence, in the main, of the prevailing mindset, sustained by the line of German physiological chemists from Liebig to Rubner. Their work on energy consumption, especially the exacting calorimetric measurements of Voit and Pettenkofer, dominated the thinking in the field. It has been argued that the 'germ theory' of disease, which emerged from the discoveries of Pasteur and Koch and swept through all of medicine, was also partly responsible, for it gave rise to the axiom that disorders could spring only from the presence of something noxious, and not from the absence of something benign. And besides, vitamins, once the hunt was on, seemed to many to be as elusive as a will o' the wisp.

The rice disease

Even though it was generally accepted by the mid-19th century that scurvy could be cured or averted by citrus and other fresh fruits and vegetables (Chapter 2), the notion that they contained a constituent essential for health, rather than some form of antidote, lay beyond the reach of men's imagination. The intellectual contortion required to arrive at such a conclusion was achieved only at the very close of the century.

When in 1867 the Emperor Matsuhito threw open the gates of Japan to the outside world, after two and a half centuries of isolation, the influx of European and American traders was soon followed by the arrival of a number of doctors. They came mainly from Germany, invited by the Emperor to introduce the methods of Western medicine into his realm. It was not long before they came upon cases of a malady that they had not encountered in Europe. This, in the words of one of them, was 'a formidable and mysterious illness, known as *kakké*'. Its first manifestations were lassitude and a numbness in the feet, accompanied by painful swelling of the ankles. This progressed up the legs and into the body. Breathing became difficult and, if there was no remission, convulsions and death from asphyxiation or heart failure would follow.

The disease was seasonal, though its incidence varied from year to year, and it respected social class. It began in late spring and reached its zenith in August. It could take either of two forms – the 'dry' paralytic, or the 'wet' oedematous variety. It spared the rich but mostly also the poor labourers, whereas artisans, servants, and the samurai were the worst afflicted. Prison inmates were particularly vulnerable. The outbreaks were generally attributed to a miasma from the damp earth, but at least one perceptive young Japanese doctor, Kanehiro Takaki, serving in the navy, was convinced that the cause lay in the diet. Takaki was chosen as a member of the first wave of his country's professional élite to be sent to the West to study and absorb the new knowledge and skills that had bypassed his own country. He had been influenced by an English doctor while a student in Tokyo, and so chose to go to England rather than (as was more or less customary) to Germany. It was at St Thomas's Hospital in London that he spent five years. Takaki's interest in *kakké*, or beriberi, had begun when he was 13. In 1906, now Baron Takaki, FRCS Eng, DCL, he delivered a lecture to the British Medical Association and related the circumstances:

> The first time I heard of the fearful disease of beri-beri was 44 years ago. At that time the [Imperial] guards were despatched by several Daimios to Kyoto to act as protectors of the Imperial Palace and my father being one of them stayed there for over a year. On his return he told me of the disease called beri-beri which killed many of these men. They attributed the cause to food and called a provision box the 'beri-beri box'. Later, in 1868, that is in the year of the Meiji Revolution, I served for eight months in the army of Prince Shimadzu but did not see any beri-beri … . I entered the navy in 1872 and began to treat beri-beri patients for the first time.[1]

The sailors, Takaki found, tended to be badly affected (which argued against the damp earth theory), their officers less so. He managed to arrange for two ships, setting out on voyages to New Zealand in December of 1882 and the following February, to be differently provisioned. The sailors in one, the *Ryujo*, received the usual rice-based diet with some fish and vegetables, those in the *Tsukuba* a similar diet, supplemented with wheat, milk, and meat. By the end of the nine-month cruise, 161 of the complement of 376 on the *Ryujo* were brought low with *kakké* and 25 died, while on the *Tsukuba* there were 14 cases and no deaths. In response to Takaki's report the naval command introduced a new standard diet, but it proved difficult to force the men to eat bread or meat. Then it was discovered that if the rice was supplemented with barley, the *kakké* subsided. A similar improvement had astonished prison governors when, for reasons of economy, barley was substituted for half the rice in the prisoners' diets. Takaki thought the cause of the disease might be protein deficiency, since the protein content of rice was found to be very low and that of barley rather greater.

In due time the connection was made between *kakké* and a similar malady, long known to missionaries in what became the Dutch East Indies, and also recognized in Ceylon (Sri Lanka), where it was known as beriberi. (The etymology is uncertain: it may derive, among other possibilities, from the Singhalese for weakness, Hindi for swelling, or from *brébis*, or ewe, referring to the staggering, sheep-like gait of the sufferers.) Then in 1868, a French naval surgeon, Le Roy de Méricourt, described what was plainly the same condition among the workers in the Cuban sugar mills and in Brazil. It was not therefore linked to any genetic predisposition of Asians. Little was done in the way of investigation until, in 1886, the sailors and the native soldiers in Dutch service, engaged in a guerrilla war against rebels opposing Dutch rule in Aceh on Sumatra, were struck down in large numbers. The Government in The Hague felt that something must be done to discover and eliminate the agent responsible for the disease. Because the 'germ theory' exercised such a powerful hold, they chose a Professor of Pathology at the University of Utrecht, Cornelis Adrianus Pekelharing, to investigate, having first sent him to Robert Koch's laboratory in Berlin to learn bacteriological techniques from the master himself. While there, Pekelharing made the acquaintance of a young compatriot, Christiaan Eijkman. Born in 1858, one of ten children of a schoolteacher, Eijkman had joined the army in order to enter the military medical school, and so spare his family the costs of a civilian medical education. He had also studied physiology and completed a doctoral thesis in that subject before receiving his posting to the Dutch East Indies. There he had his first encounters with beriberi, which was prostrating both the Dutch and the colonial troops. But after two years Eijkman was struck down by malaria, and sent back to Europe to recover. Convinced that beriberi must be caused by a bacterial infection, he decided to use his enforced leave to learn bacteriology in Berlin. Pekelharing was accompanied by another doctor, a neurologist from the University of Utrecht, Cornelis Winkler, and it is related that the two first encountered Eijkman in a café, where they were accustomed to go to read Dutch newspapers. The two older men were impressed with Eijkman's knowledge and dedication and arranged for him to join their mission.

A year's study in the East Indies served to convince Pekelharing and Winkler that beriberi was indeed an infectious disease, and that the inflammation of the nerves, which Winkler at once identified as its first manifestation, was encompassed by a bacterial toxin. So they searched for and found bacteria in the blood of patients with beriberi. Satisfied that what was now needed was a vaccine, they returned to Utrecht, Pekelharing to his Chair of Pathology and Winkler to a new appointment as Professor of Neurology. But as time went on, conflicting results came in from other quarters, and Pekelharing and Winkler began to have doubts. More work, they felt, was needed and so they prevailed on the Governor General of the Dutch East Indies to continue supporting the small laboratory on the site of a military hospital in Jakarta (or, as it then was, Batavia) and improve its facilities. Eijkman was appointed its head, and went to work to find an 'animal model' of beriberi, using the bacterial cultures prepared by Pekelharing and Winkler, as well as the blood of patients, as sources of the supposed infective agent. All efforts to infect animals failed. Although a local monkey was found to be suffering from beriberi, monkeys could no more be infected in the laboratory than rabbits or dogs. But then Eijkman had his first stroke of luck: he had turned to the study of chickens, probably because they were cheap and easily fed and housed. A laboratory assistant found that some of the birds had suddenly developed an inflammation of the nervous tissue ('polyneuritis'), very like the human victims of beriberi. Their legs splayed, they fell over and struggled unsuccessfully to get up, their combs turned blue, and they became unable to eat and soon died. This was a promising lead, especially since the disease seemed to be spreading from one chicken to another. The blood of some birds yielded bacteria – micrococci, like those found by Pekelharing and Winkler – when cultured. A new batch of healthy chickens was procured, which Eijkman divided into two groups. One he injected with his bacterial cultures and the other served as controls. All got beriberi and many died. Undismayed, Eijkman injected blood from dead chickens into healthy birds and put them into cages with his control group. All chickens in the animal house now developed beriberi, and it was plain to Eijkman that the infection was everywhere, spread probably by the faecal route. After meticulous cleaning and disinfection, a new population of birds was introduced, and after that there was no more beriberi.

Who could doubt now that beriberi was an infectious disease? Not Eijkman, who had been convinced of it all along. And yet his mind was not closed, for when he learned from an assistant that, just at the time the first chickens had fallen sick, there had been a change of dietary regime, he did not scorn to investigate. What had happened was that the cheap raw rice, which had been the chickens' staple food, was replaced by leftover cooked rice from the hospital kitchens. Again, Eijkman divided his chickens into two groups, one of which he fed on the coarse raw rice, the other on the hospital's cooked rice, and now the truth was plain: it was something in, or missing from, the cooked rice that was the cause of beriberi. A decade before, a Dutch navy surgeon by the name of F. S. Van Leent had arrived at much the same conclusion: it was something lacking in the monotonous shipboard diet of rice that brought on beriberi, probably, he thought, an insufficiency of protein and fat. (Eijkman, it should be remarked, knew nothing of Takaki's work, or of any other research on beriberi, for the small library in the Java laboratory had only a few Dutch and German journals, and none of the English-language medical literature.)

Eijkman now began a long series of tests to determine where the mysterious factor resided. The starchy rice grain is enveloped in a pellicle, a capsule with an internal 'silver skin', containing mainly protein and carbohydrate – the bran of rice. If the pellicle is removed by milling or 'polishing' the storage properties of the rice are improved, and such 'polished' rice was also often considered, like white bread compared to brown (Chapter 6), more palatable. Something in the pellicle, Eijkman believed, counteracted a toxic property of the endosperm, the starchy centre of the grains, and he showed that the bran separated from the polished rice could avert or cure beriberi. All the same, he remained uneasy about his conclusions and about the relation of the avian to the human disease. But then, in 1896, his malaria recurred, and he returned to Holland for good. Doubts about the interpretation of his results continued to prey on his mind and seemed all too justified when he started again to experiment on chickens, made available to him by the Director of the Amsterdam zoo. Nothing went right: only a few of the chickens on polished rice fell ill, and none showed signs of polyneuritis. Perhaps the metabolism of Dutch chickens differed from that of the Indonesian birds? Eijkman had chickens shipped in from Java, but the results were inconclusive. He was now seriously demoralized, and decided he had been wrong all along, and that the chickens in his laboratory in Java had all been infected with a pathogen to which inferior diet made them susceptible. He denounced the conclusions of his compatriot, Van Leent and of Takaki, and withdrew from further debate. It was Pekelharing who resolved the problems of the chickens. He went to the zoo to look for himself, and what he saw was that the indigenous chickens were simply declining the rice, and presumably feeding themselves off what they could find, as chickens will. Pekelharing suggested that they should be force-fed their rice diet, and then the polyneuritis asserted itself. Eijkman meanwhile was appointed Professor of Bacteriology and Hygiene at the University of Utrecht, and did notable work on the testing and purification of drinking water, which greatly diminished the incidence of gastrointestinal diseases. In 1929, the year before his death, Eijkman, a modest and unassertive man, was awarded the Nobel Prize for Physiology and Medicine, together with Frederick Gowland Hopkins (of whom more anon). Pekelharing had died seven years earlier.

Meanwhile, back in Jakarta, Eijkman's young successor, Gerrit Grijns, another product of the medical school in Utrecht, continued the work on beriberi and made much progress. First he disposed of a conjecture of Eijkman's, that a mineral deficiency might have something to do with the disease, a notion that had sprung from the pronouncements of Gustav von Bunge (p. 163) in Basel. Then he excluded an insufficiency of fat as a contributory factor. He also showed by arduous and meticulous experimentation that Eijkman's favourite hypothesis, of a substance in rice skin which counteracted a toxin generated in the digestive system by starches, was also without foundation. Grijns examined other foods and discovered that a locally grown vegetable, the mung bean, protected against beriberi, and that the active substance was in this case present in the seed itself. Grijns wrestled with the problem for five years. He resolved the doubts that so tormented Eijkman about whether the disease of the chickens could be equated with that of people, and, most remarkably, he prefigured in a publication of 1901 the concept of vitamins – traces of complex chemicals in food, inseparable from health. But neither this publication nor those of Eijkman's made the impact that they should have done, for they appeared in the journal of the medical association of the Dutch East Indies, or at best in a Dutch medical journal, and in the Dutch language. But in practical terms, at least, the conquest of beriberi appeared well advanced. Eijkman's friend Adolphe Vorderman, a former naval surgeon and by then Inspector-General of Public Health in the colony, inquired into the state of the inmates in the 63 prisons on the islands, and discovered that beriberi was rampant in more than half of them. These, of course, were the ones in which the diet was based on white, polished rice, rather than unmilled (brown) rice. Vorderman cleaved obdurately to the belief that beriberi was caused by an infection and cured by something in the rice bran, but he took the necessary steps. By ensuring that polished rice was eliminated from the prison diets he undoubtedly saved many lives and much misery. In 1901 another Dutch doctor, confronted with an outbreak of beriberi among a group of Chinese labourers, divided the men into two groups and persuaded one to eat mung beans, which they found disgusting. This cured the disease, which continued to torment the men in the other (control) group. The means were thus at hand to eliminate beriberi completely, but in 1902 Vorderman died, his successor had no interest in the problem, and beriberi continued to stalk the islands.

The British take a hand

It was not only in the Dutch colonies that the cause of beriberi was being sought. In what was then the Federated Malay States an Institute for Medical Research had been created, and in 1902 its director, Dr Hamilton Wright, took a look at the epidemiology of beriberi. There were four ethnic groups in Malaya, the Malays, the Chinese, the Tamils from southern India, and the Europeans. It was the Chinese who were mainly prey to the disease. They worked mostly in the tin mines and had, Dr Wright opined, dirty habits that spread disease. Their rice, he found, was contaminated with mould and bacteria, but he believed it was the fertilizer – fermented human waste – spread on the vegetable crops that was to blame. The disease, he was persuaded, was unrelated to diet. His paper on the subject instantly drew criticism for lack of rigour and insufficient research. One who rejected his conclusions was Dr Leonard Braddon, serving as a government surgeon in the Federated Malay States. He dismissed equally Grijns's surmise that an essential factor was missing from the 'best' rice, and discounted Vorderman's observations in the prisons of the Dutch possessions, because, he speculated, the skin of the 'brown' rice might contain a fungicide that preserved the rice from contamination. Braddon had, in fact, made an important observation. The four ethnic populations had different dietary habits; the staple food of the Malays was rice, milled by hand in their own kitchens; the Tamils sieved out the bran after parboiling the rice; the Chinese consumed only white rice, imported, then milled by steam-driven machines and cleaned before it was sold; while most Europeans avoided rice altogether. It was the Chinese, Braddon concurred, who suffered seriously from beriberi, which, he deduced, must therefore be due to some kind of toxin, perhaps fungal, that tended to accumulate in their rice during storage. Braddon's report inspired another English doctor in Malaya to attempt a test of the proposition, using real people as his experimental animals.

William Fletcher was the District Surgeon in Kuala Lumpur, the capital of the state, and had charge of the Kuala Lumpur Lunatic Asylum. If its name was an example of Victorian plain-speaking, Fletcher's experiment typified how the monarch's involuntary guests, whether in prisons or asylums, were regarded. In 1905 an outbreak of beriberi struck the Asylum. It began in February and peaked, as usual, in July and August, and, Fletcher reported in *The Lancet*,

> out of 219 lunatics treated in the asylum 94 persons were affected of whom 27 succumbed to the disease. The main diet constituent was uncured (Siamese) rice, and in view of the fact pointed out by Dr Braddon that beriberi occurs chiefly amongst communities with whom such rice is the staple article of diet it was decided, with the sanction of the Government, to place half the lunatics on cured (Indian) rice. The Government readily gave its consent and the experiment was commenced on December 5th, 1905.[2]

'Siamese' rice was the white rice referred to by Braddon, consumed by the Chinese population, and 'cured' rice the brownish parboiled de-husked Bengali product, preferred by the Indians.

The inmates were accommodated in two buildings at opposite ends of a quadrangle, enclosed by a high wall. Fletcher lined them up in the open and numbered them off from the left. The odd numbers went to the building on the east side, and received a white rice diet, the even numbers to the west and were given the brown rice. In other respects the diets were the same (fresh meat four days a week, fresh fish twice, salt fish once, and vegetables, along with 'curry stuff' and coconut oil, daily). The experiment was conducted with fastidious care. Of the 59 inmates in the asylum on 5 December, 30 were housed in one building and 29 in the other, and those admitted during the following year were sent to the two buildings in alternation. By the halfway stage of the experiment there were many beriberi cases in the east building, none in the west. Next: 'In view of the theory so strongly advocated by Sir Patrick Manson* that beriberi is a place disease, it was thought possible that the east ward was infected'. Therefore, on 20 June 1906 the populations of the two buildings were exchanged. But Fletcher was not a callous man, and several severely sick patients were transferred to the district hospital (where, however, 9 out of 13 died). Moreover, the other subjects with obvious symptoms of the disease were switched to a brown rice diet, and all recovered, nor did they become 'reinfected'. As the year progressed, another 90 arrivals were added to the white-rice group, of whom, by 31 December, 17 had beriberi, as well as 17 out of the original 30 occupants. The study was continued during the next year, with similar results. Throughout it all, only one patient appears to have died of beriberi in the asylum.

Fletcher owned to having been a sceptic. He knew of Takaki's study in Japan (though probably not of the Dutch work), but he had not believed that beriberi had anything to do with rice. The results of his study had come as a surprise to him, and they could not, he reasonably concluded, be reconciled with the view that beriberi was an infectious disease and nothing more. He now speculated that the lack of some nutritional substance, lost in preparing the white rice, could predispose to 'some external agency', such as a bacterial or protozoan infection. Even this, though, was too much for the adherents of the infectious disease theory, and Fletcher's report in *The Lancet* drew a stinging response from one of them, Dr C. W. Daniels, Wright's successor at the Institute for Medical Research. Fletcher's conclusions were unacceptable, he fumed; the sample had been too small, and the diagnostic criteria inadequate, especially the exaggerated knee-jerk that was taken as a sign of beriberi.

Daniels was himself presently replaced in Kuala Lumpur by a man of far greater stature and insight, Henry Fletcher from Aberdeen, a well-schooled doctor who had served his time as a medical researcher in Germany. He procured, moreover, an associate with similar qualities, a Canadian doctor by the name of Thomas Stanton, who had worked under the great Manson at the London School of Hygiene and Tropical Medicine. Together, in 1907, they undertook to confirm and, if possible extend, Fletcher's observations. Their experimental subjects were Chinese labourers, engaged in road-building in a remote area of the colony. The men were divided into two groups, each of about 150. One was allowed the much-preferred white rice, while the other was given, to high dis-

* Patrick Manson (1844–1922), often called 'the father of tropical medicine', made many important discoveries in parasitology. His most famous protégé was Ronald Ross, the army doctor who uncovered the life cycle of the malaria parasite. Ross had his own theory about beriberi: it was a reflection, he believed, of arsenic poisoning.

pleasure, the brown rice of the Tamils. In other respects the diet was the same. After three months 12 cases of beriberi had been recorded in the first group, and none in the second. The diets were then switched, and all the sufferers in the first group were cured, while the disease began to show itself in the second. Some workers left during the period of the study as their contracts expired, and others came, but in all, Fraser and Stanton reported, 20 out of 220 of those who had lived on white rice got beriberi, and none of the 273 on brown rice. They then proceeded to repeat Eijkman's experiments on chickens and took his observations a critical step further, for they found that chickens with polyneuritis from being fed white rice were cured by an alcohol extract of rice polishings. Grijns, then, had been right in his conclusion that beriberi was caused by the lack in white rice of an essential factor, and Braddon was wrong in his conjecture that it was due to a toxin. The work was recognized at home in Britain as a major contribution, which whipped Dr Braddon into a lather of indignation, for had he not shown years before that the disease was associated with the consumption of white rice? Fletcher and Stanton, he fulminated in the *Journal of Tropical Medicine and Hygiene*, 'had not contributed a single new or independent observation of *any* facts' concerning the origins of beriberi. Their chemistry, he was forced to concede, was excellent, yet 'as regards the practical, the epidemiological aspects of the beri-beri question, they have added nothing new or original to the issue'.[3] Neither, as will appear, did passions about the interpretation of experiments or about who did what subside over the years that followed.

The rearguard fights on

As with scurvy (Chapter 2), an entrenched rearguard continued to deny the evidence long after the issue was settled in the minds of the more receptive participants. In particular, it was to be several more years before the concept of essential trace substances in food came to be generally accepted. The unhappy Christiaan Eijkman, once his work saw print outside his own country, became a particular target for the scorn of the reactionaries. A Dutch medical mandarin, Max Glogner, unburdened himself thus: 'If one considers that Eijkman apparently needed six years to do this work, it must be considered the most inadequate product that can be found in the literature from the director of a scientific institute.' Nor was he any more charitable to Vorderman, whose 'attempt to secure a place for rice in the aetiology of beriberi is to be regarded as totally useless'.[4] Another Dutch doctor, who engaged Eijkman in an acrimonious exchange, had proffered the theory that white rice contained a neurotoxin; he responded to Eijkman's rebuttal with the cogent argument that opinions of colonial doctors were worthless, because, since they had all fed on infected rice during their service, their brains had been rotted by the neurotoxin.

In other parts of the world dissension also continued. The Russo-Japanese War had started in 1904 and the Japanese army of invasion in Manchuria was soon devastated by beriberi. Takaki's demonstration of the cause of the disease was set at naught, for the army surgeons remained convinced that an infection was to blame. The same article of faith prevailed in the army medical service in the French colonies, but in Madagascar an army surgeon had read Eijkman's work and kept the disease at bay in his Senegalese charges with rice bran. The American army doctors in the Philippines, where beriberi was endemic, were among the most enlightened. Two of them, Edward Vedder and Weston Chamberlain, became especially concerned with beriberi in babies, which was caused apparently by the milk of malnourished mothers and resulted in huge (up to 50%) mortality. It was as if, Kenneth J. Carpenter comments in his definitive study of the history of beriberi, the mothers' milk had become toxic. Pregnant women, it turned out, were themselves especially vulnerable to beriberi. Chamberlain and Vedder treated the sick babies with an alcohol extract of rice polishings with excellent results, and this was adopted as a routine measure. Manila was the venue in 1910 for the first international meeting on beriberi, at which it was resolved that the case for white rice as the cause of the disease was now established beyond reasonable doubt. Even so, two American researchers, one of them R. P. Strong, Director of the Biological Laboratories of the Philippine Bureau of Science, carried out another controlled study with rice diets on the inmates of the Billibid Prison in Manila, with the expected results. (The subjects, who suffered miserably, were rewarded with cigars and cigarettes.) The point was made, and it was now the turn of the biochemists. One of them, a young Pole, Kazimierz, later Casimir, Funk, was conjecturing that other diseases besides beriberi – he mentioned scurvy and pellagra – might be linked to deficiency of dietary factors unique to each. This story will unfold in Chapter 9.

The angry skin

Pellagra was, indeed is, an ailment of squalor and deprivation. In olden times it was commonly misdiagnosed as leprosy. Today's medical students learn that its symptoms are the four *d*'s, dermatitis, diarrhoea, dementia, and death. The first explicit description of the disease was published in 1735 by a Spanish court doctor, Don Gaspar Casal, who had encountered it among the poor peasants of Asturias. He called it *mal de la rosa*, the disease of the rose, from the signs by which it could be identified – an ugly pink excrescence on the limbs, buttocks, and genitals, and often in the later stages a 'butterfly rash', resembling in shape a red butterfly, over the nose. Later, in Italy, the name pellagra, meaning angry, or perhaps rough, skin, was attached to it. Casal called it 'a kind of leprosy' and thought it came from humid air, foul winds, and bad nutrition, for it struck mainly in spring, when the peasants had little to nourish them but maize meal. A later theory had it that the condition arose from excessive exposure of the skin to the hot sun. But the relation to maize was difficult to reject. In Italy pellagra became common among the poor after the introduction of maize (to

Americans, corn), brought by the *conquistadores* from Mexico (Chapter 3) and the basis of the cheap and ubiquitous polenta. That the diet of 'Turkish corn', as it was then called, lay behind the disease was also the conjecture of Goethe, who described the scene of desolation and debility that met his eye when he crossed the Brenner Pass in 1786. Ten years later a Milanese physician, Giuseppe Cerri, likewise concluded that pellagra originated from the poor diet of the peasantry. He demonstrated as much by conveying 10 afflicted rustics to the city and maintaining them on typical urban fare, when the symptoms quickly receded.

Pellagra, judged by the documented manifestations, was probably widespread in the southern states of the USA throughout the 19th century, and prevalent during the Civil War. Explicit diagnoses began to appear around 1902. The best data during this early period were collected in South Carolina, where in 1912 some 30,000 cases were reported, with a 40% mortality. Congress took note, and pellagra commissions were set up. Theories about the origins of the disease proliferated. It was linked to excessive consumption of maize (since it was on this crop that many of the poor mainly subsisted), or to fungus-contaminated maize (a conjecture that led to a precipitate drop in maize imports from the USA into Europe, and so further impoverished the south);* or was it perhaps an insect-borne infection? It was even suggested, for the disease was especially prevalent in prisons and lunatic asylums, that it was a manifestation of syphilis. Overexposure to the sun was confidently asserted by some to be the cause. And certainly, poor sanitation was a common feature of the homes in which it occurred, so perhaps, after all, it was a bacterial infection. Some insisted that it had been imported into the country by poor Italian immigrants. Malnutrition as a possible cause was mentioned only to be dismissed, for example by a Pellagra Commission funded by two philanthropists, which had reported from North Carolina in 1911. Because sufferers, either in consequence of the disease or the conditions in which they lived, were especially susceptible to bacterial and parasitic infections, the consensus emerged that pellagra was an infectious or contagious disease.

It was this sorry state of intellectual confusion and wild conjecture that in 1914 the Surgeon General of the United States asked Dr Joseph Goldberger to clear up. Joseph Goldberger was born in 1874 into a Jewish family in a small town in the Carpathian foothills, in what was then Hungary and is now the Czech Republic. In 1881 the father, a shepherd, was ruined by an outbreak of a murrain that destroyed his flock, and with borrowed money took his family to New York. There, on the Lower East Side of Manhattan, he set up a grocer's shop. Joseph

* A French physician had concluded years before that a diet confined largely to maize was responsible for pellagra in France, while in Italy Cesare Lombroso – later famous, if not infamous, as an alienist, who claimed that he could identify criminal or other asocial traits from the physiognomy – thought that a mould in maize was to blame.

worked in the shop and delivered orders, but he was a bright child and did well at school, so at 16 was allowed to enrol in an engineering course at City College, where so many poor immigrants got their start in professional careers. To pay his way he continued to work in his father's shop. Then, two years into his course, a friend who was studying medicine at the Medical College of Bellevue, the hospital for New York's poor, persuaded Joseph to accompany him to a lecture by an inspirational member of the faculty. Fascinated by what he saw and heard, Joseph abandoned his plan to become an engineer. Instead, he borrowed the money for Bellevue's modest fees from a half-brother, in whose shop he worked through the summers to pay off his debt, and eventually qualified as a doctor, coming top of his class. His attempt to establish a private practice in New York was a failure, and he tried again in the town of Wilkes Barre in Pennsylvania. There he attracted enough patients to survive, but he grew restive and after two years went in search of adventure: he applied for a position as a naval surgeon (the Spanish War having just then begun). Rejected, perhaps because of the pervasive antisemitism of the US Navy, he applied to the Marine Hospital Service (the progenitor of the US Public Health Service, into which it metamorphosed in 1912). He passed the exacting competitive entry examination and in 1899 joined the service as an Assistant Surgeon. The Service had a high reputation. Its thrust was primarily epidemiological, and by proud tradition its staff would go to wherever diseases were rife, no matter what the risks. It was not long before Goldberger had experienced in his person the effects of dengue virus, yellow fever, and typhus, the last of which very nearly carried him off.

Goldberger's first posting was to a quarantine station, responsible for inspection of ships arriving at the port of Philadelphia. There he learned to diagnose all manner of infectious diseases, and there also he made the acquaintance of the vivacious young woman who eventually became his wife, a daughter, improbably, of the southern Protestant aristocracy. When he received the call from the Surgeon General Goldberger was already a seasoned epidemiologist, who had done notable research, especially on typhus. Heading south, Goldberger found himself in 1915 in the Georgia State Sanitarium for the Insane in Milledgeville, one of three institutions that the Surgeon General had designated centres for pellagra research. Pellagra was affecting hundreds of the inmates (in many cases, no doubt, because insanity was a late manifestation of the disease), but, Goldberger observed, the asylum staff remained unaffected. Continuing on his tour of institutions, he found that this rule held true everywhere. How then could pellagra be an infectious disease? The main difference between the lifestyles of the inmates and the staff lay in what they ate. The staff were well enough off to enjoy a normal diet that included fresh meat, vegetables, and fruit. The inmates received the fare of the southern poor – some meat of inferior quality, known as fatback, but mainly cornmeal or corn-bread and molasses. To test his conclusion that pellagra must be a disease of bad nutrition, Goldberger persuaded the state of Georgia to bear the cost of a balanced diet served in two orphanages, without making any other changes in the conditions of life. It took only a few weeks for the disease to disappear.

Goldberger published a report of this study, with its obvious connotations, but they were not obvious to all the doctors who read it. Whether this was yet another case of a refusal to consider that a disease could be caused by the lack of something unknown, rather than by a noxious agent such as a mould or a bacterium, is unclear. At all events, all kinds of unlikely alternative explanations were aired, most commonly that the staffs of the institutions were both cleaner and more robust than the unfortunates who had ended up as inmates. The most acerbic of Goldberger's critics was the Chief Health Officer of South Carolina, James Hayne, who denounced the very idea that pellagra was diet-related as 'an absurdity'. The debate, in fact, was bedevilled by the tenets of the eugenic movement, which had taken powerful hold in the country. The eugenicists had no difficulty in convincing themselves that it was only genetically inferior stock that ended up in prisons, asylums, and at the bottom of the social heap generally, and they also believed that sensitivity to agents of disease was racially determined. It was therefore hardly surprising that they should have cast their own light on the pellagra problem.*

Goldberger's patience must have been severely tested by such wilful obduracy – even though his report was much better received in academic circles, and indeed he was invited to lecture on the pellagra problem at Harvard – but he did not repine. The same year (1915) he asked the Governor of Mississippi, Earl Brewer, known as a man of progressive views, for permission to carry out a nutritional study into pellagra on a prison farm. The Governor agreed – a brave decision, for a public outcry could have cost him his career. Several prisoners volunteered for the trial and 12 were selected (7 of them murderers), in return for a pardon and immediate release once the trial was over. One fell ill with an unrelated disorder and was taken out of the trial, but the remaining 11 saw it through. The choice of the farm was dictated by the absence of pellagra because of the varied food that was grown there and contributed to the felons' diets. The volunteers were given clean clothes, housed in a cleaned and disinfected building, and fed the diet of the rural poor, consisting of maize in various forms, such as hominy grits and corn bread, together with sweet potatoes, molasses, and leaves of kale. They were sub-

* The chairman of the National Pellagra Commission was Charles Davenport, founder and director of the Eugenics Record Office at the Cold Spring Harbor Laboratory on Long Island, New York, who had argued for years that diseases common in poor immigrant communities, such as tuberculosis, were a consequence of genetic susceptibility and were not to be eradicated by improved living conditions. Pellagra, he continued to assert, against all the evidence, was a similar case. Davenport had powerful patronage and a large following among the intelligentsia, and his organization had links to eugenics groups in Europe, especially in Germany. Professor Eugen Fischer, who achieved notoriety during the Third Reich as a founder of the doctrine of Racial Hygiene, was an especially admired *confrère* of Davenport's, who put Fischer's name forward as president of the international organization of eugenicists.

ject to the same arduous regime of labour as the other prisoners, who in effect made up the control group. Within a few weeks the subjects began to complain of back pains, sores in the mouth, depression, and confusion. By five months, six men had clear symptoms of pellagra. Goldberger, however, kept the trial going for another two months while he and his assistant, George Wheeler, another doctor employed by the Service, kept a close eye on their subjects. They also called in independent doctors to ensure that they could not be accused of biasing the diagnoses. There were no cases of pellagra among the other prisoners on the farm.

When the results became known, the experiment was hailed in the press as an achievement to rank with the great medical advances of the past, and Goldberger was feted as a hero. The volunteers also received praise and the Governor's action in pardoning them was for the most part applauded. Governor Brewer signed the documents at a ceremony in his office. Goldberger urged that the reprieved prisoners should remain for treatment, for the disease, he explained to them, was dangerous. Only a minority took his advice, but three of the rest returned by evening. One, whose condition was life-threatening, recovered after some months of care. There was universal scientific acclaim for Goldberger, and a professor at Harvard recommended him for the Nobel Prize. But it was still not enough to quell the ire of the sceptics, and the results were venomously disparaged, especially by the now almost hysterical Hayne. And Dr Ward MacNeal, who had led an earlier investigation into pellagra in the South, financed by two wealthy philanthropists, went so far in defamation as to accuse Goldberger of 'faking the prisoner experiments'. There was also some criticism of detail from a vastly more weighty quarter, the biochemist, of whom more will be heard later, Elmer McCollum (though he did warmly commend the report as a whole). Goldberger responded in restrained terms, but in a letter to his wife he unburdened himself: 'Let the heathens rage!', he wrote, 'blind, jealous, selfish, prejudiced asses.' All the same, he knew that still more proof was needed before the opposition would finally be silenced.

To lay the infection theory to rest once and for all, Goldberger resolved on a programme of what he called 'filth parties'. He, Wheeler, and a group of six volunteers with strong stomachs, Mary Goldberger among them, assembled at the US Pellagra Hospital in Spartansburg, South Carolina. There they systematically exposed themselves to 'blood, nasopharyngeal secretions, epidermal scales from pellagra lesions, urine and faeces' from patients. First they injected themselves with blood from severely affected victims. They rubbed secretions from their mucous sores into their nose and mouth, and after three days swallowed pellets consisting of the urine, faeces, and skin scabs from several diseased subjects, bound with flour paste. In Goldberger's case, quite severe diarrhoea resulted but no pellagra. Astonishing to relate, the results still did not satisfy the medical profession in the southern states, nor even all the federal organizations, and the victims of pellagra continued to be treated by such lethal nostrums as the arsenical 'Fowler's Solution'. But in the great centres of medical research, at Harvard, Yale, and elsewhere, and in Europe, the import of Goldberger's work was understood, and he was twice more nominated for the Nobel Prize. He shared the view, first articulated in 1911 by Casimir Funk (p. 166), that pellagra, like scurvy and beriberi, was caused by the lack of a trace substance in the diet.

How, then, to eliminate the curse of pellagra? One of his 'filth' volunteers, Edgar Sydenstricker, was a statistician and demographer, who eventually became, and remained for 20 years, the Chief Statistician of the US Public Health Service. Sydenstricker (1881–1936) was developing improved techniques of epidemiological research and was especially concerned to relate the incidence of disease to social circumstances. He and Goldberger formed a close partnership, and together started an exacting epidemiological study of the distribution of pellagra in the impoverished mill towns and villages in the southern states. They discovered that the incidence of the disease was patchy and strongly linked to diet. The message in the end was that three possible causes of pellagra should be considered; the first was deficiency of protein, or perhaps one of the amino acids, 20 in number, from which proteins are constructed but which differ in their relative abundances in different proteins. The second was a deficiency in intake of mineral substances, and the third, the lack of a trace factor – a vitamin – in the diet, as Goldberger had already hazarded. For the moment, the best hope lay in improved economic conditions in the South, for it was pure poverty that condemned the populace to a monotonous and inadequate diet. But the destitution of the region had been aggravated by the drop in world cotton prices, and Goldberger and Sydenstricker loudly urged an end to the cotton monoculture, with diversification of agricultural production. It was decades before this came about.

At that moment, in 1918, the deadly Spanish influenza epidemic, which was to kill many millions around the world, struck America. The Spartansburg hospital was turned over to accommodating the victims, and Goldberger was instructed to confront the crisis. It was two years before he was able to return to the study that by then obsessed him. When the influenza epidemic had subsided he installed himself in the Hygiene Laboratory in Washington, DC (the precursor of the National Institutes of Health, which now constitute a veritable laboratory city in Bethesda, outside Washington). From there he supervised a nutritional study in Milledgeville in Georgia, and in the Hygiene Laboratory began a series of animal experiments that might have led to the isolation of the mysterious factor. Some years before, Russell Chittenden, the physiologist at Yale who had toiled all his life at protein nutrition (Chapter 5), had found that a condition with pellagra-like symptoms could be induced in dogs, which also displayed a sign sometimes seen in human pellagra patients – a black tongue. This was originally taken as the mark of an infection, and, if not treated, the dogs died. But all symptoms vanished when the animals were fed fresh meat or brewer's yeast. This last, indeed, both cured and prevented black-tongue and the associated effects. The study in Milledgeville meanwhile came up yet again with the result that animal protein in the diet averted pellagra. What might there be in animal protein that was missing from plant protein? Tryptophan was an amino acid of generally low abundance in proteins, and Goldberger wondered evidently whether this might be deficient in a bad diet. A woman in Georgia suffering from pellagra responded well when given tryptophan, and Goldberger asked for proper trials, which proved quite successful. But he was no biochemist and was apparently at a loss as to how to take the matter further.

In the wet spring of 1927 God smote the god-fearing southern states with another natural disaster. The land was devastated by floods, food became scarce, and pellagra was suddenly rampant. Goldberger was summoned, and he and Sydenstricker toured the stricken lands to assess the extent of the problem. Goldberger asked that some 12,000 pounds of yeast be distributed among the affected communities, and advised that it should be taken in milk or molasses. The measure was spectacularly successful: the recommended dose was two teaspoonfuls three times a day for adults, and half that amount for children. Goldberger was hailed as a saviour. Two years later he fell sick. It was a rare form of cancer, identified only at autopsy. Joseph Goldberger died in 1929, greatly lamented by all who had known and worked with him.

Much of the opposition to Goldberger's irrefutable results was undoubtedly due to resentment at the incursion of northerners into the affairs of the South, and the implicit criticism of its ways, although it was also true that the very concept of a disease without a pathogen remained inconceivable to many doctors. Goldberger's friend, admirer, and closest associate, George Wheeler, himself a southerner, did his best to convince the sceptics. He now had to contend with assertions that pellagra was caused by a virus (for viruses had recently become fashionable in the healing profession), and with a publication by a professor at Tulane University in New Orleans purporting to prove that it was a disease of iron deficiency. At all events, hardly any measures were taken to improve the diets of the poor farmers and mill-workers, and pellagra continued to inflict immeasurable suffering and a painful death. Even as late as 1930 some 6000 people died of it in that year. Reform came partly as a result of the devastation caused by the boll weevil, which ruined the cotton crops and forced farmers to diversify, and more especially from federal help after the stock-market crash of 1929, in the form of soup kitchens, and the provision of vitamin-enriched flour. Pellagra, it should be remarked, has not disappeared from the world. In 1990 doctors who examined refugees in Malawi fleeing from the civil war in Mozambique discovered that some 18,000 people out of 286,000 were suffering from the disease.

The sickness of the older child

Kwashiorkor is a deficiency disease of a quite different kind. Its story begins much later – in 1933, when there appeared in a paediatric journal a brief report describing a dire disorder that was afflicting children in the West African colony of the Gold Coast, now Ghana. Its author was a British doctor, Cicely Delphine Williams. This remarkable woman was born, the fourth of six siblings, in 1893 in Jamaica into a Welsh family. Her father belonged to the 14th generation of the Williams line established on the island since the 17th century. Cicely wanted to study medicine and win a place at Somerville College in Oxford, but an earthquake ruined the family estate and she was denied her opportunity. It was only her brother who was to be allowed to enter into a medical career in England. Cicely was sent instead to the local Montessori school to train as a teacher. This gave her little satisfaction, and after two years of cajoling she managed to raise enough money to fulfil her ambition. At the age of 22 she belatedly began to read for a degree in medicine at Somerville. It was the first year that Oxford permitted women to take a medical degree. After graduating she moved to London for clinical work at King's College Hospital, where she was awarded the prize for the best first-year student in neurology. Cicely Williams found a distinguished patron in Sir George Frederick Still, one of the leading paediatricians of the day. But then the family money ran out once more and, destitute in London, she had no option but to return to Jamaica. Two months later she was back at King's, and began the laborious climb up the academic ladder as a Junior House Doctor in a hospital in south London. It was then that she made up her mind about the direction her career would take, and enrolled in the diploma course in Tropical Medicine and Hygiene. With this behind her, she applied for a position in the Colonial Medical Service in the expectation of a career in Jamaica, among the children of the rural poor. She waited more than a year for a reply, but then, through a friend, she was introduced to the Head of the Service, Sir Henry Stanton (whom we last met when, as a young doctor and researcher in Malaya, he was investigating the causes of beriberi). A posting to the Gold Coast swiftly followed.*

* Cicely Williams became a heroine of the Second World War. Her work had by then taken her to Singapore, and after the Japanese invasion and the British surrender in 1941 she strove to keep her hospital functioning in the face of appalling shortages and an endless flood of desperately ill patients. Some telegraphic case notes that she exchanged with another doctor were found by the Japanese, who were unable to interpret the clinical shorthand and arrested both as spies. Dr Williams was incarcerated in the notorious Changi jail, where she was elected by the women prisoners to organize their diet with the meagre supplies that their captors made available, and she did her best to treat malnourished and often dying inmates. She negotiated with her jailers and was able to wring some concessions from them. Through all this, she managed to produce a report on her work among the deprived, *An Experiment in Health Work in Trengganu in 1940–1941*. By the end of the War, Cicely Williams was herself close to death. She was restored to health and very soon resumed work. She urged that nutrition in poor parts of the world should be based on local customs and resources. In 1949 she was appointed Head of the section on Maternal and Child Health of the World Health Organization, but in 1951 moved on to academic appointments at the American University in Beirut, and then in the United States. In retirement in Oxford she wrote, lectured, and campaigned into old age. Cicely Williams died in 1992 just short of her century.[5]

It was in 1933 that Cicely Williams published a remarkable paper in which she described a disease, confined almost entirely to young children, generally between the ages of six months and four years, who had been weaned early. She noted that it differed in its symptoms from the known deficiency states, including hunger disease, marasmus. The effects, graphically described, were oedema (swelling, or dropsy), especially of the hands and feet, followed by wasting, diarrhoea, sores, especially on the mucous surfaces, and general physical irritability. The skin developed a 'crazy-paving' appearance and peeled, leaving raw red patches. The child was by then in a pitiful state and would soon die. Cicely Williams found time to conduct a few makeshift autopsies and discovered a fatty liver as the only abnormality. The diet of such children, she reported, consisted of (besides breast milk) a gruel or a dough, made from partially fermented white maize. If the condition was caught in time, it responded to condensed milk and cod liver oil.

This paper, now considered a classic in the history of clinical medicine, drew a cutting response from one of the panjandrums of British tropical medicine, Dr H. S. Stannus (or in his full majesty, Hugh Stannus Stannus), who wrote to the journal with the stark assertion that the condition Cicely Williams had described was 'clearly' nothing more than infantile pellagra. But Cicely Williams persisted. In 1935 a full report was published under her name in *The Lancet*,[6] consisting of a study of 60 cases. Here the name, *kwashiorkor* appeared for the first time. An indigenous nurse had explained that the term meant 'the sickness that the older child gets when the second child is born'. The paper reiterated the conclusion that this was a previously unrecognized deficiency disease, associated always with inadequate feeding. It commonly struck when the mother was sick or undernourished, weakened perhaps by a new pregnancy, or had died, because then breast-feeding might devolve on 'an unsuitable foster mother, very often a senile grandmother, for among these African women some mammary secretion may be present in a woman who has not had a child for 20 years'. If caught very early, condensed milk, cod liver oil, and malt could effect a cure, but by the time a skin rash has appeared the outlook was generally hopeless. The paper featured a table contrasting the symptoms of kwashiorkor and infantile pellagra. All this excited Dr Stannus's wrath anew, and he again published a reprimand in the next issue of the journal. Why had Dr Williams not bowed to his rebuke? Why introduce a barbaric name for what was still infantile pellagra? 'How will it be possible', he demanded, 'for anyone later looking for references to nutritional disorders to recognise one with such a designation?' Diseases without Latin stems, were evidently, to Stannus's mind, uncouth. He went on to dismiss the discussion of symptoms: 'Dr Williams has now again repeated these misstatements in regard to pellagra, presumably based upon a failure to realise the course and symptomatology of that disease, and tends to confuse the problem at issue.'[7] It must be conceded that Stannus's point of view was not totally unreasonable, for pellagra had of course also been associated with a maize diet, and other authorities, too, were sceptical, although they had not seen the patients. Oedema was equally a symptom of pellagra and beriberi (though elsewhere in Africa kwashiorkor was often diagnosed as congenital syphilis).* The disease, in fact, was no small problem, for 90% or so of children who contracted it died. Cicely Williams, however, received some support from other medical officers in Africa, who published descriptions of cases.

The argument dragged on for years. A leading protagonist was another British physician, Hugo Trowell, who 20 years later related that the disease had been familiar to medical missionaries in Africa. They had wondered whether it might not have been a manifestation of chronic protein deficiency. Trowell, though, had been persuaded by Stannus's strictures, and in 1940, while still working in Kenya, surveyed the field, under the title 'Infantile Pellagra'. He thought that the oedema did arise from protein deficiency, but that the other symptoms were those of pellagra. All the same, Trowell found that the children did not respond to nicotinic acid, the vitamin already identified (p. 180) as the deficient factor in pellagra. Cicely Williams sent a rejoinder from her post in Singapore: kwashiorkor was nothing like the pellagra that she had herself seen during a stay in New Orleans. Trowell gradually came round to her point of view, but he did not care

* Kwashiorkor was not the only malady misdiagnosed as pellagra. Towards the end of the 19th century reports began to appear in Britain and the United States of a new syndrome. It afflicted babies between about six months and two years of age, and was called pink disease. Its effects were distressing: the babies turned bright pink, even red, they became painfully sensitive to light, and terrible rashes with peeling skin ('raw beef') appeared on their hands and feet, not, indeed, unlike the 'red gloves' of pellagra patients. The death rate was estimated at about 7% on average, although in some places it was much higher. The survivors generally recovered spontaneously, but were sometimes left with severe kidney or other organ damage. It was characterized as a type of infantile pellagra, though some doctors were convinced that it was a bacterial infection, and others again that it was due to a virus (true to the generally fashionable explanation for the inexplicable at that time). In 1922 an American doctor noted the similarity between the symptoms of pink disease and those of mercury poisoning, but he did not pursue this insight because some of the sufferers had no known record of calomel treatment. Calomel, a white chloride of mercury, featured in the pharmaceutical armoury from early times as a remedy for bacterial infections and many other conditions, notably syphilis, and its toxic effects were well known. (In the 'lock hospitals' in which syphilis was treated the patients often found the cure more cruel than the disease. Among its effects were suppurating gums, loosened teeth, and profuse and uncontrollable sweats.) It was not until 1945 that a paediatrician in Cincinnati, Josef Warkany, discovered high concentrations of mercury in the urine of babies with pink disease. The culprits, it turned out, were teething powders, which contained calomel as a cure-all, and especially a prophylactic against infections. These formulae appeared on the market in the late 19th century after opiates to quieten babies (p. 113) had gone out of use. Aggressive advertising was aimed at anxious mothers, who were instructed that teething was a hazardous process, during which the baby was especially prone to infections. Even after it was published, Warkany's evidence did not become widely known, nor was it universally accepted by those paediatricians who were made aware of it. Up to 1960 textbooks and guides published by organizations like the Royal College of Physicians in Britain listed pink disease under such categories as skin and nerve disorders. It is difficult to estimate how many babies endured pink disease and how many died.[8]

for the name, which he had been told meant 'red boy' in the principal Gold Coast language, alluding to the familiar pallor and lack of pigment in the hair. It was not, he felt, a valid general description of the appearance of the children. This etymology is still everywhere cited as the origin, but is evidently wrong, and Trowell, too, conceded later that 'the deposed child' was probably the best translation; his own preferred term, 'malignant malnutrition', never caught on. In Uganda, in 1945, Trowell found the disease in the children of one tribe, neither weaned early nor fed on corn, but rather on plantains, sweet potatoes, and tea. It also made its appearance among adults, especially those who had made the long trek north on foot from Rwanda–Burundi, in search of work. They had subsisted on a dismal diet which gave them only 2000 calories each day, and their symptoms resembled those of the children, although their hair colour did not fade in the same way. On a diet that included meat and milk they would gradually recover.

Also in 1945, a publication by doctors from Johannesburg reported a controlled study of what they identified as infantile pellagra. Seven of the children were treated with vitamins and all died, whereas another six, administered a commercial preparation from pigs' stomachs, made a rapid recovery. Autopsies on children of the first group revealed fatty livers, just as Cicely Williams had described. The next year brought a report from Southern Rhodesia (Zimbabwe) of more cases of what this time was recognized as kwashiorkor, but because the condition was not seasonal, the author concluded, it could not be due to any dietary deficiency, and in any case, many children did not develop the symptoms. The theory that a toxin in the maize was responsible was accordingly dusted off and brought out. In 1950 medical scientists of the World Health Organization and the Food and Agriculture Organization took a grip on the problem. The disease, they found, was common everywhere in Africa except in regions where animal protein was in good supply. Skimmed milk proved to be an excellent curative agent, and so there could be little doubt that kwashiorkor was a disease of protein deficiency. Analysis showed that the staple crops in much of Africa – sweet potatoes, cassava, plantains – were all very low in protein, so plant proteins were not in general making good the deficiency of meat, dairy products, or fish. Cicely Williams, who had travelled across Africa with the British Committee of the Food and Agriculture Association, made a film about kwashiorkor. It emphasized the health of the meat-eating Masai, with their herds of cattle which provided them with beef and milk, often mixed with blood taken from the jugular vein, compared to the sickly agrarian Kikuyu, scraping a precarious living from the parched soil. Cicely Williams, then, has been abundantly vindicated, but the problem of finding protein-rich foods is still acute in Africa and other impoverished parts of the world. Kwashiorkor indeed remains common throughout the tropical belt.

Hunger disease

What happens to the human frame when it is deprived of food altogether? Famines have occurred throughout recorded history and are with us still. Their effects have been abundantly documented, but probably the first scientific description pertained to the Irish potato famine in the years from 1840 onwards (p. 34).[9] The causes of death were meticulously, if by today's standards not altogether accurately, recorded by the doctors (more numerous, to be sure, in the cities than in the poor counties in which the worst conditions prevailed). The most striking feature of the assembled data is that infectious diseases, such as dysentery, typhus, and cholera, were the commonest causes of death, and seldom the lack of food – of the protein and carbohydrate needed to prevent the body from wasting away. Spoiled and dirty food, consumption of plants incompatible with human metabolism, neglect of hygiene, all these contributed, but diseases took hold primarily because the immune system is one of the first casualties of malnutrition. In 1847, according to contemporaneous records, only about 6000 people died from starvation out of a total of close to a quarter of a million. A further minority was supposed to have succumbed to the effects of hunger oedema (the accumulation of water in the tissues and the resulting strain on the heart, aggravated by muscle wastage) and, in the case of babies and children, to what was termed marasmus, a starvation-induced state, probably resulting from a vitamin deficiency of some kind. Because of rampaging infections, it was not only the most deprived who were at risk, and many doctors also died.

The first rigorous observations of hunger were probably those of Jean-Baptiste Dumas (Chapter 5) during the 10-month siege of Paris by the Prussian army in 1870–71. Food quickly ran short. Attempts were made to smuggle cattle through the Prussian lines by night (and indeed Claude Bernard suggested that their vocal cords should first be severed so that they would pass by in silence), but without success. The Parisians ate the horses, then the zoo animals, including the two much-admired elephants, Castor and Pollux, and then dogs, cats, and sewer rats (but the evidence is that, even though two books of recipes for dog and rat dishes were published, few had the stomach for such unaccustomed fare). A diary of the time alludes to a gourmandizing citizen who was fattening up a large cat for Christmas, which he proposed to serve surrounded by mice, like sausages. An English visitor, the Member of Parliament Henry Labouchère, rebelled at the notion of feasting on dog steaks, but recorded how different breeds of dog were rated by the Parisian epicures (bulldog tough, spaniel best). Dumas foresaw the ultimate inevitability of capitulation. He was particularly concerned by the high infant mortality. As milk and eggs became vanishingly scarce, he began to think about creating a milk substitute, and attempts were soon set in train: fats were emulsified with sweetened solutions of 'albumens' (proteins of various kinds) but the effects on infants were uniformly pernicious. This was when Dumas concluded, as we have seen, that a diet of the kind that Liebig had prescribed, containing only his vaunted 'trinity of nutrition' – protein, carbohydrate, and fat, with added salt and water – would not serve to support life, and that natural foods must contain something else essential besides. His paper on the subject (written in English and published in 1871 in the *Philosophical Magazine*, a respected journal) attracted little attention.

Improvements in the treatment of infectious diseases, and especially rehydration, as well as the eventual arrival of antibiotics, have mitigated the effects of famines in more recent years, but where hunger has struck, doctors and life-saving drugs have all too seldom been available. In terms of the total number of deaths (if not the proportion of the population exterminated), the great Irish famine was trifling compared to more recent disasters – to the apocalyptic consequences of Stalin's collectivization policy in the Ukraine in 1932, or to the famines that followed from Mao Dzedong's deranged Great Leap Forward in China in the three years from 1959, or indeed those within the last decade in Sudan and the Horn of Africa and probably North Korea.

Charged with finding out how German workers during the starvation blockade in The Great War (p. 112) could best be preserved from prostration, the physiologist Nathan Zuntz made a study of the changes that starvation visits on the body. He discovered that the essential organs, on which the maintenance of life depends, are the last to atrophy. Thus metabolic activity continues and workers can go on working until the vital organs start to fail. At that point only a return to a nourishing regime can stave off death. This discovery allowed industry and basic social services to continue functioning for much longer than they might otherwise have done. The same held true for the slave-labourers who were systematically worked to death in the Third Reich. Worse even than that, the lessons that had been learned in the 19th century were applied by German doctors in the euthanasia programme, initiated in 1940.[10] When the law of 'life unworthy of life' was promulgated, condemning to death those with hereditary or mental illnesses or indeed any debilitating conditions, the regime found little difficulty in recruiting doctors to implement it. Dr Valentin Falthausen, the director of a large mental hospital, suggested that the best means of eliminating these unproductive members of society, mainly young children, was to starve them to death. The advantage of this method was that the doctors were not, so they felt, actually killing their charges, merely allowing them to die. Strengthened by such sophistry, Falthausen himself devised a diet devoid of fats and very low in protein. It consisted of potatoes, turnips, and boiled cabbage. The effect, he told his colleagues at a meeting called to discuss the plans for the new policy, 'should be a slow death, which should ensue in about three months'. The proposal was eagerly espoused by Dr Hermann Pfannmüller, who, in his large hospital in 1943, erected two new buildings, designated the *Hungerhäuser*, the starvation blocks, wherein patients were starved to death on the Falthausen diet. It was only later, when this method of execution proved too tedious and uncertain, that more efficient measures were adopted.

The Second World War brought all too many opportunities to study the effects of starvation. In November 1944 the western part of Holland was isolated by the German occupiers in reprisal against a heroic strike by Dutch transport workers. This had been aimed at impeding the transport of German reinforcements to the south, where British forces were trying to reach parachute troops cut off in Arnhem. The acute food shortage, affording as little as 400 calories per day in some towns, lasted through the coldest winter for many decades. This became known as the Dutch Hunger Winter.[11] Some 10,000 people had died by the time the siege abruptly ended in May 1945, when the country was liberated. Studies of the population over a period of 50 years revealed that it was the babies *in utero* that were most affected by this relatively brief but severe period of malnutrition. Children born to women whose pregnancies had lasted out the siege were smaller. Stillbirths increased nearly twofold, deaths of babies during the first three months of life by fourfold, and birth defects were common, in consequence probably of deficiencies of B vitamins or folic acid (Chapter 9). Most severely affected were the babies of mothers malnourished during the third trimester, the time of most rapid growth. It was clear, moreover, that the mother's fat and muscle reserves would be depleted first, and only when they were effectively exhausted did the fetus begin to suffer. The most striking conclusions emerged many years later, when the 40,000 or so offspring of the pregnant women had grown to middle age. For the most part the survivors developed normally into adult life, but those who had been deprived in the womb during the first two trimesters of the pregnancy had an 80% higher likelihood of becoming obese as adults, and were more prone to diabetes and some other conditions. Those starved during the third trimester were less likely to grow obese. This seems to show that metabolic character evolves in the womb. Another, and seemingly paradoxical outcome was the increased risk of obesity and late-onset diabetes in the *granddaughters* of women who had been pregnant during the Hunger Winter (a phenomenon to which we will return).

The worst hunger episode of the war occurred during the siege of Leningrad,[12] imposed by the German army in its attempt to starve the city into capitulation. The siege lasted 900 days, from September 1941 to January 1944, and from October 1941 until May the next year the city was encircled and practically no supplies got in. But Leningrad did not capitulate. Of the population of 3.2 million, between 1 and 1.3 million were estimated to have died. Almost from the outset savage rationing was imposed. Children and those not engaged in heavy work were allowed each day one-third of a loaf of bread, baked with all manner of unwholesome additives, and, for an entire month, one pound of meat, one-and-a-half pounds of cereal, three-quarters of a pound of oil or butter, and three pounds of pastry or confectionery of a sort. Some allowances were made for the old and for the youngest children. Those performing essential heavy labour received about twice the amount that had to suffice for the rest of the population, and soldiers manning the front lines about twice as much again. Almost at once the supplies fell even below this desperate level, and people began to die. Dmitri Pavlov, the young civil servant in charge of food distribution, scavenged for the remains of potatoes and other root crops in the hinterland, often under fire, collected malt and yeast from breweries, and recovered cottonseed cake from the ships on which it was destined to be burned for fuel. This contained a dangerous toxin, which Pavlov found could be rendered harmless by heating. This repulsive material was mixed into the bread dough. After a month rations were cut, and by December the daily ration of flour was down by 75%, and thus below starvation level.

The work of the research departments in hospitals and the university never ceased, and much information on the effects of starvation was gathered. The first symptoms were swelling of the limbs and pains in the joints. The youths and young adults – those between about 14 and 18 – died first, in part at least because they tended to receive the smallest rations; men died before women, and strong, healthy people generally before chronic invalids. All animals were eaten – horses and then dogs and cats, although the practice was frowned on. Cannibalism, which was illegal and severely punished if discovered, became common. The Russians remained sensitive on the subject and all mention of it was forbidden until after the collapse of the Soviet Union. Recent trawls through the police archives by historians have brought to light stories such as that of the mother who smothered her six-week-old baby in order to feed her three older children. Blood donors continued to come forward to meet the needs of the many wounded, and were given a little extra food.

The indefatigable pathologists, stepping over mounds of corpses, established that it was disease that had carried off the majority. In the early part of the Siege it was most often pneumonia, and a little later gastrointestinal infections. As typhus and especially dysentery took hold, fewer and fewer of the victims were found to have succumbed to what the Russian doctors termed 'nutritional dystrophy', and then in the last phase tuberculosis and scurvy asserted themselves. Starvation was accompanied by oedema (that is, swelling of the tissues through accumulation of water) in some cases though not in others, but pathologists decided that 'oedematous' and 'cachexic' (wasting) hunger disease were one and the same. The oedematous form became more prevalent in the later stages of the siege, and was ascribed to intestinal failure.

In December, when even the drastically reduced food supplies were within days of running out, Lake Ladoga to the north-east of the city began to freeze. At once soldiers were detailed to probe for routes that would support a horse and sledge, and so supplies by this 'ice-road' began to trickle into the city. As the ice thickened, heavier transport, finally trucks, travelling slowly by night, brought in food. By the summer of 1942 the German armies were being held, and the siege was gradually breached. Food began to arrive by barge across Lake Ladoga, by rail and by air, and an estimated million people were evacuated by all these routes. Relief came too late for many Leningraders, fatally weakened by malnutrition. Doctors noted that throughout the next year or two people were dying from heart disease brought on by elevated blood pressure, but not from strokes. This accorded with the weakened heart muscle that pathologists had observed in autopsies. Kidney disease became common and so did atherosclerosis and spinal and joint disorders. The development of children, especially those aged around eight years at the time of the Siege, suffered, but curiously the longevity of the survivors was not greatly affected.

There was no record of the manifestation during or after the Dutch Hunger Winter of a phenomenon seen among British prisoners in Japanese captivity. Some malnourished men experienced galactorrhoea, or secretion of 'witches' milk'. The normally dormant milk glands of baby boys, as of girls, are occasionally activated in the womb by circulating maternal hormones. A small amount of a milky secretion, though not necessarily true milk, is exuded by the baby's nipples, an effect that normally ceases a week or so after birth. Some form of endocrine derangement, occasioned by poor diet, in combination perhaps with other forms of physical stress, would provoke the effect in adult men, sometimes during and sometimes following captivity, when the diet was abruptly changed. Certainly endocrine disturbances have been noted in starvation, nor is this surprising, since the synthesis of hormones depends on the intake of dietary precursors.

The camps

The slave-labourers brought to Germany from all over Europe were not deliberately starved to death, but fed a minimal and grossly inadequate diet, low in protein and vitamins, that would allow them to work for a period until malnutrition or disease killed them or rendered them incapable of physical activity. They would then be replaced by fresh victims. As in earlier episodes of starvation, disease was responsible for the overwhelming proportion of deaths. Typhus was an especial danger, and much feared by the guards in labour and concentration camps. Deaths from starvation per se ensued only when the body's fat reserves were exhausted or from atrophy of muscle tissue and especially of the heart muscle.

When the camps were liberated, Allied army doctors strove to cope with the paralysing problems.[13] Attempts were made to treat the universal protein deficiencies of the emaciated prisoners with an acid hydrolysate of protein, that is to say, an effectively predigested protein solution. This was an American proprietary product called Amigen, and was only partly successful. Janet Vaughan, a distinguished British doctor (later Principal of Somerville College, Oxford) arrived at Belsen concentration camp with a Medical Research Council team two weeks after the liberation. She recorded that the inmates suffered from a diarrhoea which was refractory to treatment, and that they were severely dehydrated and protein-deficient, as evidenced by the low level of proteins in the blood serum. The doctors found by trial and error that injection of huge volumes of concentrated serum – as much as two litres – offered the best hope of survival. The best-tolerated diet was skimmed milk and glucose. Descriptions of the effects and treatment of starvation in Japanese prison camps appeared after the War, and in general, concurred with these conclusions.

Other doctors noted the difference between the symptoms of long, slow starvation and the episodes of relatively brief, acute starvation, experienced during the Second World War. Only in the latter was oedema generally observed. In the concentration camps, the children suffered particularly badly, some from the hideous disease of noma, a bacterial infection that starts in the mucous surfaces and eats away lips, cheeks, and muscles. It never afflicts healthy children and is fatal unless quickly treated. A Dutch woman doctor, who had survived incarceration in Auschwitz, described how children found at the time of liberation to have the disease were cured by injections of nicotinic acid (Chapter 9) and antibacterial agents.

Warsaw

Perhaps the most poignant scientific study of the effects of starvation was that of the heroic group of Jewish doctors, tending the sick and dying in the Warsaw Ghetto in 1940, when the Germans had decided to exterminate its occupants by starvation. In these desperate circumstances, five specialists gathered around Dr Israel Milejkowski, who had been elected as head of what health services the Ghetto could assemble. He had decided that if, as he supposed, they were all to die, the facts should be made known to the post-war world, and also that the opportunity of making a scientific study on an entire malnourished community should not be lost, so that at least something good might come from all the pain and death. It was indeed the first study of the kind ever made.[14] The diet provided by the Germans yielded 800 calories per day, often in practice less. It was grossly deficient in vitamins and minerals. Weight loss, commonly to half the original body mass, the doctors found, developed in three stages: depletion of the body fat, withering of skin and tissues as if by ageing, and terminal cachexia, or wasting away. The effects on muscle and eyes were the most severe. Osteomalacia, the weakening of the bones, often leading to spontaneous fractures, was common. Children were the most vulnerable. Meticulous measurements were made of carbohydrate, nitrogen and mineral metabolism, blood pressure and volume, circulatory disturbances, water and electrolyte balance, blood composition, and so on. All the doctors were well-known specialists, some with a training in research. A famous pathologist performed autopsies. The studies on the effects of starvation on the eyes are considered unique. They revealed that, against all expectation, there was no night-blindness, therefore no vitamin A deficiency. There was clouding of the lens (cataracts), especially in younger people, reduced pressure in the fluid under the cornea, the aqueous humour, and degradation of the cornea itself. The whites of the eyes became bluish, and the response of the pupil to light was sluggish. Through all this the doctors, themselves starving and debilitated, worked ceaselessly to give succour to their patients. One of their great concerns was the extreme anaemia of the Ghetto's inhabitants, and they tried many ways of relieving it, none of them successful.

The results of the study, to which in the end 28 doctors contributed, were recorded in a series of scientific papers, which of course there was no way to publish. The programme was terminated when the deportations to the death camps began in 1942. About half of the reports were lost. The remainder were smuggled out and entrusted to the non-Jewish Director of the Department of Medicine of Warsaw University, Professor Orlowski. He buried the papers, and managed to retrieve them when Poland was finally liberated. Few of the participants in the study survived. Of the original group of six doctors, two died in the Treblinka extermination camp, two, including Dr Milejkowski, committed suicide, one, a noted woman paediatrician, former director of a children's hospital, was killed during the Ghetto Rising, and one survived to recover some of the papers, but died of a heart attack in a Warsaw street the next year at the age of 56. The surviving manuscripts were published in 1946 in Polish and in French, but made little impact until they were translated into English and published in the United States in 1979. The commentary by one of the American editors observes: 'One is struck by the clarity of thought, by the rational, critical analysis of the observations, and by the persistence and tenacity of the observers.' By this time much of the research had been bypassed by more recent progress, though by no means all. Dr Milejkowski's epigraph in his foreword stands as a memorial: *non omnis moriar* – not everything of us will die.

9

The Quarry Run to Earth

The paradigm shift

The physicist Max Planck, who discovered that radiation travels in packets, which he called quanta, and thereby shattered the smug perfection of classical physics, is remembered also for a much-quoted utterance. New concepts in science, he reflected, gain acceptance not when the sceptical majority is at last won over by the evidence, but when they die off and are replaced by a new generation whose opinions have not yet hardened. This is really only a half-truth, but certainly scepticism is deeply embedded in the scientific process. A departure, therefore, that threatens to overturn the prevailing intellectual structure of a field of knowledge – what the philosopher Thomas Kuhn termed 'a paradigm shift' – must pass through the fire of doubt, even hostility, before it can supplant the accepted wisdom. So it was with the vitamin concept, for the very notion of a disease caused by no pathogen or other noxious agent ran counter to all 19th-century medical teaching. Albert Szent-Györgyi, whom we shall meet again shortly, was famous for his gnomic asides, and he encapsulated the matter thus: 'A vitamin is a substance you get sick from if you don't eat it.' To conceive of such enigmatic essences required an intellectual leap of which few physiologists, much less doctors, were capable, yet when the time came several remarkable researchers achieved it more or less simultaneously, as we have seen. But it needed biochemists and chemists to run the elusive factors to earth.

Vitamin A revealed

First came vitamin A. Its existence was foreshadowed in Hippocrates's time, perhaps indeed earlier, for night-blindness, the first manifestation of a deficiency, and also its treatment, were familiar to many. The remedy – liver in one form or another – is prescribed by Hippocrates, Galen, and by Chinese and medieval European scholars. Eduard Schwartz, a doctor in the Austro-Hungarian navy, described in 1861 an outbreak of night-blindness in the course of a long voyage, and the 'miracle' recovery that resulted when he fed the ship's company boiled liver. He was savagely attacked in the Vienna physicians' journal, perhaps because the liver cure had been regarded as an old wives' remedy. Reports, however, continued to flow from travellers and doctors (David Livingstone was one) throughout the 19th century. But one of the great missed opportunities occurred in 1880, when Nikolai Lunin, a Russian research student at the German University of Dorpat (now Turku) in Estonia, found that mice fed on what was then regarded as a complete diet, consisting of carbohydrate (sugar), protein in the form of milk curd, fat, and mineral salts, all died in two to four weeks. By contrast, mice on the same diet, but with a little added milk, thrived. Lunin and his supervisor, Gustav von Bunge, were astonished, for this went against the nutritional precepts to which von Bunge was wedded. Lunin, in his publication in 1881, though, drew what now seems the obvious inference – that one or more trace substances in the milk must be required to support life. The professor, however, would have none of it: he did not believe in 'accessory factors' and thought that the milk-free diet must simply have been deficient in certain minerals, lost in the course of the preparation.

Von Bunge was in fact fixated on the role of inorganic (mineral) substances, which must, he thought, exist in foods in something like their functional states. For he further believed that the chemistry that went on in the body was rudimentary when set against the skills of an organic chemist. Thus the phosphorus-containing lipids (fats) of cell membranes could not, he declared, be generated from simple inorganic phosphates, nor simple iron compounds be turned into the red blood pigment, haemoglobin. He continued to insist that all Lunin had proved was the inadequacy of milk components in milk that had lost essential mineral-derived substances. It did not help that two other students in von Bunge's laboratory both reproduced Lunin's result: the professor discouraged them all from searching for an active substance.

Von Bunge (1844–1920) was by then Professor of Physiological Chemistry at the University of Basel. Born and educated in Dorpat, where he gained doctorates in both chemistry and medicine, he was highly regarded, so much so that he was invited to succeed the famous German physiologist Emil DuBois-Reymond in Berlin, a call he declined. Von Bunge's accomplishments were far from trivial. He showed that sodium and potassium were present in cells and in the serum of the blood and to differing extents in plants, and that iron in some form was required for haemoglobin formation. His prestige and influence were considerable, and his *Textbook of Physiological Chemistry* was for many years a standard work. He was at the same time something of a crank, obsessed with abstinence and prey to a vague mysticism about the nature of life. He may in the final reckoning have held back more than helped progress on nutrition.

Other workers early in the 20th century rediscovered what Lunin in Dorpat and his successors in Basel had first observed. In 1905 Cornelis Pekelharing in Holland found that while animals offered only a 'basic' diet of protein, carbohydrate, and lard deteriorated and died, addition of a very small amount of milk allowed them to develop normally and survive, and he fully grasped the implications. But he published his account in a Dutch medical journal, which no English- or German-speaking researchers would have read. Frederick Gowland Hopkins in England, who made broadly similar observations in 1906 without pursuing them further at the time, noted as much in his Nobel Prize lecture in 1929. 'It is indeed astonishing', he observed, 'that the results of such significant work as his, though published in the Dutch language alone, should not have become rapidly broadcast.' The work had not even appeared in the regularly published abstracts on which researchers in the field were accustomed to rely. And, Hopkins continued, 'I shall never cease to regret that, in common with so many others, I was then [in 1906 and later] completely ignorant of [Pekelharing's paper].'[1] Here is how Pekelharing had summed up his results:

> ... there is a still unknown substance in milk which even in very small quantities is of paramount importance to nutrition. If this substance is absent, the organism loses the power properly to assimilate the well-known principal parts of food, the appetite is lost and with apparent abundance the animals die of want. Undoubtedly this substance occurs not only in milk but in all sorts of foodstuffs both of vegetable and animal origin.[2]

The essence of the matter could hardly have been more clearly put.

Hopkins had independently come to a similar conclusion, but felt as Pekelharing perhaps had, that to publish a mere speculation about the existence of such substances would not do. The physiologists engaged in calorimetry and other whole-body experimentation would scorn what the biochemists had to say unless a factor was actually identified. This Pekelharing had failed to achieve, despite strenuous efforts, and Hopkins had only then resolved to start along this arduous trail. Then, in 1907, a young man, who was to become one of the giants of the subject, made his appearance on the historical stage. Elmer Verner McCollum was born in 1879 on a farm in Kansas.[3] He attended the one-room local village school, and from there won a scholarship to study chemistry at the University of Kansas. Having graduated, he enrolled at Yale University as a PhD student in organic chemistry, but did his research at the Connecticut Agricultural Experiment Station, under the noted chemist, T. B. Osborne, whose interests lay in the chemistry of foods, especially proteins. McCollum was awarded his doctorate in 1906, and the following year found a position at the Wisconsin Agricultural Experiment Station. As a junior member of a research group engaged in the study of nutrition of farm animals, McCollum was put to work applying the latest methods of chemical analysis to the food that went into dairy cattle and to what came out. The project was of very practical importance, for the leader of the group, E. B. Hart, had found that cows did well on a diet of maize and less well on oats, while on a wheat diet their health and fecundity declined. They weakened, often went blind, and produced only stillborn or sickly calves. The initial conclusion, of course, was that wheat contained a factor toxic to cattle, but this hypothesis was soon discarded, and a search began for whatever beneficial substances might be present in maize, yet lacking in wheat. It was not long before young McCollum began to show his independence of mind. The health and productivity of livestock might be the institution's concern, but cows were less than ideal laboratory animals, and McCollum was finding the work tedious and progress frustratingly slow. Rodents, he thought, were the answer, and he set up a breeding colony of albino rats, in the face of the Dean's disapproval and the rudimentary facilities. McCollum's rationale was that rats took up less space than cows, so many more animals could be accommodated, they mature quickly and breed rapidly, and being small they were cheap to feed. Like Osborne and his colleague at Yale, Lafayette Mendel, McCollum found that rats raised on a diet of the purified milk protein casein, together with starch, lard, and salt, died young, having developed a number of symptoms, including a form of blindness. This indeed had already been noted some three years previously by an ophthalmologist in Basel, who had characterized not only this but other eye disorders, including xerophthalmia, a desiccation of the cornea, and corneal ulceration, in deprived rats. (In human subjects, night-blindness* was later recognized as one of the first manifestations of vitamin A deficiency.)

Osborne and Mendel had also by this time (1912) discovered that the part of milk from which protein had been eliminated would ensure survival when added to the basic diet. And from 1911 papers in German journals had begun to appear under the name of a young doctor called Wilhelm Stepp, working at what was then the German University of Strassburg (that is to say Strasbourg). His professor was one of the leaders in physiological chemistry, or as we would now say, biochemistry, Franz Hofmeister. Hofmeister ran his department in a liberal style, highly unusual at the time, especially in Germany. Stepp, therefore, was allowed the freedom to develop his own line of research: he wanted to devise a complete diet on which his experimental mice could thrive, and then try to extract from it trace substances needed for healthy survival. The first of the foods that proved satisfactory was bread soaked in milk and then baked to dryness. Stepp prepared an extract from this with boiling alcohol, and a second one with boiling ether (in what chemists know as a Soxhlet apparatus), and found that after such treatment the residual fodder no longer sustained the life of mice, which lost weight and died. He then added the extract back to the depleted food, and the mice once again flourished. The conclusion seemed clear: a fatty ('lipoid') substance, soluble in ether, was the vital essence. Stepp also showed that a similar life-giving extract could be prepared from egg yolks, and that whatever its nature, it did not

* Night-blindness, and its association with impoverished diets, was, as has been remarked (Chapter 3), recognized even in ancient times. Hippocrates of Cos recommended raw liver steeped in honey as a cure, and beef liver continued to be prescribed until the advent of cod liver oil.

consist simply of the fats contained in eggs and dairy products, for purified known fats* added by themselves to the extracted milky bread were inactive.

All this was known to McCollum when he began his search for the 'lipoid' substance. He found it in ether extracts of milk, butter, and egg yolks, but not in olive oil, any more than in lard. To demonstrate to the world that the life-sustaining component was not a hitherto undiscovered fat (triglyceride), he did a clever experiment: he treated his ether extract† with alkali. This destroyed the triglycerides by a reaction known as saponification, because it is used to turn fats into soaps.‡ The resulting material was insoluble in ether, and McCollum then showed that something had been left behind, untouched by the saponification reaction, and this something possessed the activity that he was seeking. Similar experiments showed that the activity could also be recovered from some other types of food, notably liver. McCollum called the factor 'fat-soluble factor A'. This distinguished it from a water extract of rice polishings that prevented beriberi symptoms, which McCollum called 'water-soluble factor B'. Such was the origin of the later designations of the vitamins. Stepp, in fact, had got into a muddle over that very issue, for having found that the fat-soluble substance in milk, egg yolk, and also later butter was indispensable to life, he was dismayed to discover that mice also died if they were not given an alcohol extract too. (A water extract would have done equally well.) It evidently did not occur to him that there might have been more than one essential factor. Stepp thereafter surfaces only occasionally in the scientific literature, for he largely abandoned research in favour of a career in clinical practice, finishing as a professor at the University of Munich, the city in which he died in 1964.

In 1911 a new comet appeared in the firmament. Casimir (originally Kazimierz) Funk[4] was the son of a Polish doctor. At the time of his birth in 1884 Poland was a brutally repressed Russian satrapy, where there was little prospect of fulfilment in a professional career, and so his father sent Casimir, at the age of 16, to study at the University of Geneva. At 20 he had a doctorate in organic chemistry from the University of Berne, and the same year, being already interested in physiological chemistry, he headed for Paris and the Pasteur Institute, and two years later for

* The fats in question, known to biochemists of the time, were triglycerides, lecithin, and cerebroside, and Stepp also included cholesterol.

† Solvent extraction is one of the first steps along the route by which biochemists seek to prepare a pure substance from biological materials. The solid matter is shaken up with water, alcohol, ether, or some other solvent. Some substances (those that are fatty in nature) would be extracted from the solid into ether, while others would be left behind, but might instead be extracted by water or alcohol for instance. Many substances show fine discrimination between their solubilities in different solvents.

‡ The triglycerides fall into a class of chemicals called esters (p. 253). When heated with alkali, the esters, which are insoluble in water, are split into their constituent parts, comprising alcohols and acids, but acids in this case with long chains of carbon atoms attached. Such molecules have detergent properties, and are indeed soaps.

Berlin. There he spent four years in the laboratory of Emil Abderhalden (p. 170), engaging in research on proteins, and their constituents, the amino acids, and in 1910 moved again, this time to London and a position at the Lister Institute, then the centre of research into nutrition. His plan was to isolate and identify the anti-beriberi factor in rice husks, which he and others at the Institute believed would turn out to be an amino acid. The substance was water-soluble, he soon found, but it did not appear to share any of the other characteristics of amino acids. He thought, though, that it was a base (a substance that neutralizes, or is neutralized by, acids). The only type of base then known to physiology was an amine (a compound containing a nitrogen atom with attached hydrogens). In 1912 Funk wrote an article that attracted wide attention, a *tour d'horizon* of the subject of deficiency diseases. In it he conjectured that scurvy and pellagra, like beriberi, would prove to arise from a lack of one member or other of a class of trace substances. It was in this review that he introduced a generic term to describe these compounds: *vitamine* was a conflation of 'vital amine', and of course a misnomer. Not everyone approved of such presumption – McCollum, for one, hated the word – and when in due course it emerged that the 'accessory food factors' (Gowland Hopkins's usage) were for the most part not amines at all, *vitamine* was, apparently by common consent, shortened to vitamin. Alternative coinages, such as nutramine and food hormone, did not catch on. In any event, Funk's paper was influential, largely by concentrating the minds of the nutritionists and focussing attention on the importance of trace substances. The primacy of the whole-body physiologists with their calorimeters was drawing to an end.

It was to be some years, all the same, before McCollum's 'fat-soluble factor A' was finally run to earth. One of McCollum's students, who achieved high distinction, was Harry Steenbock (1886–1967). Like his mentor, he came from Wisconsin, and it was at the University of Wisconsin that he spent his entire career. His work on vitamin D (see below) led to a patent, the profits from which Steenbock used to found the Wisconsin Alumni Research Association. This body, which still flourishes, has disbursed funds to research workers in the state and beyond, amounting since its inception in 1925 to some $800 million. Steenbock first attracted notice in 1920 through a curious observation. He had been trying ether extracts of a variety of plant and animal tissues for their efficacy in supporting the growth of young rats and averting eye disease. What struck him was that yellow extracts from yellow plants such as carrots or sweet potatoes worked, while those from white plants (parsnips, potatoes) or red (beetroot) were inert. Now there are many substances in cells that will impart a yellow tinge to solvent extracts, but Steenbock guessed that the active principle might really be a yellow pigment. He cleaved to this idea even though liver extract, which was undeniably effective, was colourless. Were there then two different forms of 'factor A', one yellow, one colourless? It was to be another 10 years before a biochemist, Thomas Moore at the Dunn Nutritional Laboratory in Cambridge, recovered a yellow compound, carotene, from carrots and showed that it had the requisite activity. This laboriously purified substance was as effective as cod liver oil (but no more so) in supporting the growth of young rats. Then Moore kept a cohort of rats on a diet lacking vitamin A (and carotene) for several weeks, killed some of them, and analysed the contents of their livers. The remainder were fed carotene-rich diets and after an interval their livers too were examined. In the livers of this group, but not of the first, Moore found a high concentration of the colourless 'fat-soluble factor A' (already designated vitamin A), as extracted from milk and egg yolk. The yellow carotene, then, was converted into colourless vitamin A in the liver. The two are now known to be closely related to one another chemically, and also to retinene, the molecule that, in association with a protein, forms the visual image on the retina. The carotene is the chemical precursor from which vitamin A and retinene both derive. Carotene is plentiful in maize but not in wheat, and this was why the cows with which McCollum had begun his career did well on the one but not on the other. The year after Moore's discovery, the Swiss organic chemist Paul Karrer, with two of his army of students and assistants, obtained the vitamin in pure state and determined its structure, and a little later achieved its synthesis.

Change and reaction

The first decade of the 20th century saw 'the dietetic trinity' finally buried. An important milestone was the proof by Frederick Gowland Hopkins that proteins are not all alike. It had of course been shown by Magendie (p. 74) that gelatin was a protein with poor nutritional value, but there the matter had rested for a hundred years. Hopkins was a true pioneer. Born in 1861 into a solid middle-class English family, he made the most hesitant of starts to a scientific career. His father died when the boy was still young and he was brought up by his mother. At school, though poorly taught, he did well in chemistry and developed an interest in insects. He left school for a miserable spell as a clerk in an insurance company, but then apprenticed himself to a consulting analytical chemist in London. He attended indifferent evening lectures in chemistry, but then managed to get into University College where he qualified as an Associate of the Institute of Chemistry. So well in fact did he do in his examination that he came to the attention of the Medical Jurist at Guy's Hospital, and Home Office Analyst, Sir Thomas Stevenson, who offered him a position as his assistant, to perform forensic analyses. Stevenson encouraged young Hopkins to better himself, and enrol at Birkbeck College, an evening school then as later, to read for an external London University degree. By the time he had finished Hopkins had gained confidence and entered Guy's Hospital Medical School to train as a doctor. During his time there he met and learned from one of the towering figures in the history of academic medicine. A. E. (later Sir Archibald) Garrod was working at the time at St Bartholomew's Hospital, on the other side of the river from Guy's. There, in his laboratory, he founded a new field of scientific inquiry, that of inherited metabolic disorders. Garrod, unlike other doctors, thought about diseases in terms of chemistry, and seized on the discovery that the reactions that occur in the life processes of the body require enzymes. From that came his most profound insight, for he postulated that a defective enzyme would cause a metabolic step to fail, and that the defect in the enzyme must be the result of an answering defect (or as we would now say, a mutation) in the patient's genetic make-up. This was the famous 'one gene, one enzyme' hypothesis. Garrod's first exemplar was a disease, which he himself first identified, called alkaptonuria, marked by a darkening of the urine and unpleasant attendant symptoms (p. 259). He predicted that the fault would be found to lie in an enzyme involved in protein breakdown, and so (50 years later) it proved. The course of Hopkins's later work was deeply influenced by Garrod's style of thinking.

But for the present, Hopkins, already 32 when he qualified, taught physiology at Guy's for the next four years, indulging himself in experimental work in the little time that remained. By this tortuous route he eventually found his way to Cambridge, where he again taught physiology and did what research he could. He advanced by degrees, and finally in 1914 was appointed the first Professor of Biochemistry, though it was to be another decade before the newly established department was favoured with a building of its own. Before that, most of Hopkins's research work was done in a small laboratory in the Physiology Department.

Hopkins commented in his memoirs that in none of his research projects had he ever worked with anyone senior to himself. He had had to plough his own furrow, and he felt that in this and in the varied experience of his earlier life he had been fortunate. 'I was led', he reminisced, 'at a right moment to follow a path then trodden by very few and where each wayfarer was conspicuous. It is now a crowded path on which individuals cannot fail to jostle each other.'[5] (This in 1937, when there were a hundred times fewer biochemists to jostle each other than today.)

As early as 1900 Hopkins discovered the amino acid (p. 257), tryptophan, in proteins. This is one of the least abundant of the 20 natural amino acids from which the proteins are built, and it is one of the 'essential' amino acids, that is to say, one that cannot be synthesized in the body, and must be provided from the diet. Some proteins contain no tryptophan at all, and among these are gelatin and the maize protein, zein (which takes its name from the maize plant, *Zea mays*). Hopkins fed mice on a regime of zein and dog biscuit, with ashed oats to provide minerals, and found that the animals all died after two weeks, whereas fresh oats and dog biscuit kept them alive. When tryptophan was added to the zein diet the mice survived longer. This was the first clear demonstration, complete with explanation, of nutritional differences between one protein and another. Hopkins went on to study the beriberi factor, and eventually shared the Nobel Prize with Christiaan Eijkman. He was the first (though challenged by the pugnacious Casimir Funk for the privilege) to articulate the 'accessory food factor' or vitamin principle of nutrition, and in a much-cited lecture, published in 1906, he stated it like this: 'No animal can live on a mixture of proteins, carbohydrates and fats, and even when the necessary inorganic material is carefully supplied, the animal still cannot flourish.' He had, as he later wrote, been especially struck by the strange circumstance that when he used the purified milk protein, casein, in feeding experiments, there were enormous differences between the capacities of one preparation and another to support the life of mice. When washed with water–alcohol mixtures the efficacy of the active preparations vanished, and Hopkins at once deduced that a very minor factor essential for life had been present as an impurity – in short, a vitamin. Much later, in 1920, he began a lecture to the British Medical Association with the bold declaration: 'In what I am about to say I refuse to speak about the vitamine "hypothesis". Vitamines, though still of unknown nature in the chemical sense, are not merely hypothetical.' And later: 'But I have found that there is at the present moment some scepticism about the whole question, particularly perhaps among certain members of the medical profession.' Hopkins was not exaggerating. Here is what one of the luminaries of that profession, Sir James Barr, had to say on the matter: 'Vitamines, so far as their composition is concerned, seem to be a figment of the imagination .... All these observations are easily explained without invoking any recondite influence of "vitamines".'[6] It was not, in fact, only among the doctors that doubters were to be found, for they also inhabited the highest reaches of academia. That same year, Gustav von Bunge, not long before his death, wrote a letter to his brother in which he voiced the following sentiment: 'Further, vitamins remain a mere name; no-one has isolated them. Modern physiological chemistry threatens to become physiological economics, consisting in the invention of Latin names for unknown substances.'[7] And only a few years earlier the influential, malevolent (even fraudulent) German physiological chemist, Emil Abderhalden, had asserted in a publication that 'we arrive at the conclusion that at present there is no convincing proof for the hypothesis that there exist entirely unknown substances essential for survival.'*

* Abderhalden was an unappealing personage, with a penchant for being influentially wrong. He had studied initially under von Bunge and then with the great organic chemist Emil Fischer, who was determining how proteins were constructed. In Fischer's laboratory Abderhalden learned the craft of synthesizing peptides – amino acids linked together as in proteins (although the technology of the time limited the lengths of the chains that could be assembled to little more than two or three amino acids, whereas proteins comprise chains of often many hundreds). Abderhalden used these peptides in feeding experiments and in many other investigations, mainly misconceived. In the worst of these undertakings he discovered, as he thought, a new class of enzymes that break down peptides and proteins into their constituent amino acids. These, he asserted, acted on alien proteins in the body, and he termed them 'defence enzymes'. Among the intruders, naturally produced, were proteins released by the fetus into the bloodstream of pregnant women. Therefore, he claimed, his defence enzyme preparations could be used for early pregnancy tests. A little later they emerged as detectors of cancer and then of other diseases. The entire scheme was an illusion, as quickly became clear to

Hopkins was able to ignore the background noise. He busied himself profitably with the pursuit of the anti-beriberi factor and with many other areas of biochemistry. The laboratory in Cambridge, under his benign rule, drew visitors from all parts of the world, and many of the most illustrious members of the next generation of biochemists trained there. In the years before the Second World War Hopkins and a small group of colleagues founded the Academic Assistance Council to give succour to displaced Jewish and other scholars, fleeing persecution in Germany. Hopkins was showered with all the honours his profession and his country could bestow, and died at the age of 86 in 1947.

Minerals and the inorganic organism

Gustav von Bunge was not always wrong. We cannot do without mineral salts. He was not, to be sure, the first to proclaim as much, and indeed it was the alchemists of the 17th century who discovered an inorganic substance, phosphorus, in bodily materials. Phosphorus is an element that looms large in the chemistry of life. In combination with calcium it forms the substance of the bones and teeth, it is in our genetic material, the deoxyribonucleic acid (DNA), and in the other natural nucleic acid, ribonucleic acid (RNA), and also in the lipids of cell membranes. One of its most ubiquitous compounds is the molecule that provides the energy of muscle and movement and much else. This is adenosine triphosphate, or ATP. ATP releases energy when it is broken down to adenosine diphosphate, ADP, and the amount of ATP that we utilize daily amounts to our entire body weight. A traveller in South Africa had noted in 1785 that cattle on exhausted land would chew the bones left by dogs, and even each others' horns as their craving became more insistent. Many died. The same aberrant behaviour recurred early in the 20th century, when cattle were driven from grazing land and the soil was not manured, and thus depleted of phosphorus. The farmers then found that this strange form of cannibalism ceased as soon as phosphate-containing fertilizer was made available to the animals.

The metal, calcium, was discovered simultaneously in 1808 by Berzelius and Humphry Davy, but 1804 had seen the beginnings of the study of metal compounds in living organisms, when the Swiss Théodore de Saussure incinerated

biochemists who tried to reproduce the effects, but Abderhalden used his position and his connections to persecute all who dared to voice their doubts. Not merely was the saga of the defence enzymes atrociously bad science, but it also acquired a particularly unsavoury political aspect when Abderhalden suggested that there might be variations between the defence enzymes of people of different races. By this time the Third Reich was approaching its zenith, and the pseudo-scientific experiments on prisoners in concentration camps were in full swing. Human material could be readily obtained with the help of such would-be researchers as Josef Mengele in Auschwitz. Abderhalden seems to have connived in all aspects of the programme. Research on his imaginary enzymes continued even after his death in 1951, conducted by his son and other aficionados.

soil and found that the residue, which could have contained only mineral matter, encouraged the growth of plants in exhausted soil. Plant material itself, for that matter, contained mineral substances. It was rich in potassium but much less so in sodium. Liebig, in his book of 1842 (Chapter 5), cited this work as proof that minerals are essential for plant life. That animals were drawn to the salt lick had of course been known for centuries, and must have been seen as evidence that they, too, needed salt, but it was Boussingault yet again who did the critical experiment on his farm: he fed two groups of oxen on identical diets, except for the lack of salt in one. The salt-deprived beasts developed a staggering gait, and lost their hair. Soon Liebig and other chemists began systematic analyses of animal tissues and found sodium phosphate in blood and lymph, but potassium and magnesium phosphates in other fluids. Was sodium phosphate then essential for formation of blood? Salted meat oozed fluid rich in phosphate, and this led Liebig to one of the many false inferences concerning the cause of scurvy (p. 94): it was, he thought, a consequence of a deficiency of phosphates, lost from the salted meat on which seamen relied.

This was how matters stood when Gustav von Bunge began his researches into mineral salts. The problem was that he and the other chemists of the time were ignorant of the existence of ions. Common salt, sodium chloride, when dissolved in water splits into its constituent parts, not the elements, but their ions – sodium ions with a positive charge and chloride ions, which are negatively charged. This is equally true of other salts, so a solution made by mixing, say, sodium chloride and potassium phosphate is the very same as one made with potassium chloride and sodium phosphate. Von Bunge supposed that one mixture could be changed into the other only by a chemical reaction, and so he deduced that potassium compounds taken into the body when plants are eaten would displace sodium from its compounds, with formation of foreign sodium compounds. These would be eliminated by the kidneys, so that replacing sodium in its acceptable form would require an intake of common salt. To provide experimental proof, he ate a diet laced with potassium instead of sodium chloride and analysed his urine, which he found to be rich in sodium. His theory – not unreasonable in the state of knowledge of the day – became the accepted orthodoxy. Von Bunge, as we have seen, rejected the very concept of nutritional factors other than minerals, which, he tenaciously held, sufficed to meet all dietary needs, besides of course proteins, carbohydrates, and fats. Gradually, as chemistry advanced, the action of salts, too, became clearer, and it was recognized that calcium and magnesium, as well as sodium, potassium, and phosphorus, were indispensable for life.

The beriberi factor

Robert Runnels Williams was an American chemist. Born in 1886 to a Baptist missionary couple, he passed his childhood in India and was eventually sent to study chemistry at the University of Chicago. In 1909 he found himself teaching in a school in the Philippines, and within months he had made the acquaintance of Edward Vedder, the American army doctor who had deduced that babies with beriberi were suffering from deficiency of an essential food factor (p. 144). Vedder needed the skills of a chemist, and Williams left his teaching job for a position in the Philippine Bureau of Science, where he set out on what was to prove a long road to isolate the anti-beriberi vitamin. Williams extracted rice husks with a hot water–alcohol mixture and showed that the extract contained the anti-beriberi activity. At the same time a team of researchers at the Imperial University in Tokyo under Professor Umetaro Suzuki had been proceeding along much the same lines and had recovered crystals of what they called oryzanin (from *Oryza*, the botanical name for rice), which prevented polyneuritis in chickens. They believed they had isolated the factor itself, but were soon enough disabused when the crystals turned out to contain a mixture of compounds. The hunt for what McCollum had dubbed factor B now became an epic, for it engaged the skills of many of the world's leading biochemists and chemists. The difficulty of course was its enormous activity, which meant that it was needed – and present in food – in only minute amount. Casimir Funk was one who, like Suzuki, made a crystalline preparation from rice husks – the starting material was 836 pounds of rice – and believed it to be the vitamin, and indeed 1 mg sufficed to cure pigeons of polyneuritis. But then, assisted by the young Jack Drummond, at the start of a notable career (p. 14), he found that his material could be separated into two fractions (though there must in actuality have been many more), and he inferred that two substances acted in unison to suppress beriberi. This, too, turned out to be wrong. Then the First World War began, and Funk felt himself unwelcome in England and so migrated to the USA, where he joined a pharmaceutical laboratory and redirected his energies to the study of hormones.

Williams, meanwhile, who had briefly returned to the USA to learn about the newer methods of preparative biochemistry, was back in Manila trying again. After four years of toil he had some tolerably active preparations, which he made available to doctors treating beriberi around the Philippines. Then his contract ended and he had no option but to return home, to a position with the Bureau of Chemistry of the US Department of Agriculture. He was allowed to devote only part of his time to the beriberi problem, but made no less than 40 attempts, all of them unsuccessful, to isolate the tantalizing factor, starting each time from sacks of rice polishings or of yeast. The chase continued for another decade, but in the interim an important fact had come to light. Eijkman and his colleague and later successor in Java, B. C. P. Jansen, began to suspect that the 'water-soluble factor B', which rats needed for growth, might not be the same as the substance that prevented beriberi. They were in due course proved right, when two American workers found that one of these activities in yeast preparations was destroyed on heating, while the other was unaffected. So when Jansen and his colleague, W. F. Donath, set out to isolate the vitamin, it was only the anti-beriberi activity that they allowed to guide them. Theirs was a heroic effort, and it took them eight years. For their assay they used a bird, the bondol, which lived in the rice fields of Indonesia, so small that it could be sustained on no more than 2 g of rice per day – far less than a chicken, or even the pigeons that Funk had used. On polished rice they developed the avian form of beriberi, polyneuritis, which a minute amount of polishings or of their extracts would avert. The account of the purification is almost painful to read, but starting from 100 kg (one-tenth of a ton) of rice polishings, Jansen and Donath ended up with 100 mg (one-millionth of the original weight) of the pure vitamin. A sample of this they sent to Eijkman in Holland, so that he might have the pleasure of assaying it and observing its potency for himself. They themselves performed an elementary analysis for the content of carbon, oxygen, nitrogen, and hydrogen, and obtained the composition $C_6N_2H_{10}O$. In this they were subtly wrong, for they had slightly overestimated the hydrogen, and the molecule also contained an atom of sulphur. This was enough to thwart Jansen's attempts to synthesize the vitamin, and a compound that he identified with the supposed atomic formula bore no relation to it.

Meandering down the same valley of tears was the dogged Williams, who in 1920 had left US Government employ for a position at the Bell Telephone Laboratories in New Jersey (then, and for another three or so decades to come, a celebrated centre of basic research in many branches of science). As before, Williams could pursue the anti-beriberi factor, which was by now an obsession, only in his spare time. Jansen sent him some of his crystals, which Williams assayed. But he found difficulty in reproducing Jansen's preparative procedure, which was indeed very complex. He decided instead to develop a method of his own, starting from an even greater mass of rice husks. He found laboratory space in the basement of a hospital and conducted his assays on pigeons in his garage at home. Working evenings and week-ends with the help of a colleague from the Bell Telephone Laboratories, Williams eventually prevailed: pure material emerged, and a corrected elementary formula was determined. He had managed to improve his circumstances by way of small grants from a company with an interest in yeast and from the charitable Carnegie Foundation, and better laboratory space, made available by Columbia University. Williams also remarked that he was much helped by the Great Depression, which forced the Bell Telephone Company to reduce its working week to three days. The next challenge was of course to determine the structure of the vitamin, now designated B_1, and for this Williams sought the help of the chemists at the Merck pharmaceutical company. In the event the structure was determined simultaneously by them and by the great organic chemist Adolf Windaus, in Germany (a Nobel Laureate for his work on steroids), who had obtained the pure vitamin from the mighty chemical and pharmaceutical concern, I. G. Farben. Williams and the Merck chemists named the substance aneurin (a contraction of antineuritic vitamin), but this was later changed to thiamin (from *thio* for sulphur), and then again to thiamine (for it is indeed an amine), by which name it is now known. The biochemical route by which the vitamin acts in reversing within minutes the symptoms of beriberi in children or of polyneuritis in birds was brought to light not long afterwards by R. A. (later Sir Rudolph) Peters in Oxford.

The antiscorbutic substance

In 1895 a researcher in the US Department of Agriculture, one Theobald Smith, who was studying bacterial infections in pigs, fed various diets to guinea pigs and found that the animals died when restricted to oats and bran. The symptoms were those of scurvy, but since this was not what Smith was after, the explanation never crossed his mind. This, though, was the first 'animal model' for the human disease, and it probably gave impetus to later work, in particular that of Axel Holst and his collaborator, Theodore Fröhlich, in Oslo. From 1907 to 1912 they looked for a source of an anti-beriberi factor in different foods, which they tested on guinea pigs. The animals remained healthy on a diet of cereal grains and fresh cabbage, whereas cereal alone resulted in death from what Holst and Fröhlich identified, to their surprise, as scurvy. As expected (Chapter 2), fresh fruit or fruit juice averted this fate. The most acidic juices retained their antiscorbutic activity longest, and, more surprisingly, when the cereal seeds were soaked in water and allowed to germinate they too acquired activity.* The report of this work was not uniformly well received, and Elmer McCollum, then at the height of his career and prestige, for once went seriously astray when he insisted that scurvy was not a disease of vitamin deficiency at all. He was led to this unfortunate conclusion by his faith in the rat as the prime subject for nutrition experiments, for the rat, unlike man and most other animals, has the capacity to synthesize its own vitamin C, so does not depend on an external source. The observations of Holst and Fröhlich were taken more seriously when similar results on monkeys emerged from English laboratories. The more sagacious researchers were by this

* A puzzle that had given rise to scepticism about the existence of an antiscorbutic vitamin was the well-known circumstance that the Inuit people of the Arctic lived healthy and robust lives on a diet seemingly almost entirely devoid of vitamin C, and consisting of meat and offal, much of it barely cooked. An anthropologist and explorer, Vilhjalmur Stefansson had spent long winters in the Arctic and had developed a taste for raw seal blubber and putrefied whale oil, while eschewing all vegetable foods. He stayed strong and healthy on this diet. Stefansson's book about his experiences, gastronomic and sociological, caused a stir, but from the medical profession he encountered nothing but scepticism. Exasperated, he resolved on a demonstration and in 1928 he and a colleague began a year on an exclusively meat diet, supervised by doctors at Bellevue Hospital in New York. They started on low-fat rabbit meat, and within days began to suffer from digestive problems. But Stefansson knew about the condition of 'rabbit starvation', which afflicted his friends, the Inuit, when fatty foods were scarce. A cure was effected when the rabbit meat was supplemented with calves brains, cooked in bacon fat, and from then on Stefansson and his friend subsisted on a diet of nothing but meat, and thrived. The vitamin C content of meat was later established.

time fully attuned to the concept of vitamins, and Jack Drummond (p. 14) went so far in 1920 as to suggest that the still unknown antiscorbutic factor should be recognized as the third vitamin, therefore vitamin C. This undoubtedly stimulated attempts to isolate and identify the substance. Several laboratories entered into a race, and more than one produced highly active concentrates from lemon juice and other sources. The best was probably prepared at the University of Pittsburgh by Charles Glenn King and his assistant, a young man of Hungarian parentage, Joseph Svirbely. From chemical studies they inferred that the vitamin was probably a hexuronic acid – a class of molecules derived from the oxidation of common kinds of sugar. Into the human drama now strode one of the most flamboyant figures in the history of biochemistry.

Albert von Szent-Györgyi Nagyrapolt (later plain Albert Szent-Györgyi)[7] was a product of the minor Hungarian aristocracy, born in Budapest in 1893 into a landed family. But a maternal uncle and grand-uncle were both anatomy professors at Budapest University, and it was there that Albert was sent to study medicine. He emerged a specialist in the arcane discipline of proctology, wished on him by his uncle, hoping apparently for a compliant nephew who would minister to his chronic haemorrhoids. This unpromising beginning was interrupted by the First World War. Szent-Györgyi was lucky to survive the ill-fated campaign on the Ukrainian front. He was decorated for bravery, saw the army decimated by typhus and cholera, and decided that he had had enough. He shot himself through the arm, was invalided home, and, on recovering, narrowly avoided the slaughter in the mountains of north Italy. By then he had resolved on a career in science and, the War over, his *Wanderjahre* began. With his wife and young daughter in tow, he migrated from one place of scientific pilgrimage to another. He at length found a position at the Hungarian University of Pozsony, the ancient Pressburg of the Austro-Hungarian realm, but no sooner was he installed than the Treaty of Versailles redrew the frontier, Hungarian Pozsony became Slovakian Bratislava, and Szent-Györgyi, as a Hungarian citizen, was expelled. Peripatetic once more, he spent a formative period in Gowland Hopkins's laboratory in Cambridge, but it was in Gröningen in Holland that Szent-Györgyi enjoyed his first great slice of luck. His interests had settled on the chemical reactions, then a matter of wide debate amongst the world's biochemists, by which energy is generated in the cell. He found in the adrenal gland a strong reducing agent (a substance that abstracts oxygen, or simply electrons from other substances, and is thereby itself oxidized). For no good reason he conjectured that this might be the selfsame material, with reducing as well as antiscorbutic properties, that had been isolated from lemon juice. In Hopkins's laboratory in Cambridge he established that his reducing agent was indeed antiscorbutic. He supposed the substance (incorrectly) to be a sugar, and in his publication on the subject proposed the name *ignose*, in accordance with the nomenclature for sugars (the names of which – glucose, sucrose, fructose, etc. – all end in *ose*). When the editor of the learned journal demurred, Szent-Györgyi tried the still more frivolous *godnose*, but the editor would have none of it and insisted on hexuronic acid, since the factor was indeed acidic. Later Szent-Györgyi called it ascorbic acid, by which vitamin C is still known to chemists.

When Szent-Györgyi was finally accommodated in Hungary, with a position of his own at the University of Szeged, he had two further strokes of luck. First he received an application from Joseph Svirbely, who had been awarded a travelling fellowship and was looking for a laboratory in the country of his ancestors. This could not have come at a better time, for Svirbely had learned the tricks of assaying for antiscorbutic activity in Glenn King's laboratory in Pittsburgh, and much else about the properties of the vitamin. And then it transpired, by way of another wholly serendipitous discovery, that red paprika peppers, for which Hungary and especially the region around Szeged was famous, were the most abundant of all the sources of vitamin C. Many grams of the pure substance were now available, and Szent-Györgyi sent some to Norman Haworth, a distinguished organic chemist specializing in the chemistry of sugars at the University of Birmingham in England. There were already clues about the structure, but it was Haworth who arrived at the complete answer. He and his colleague Edmund Hirst later achieved the synthesis of the vitamin, as also, simultaneously, did Tadeusz Reichstein in Switzerland. It was Haworth who walked off with the Nobel Prize for chemistry in 1937.

King's paper on the isolation and nature of the vitamin appeared two weeks before Szent-Györgyi's, but King and his students had actually obtained the material in crystalline form the previous year. Moreover, King and Svirbely had demonstrated its antiscorbutic potency in guinea pigs, but King, a cautious man, had held back his publication to look into a spurious claim from another laboratory of huge antiscorbutic activity of an opiate-related substance, purportedly the true vitamin C. When the two reports, the one from Hungary, the other from the United States, appeared in effect simultaneously, accusations flew of plagiarism by Szent-Györgyi. How much had Svirbely been able to divulge in Szeged of what had been discovered in Pittsburgh? King (1896–1988) was an altogether less assertive and extroverted character than Szent-Györgyi, and of very different background. He had been born in the Midwest and attended a one-room school before finding his way into university. His studies, like Szent-Györgyi's, had been interrupted by the First World War, and he had served on the opposite side, with the US infantry in France. On his return he had embarked on a research career at the University of Pittsburgh, where he spent his working life. His interests were in nutrition, on which he did much valuable work. That it was Szent-Györgyi alone who emerged in 1937 with the Nobel Prize in Physiology and Medicine caused outrage in the United States. The press accused him of cheating and plagiarism, and the affair continued to haunt him throughout his long life.

But Szent-Györgyi did other notable work, most famously on the nature of muscle contraction. In 1939 he was elected Rector of the University of Szeged.

When Hungary joined the War on the German side he bravely stood up against the fascists who were seizing control of the country's institutions. He was active in the underground movement and, when the dictator, Admiral Horthy, whom he had resolutely opposed, perceived that the time had come to change sides, it was Szent-Györgyi who was chosen to make clandestine contact with the Allied command. In 1945 he was hailed as a hero and offered the presidency of the liberated Hungarian Republic, which he declined. He had been on friendly terms with the invading Russians, and had secured their help in establishing a new laboratory in Budapest, but when he realized that Hungary was to be assimilated into the Soviet imperium he made his objections known. Once more suddenly *persona non grata*, he departed for a new life in the USA. He was not universally welcomed, and it took some time for him to re-establish himself in research. As he aged his ideas grew increasingly extravagant and controversial. He resented having to justify his demands for funding, and soon was able to attract support only from private foundations on which he exercised his enveloping charm and powers of persuasion. His vociferously articulated theories about cancer have not survived scrutiny. He continued to work in the laboratory with undiminished energy and an unshakeable conviction that he was right until a few weeks before his death at 93. It is a tribute to his vigour of mind and body that he was four times married, twice to women 50 years younger than himself. As to vitamin C, or ascorbic acid, there have been many attempts to promote it as an elixir, associated most of all with the mighty name of Linus Pauling, but more of that anon.

Vitamin D, the anti-ricketsial factor

The hunt for the anti-ricketsial substance in dairy products and cod liver oil was set in motion in the first decade of the 20th century by another accidental discovery in Elmer McCollum's laboratory at Johns Hopkins University in Baltimore. Infant rats on certain of the restricted diets that McCollum and his assistants were exploring developed gross deformations in their bone structure: they had rickets. The effects could be moderated by the addition to the diet of calcium compounds, vitamin A preparations, or animal proteins. These contain amino acids sparsely represented in the proteins of the grain that made up the bulk of the animals' feed. Then McCollum tried butterfat. This prevented xerophthalmia, the eye disease engendered by vitamin A deficiency, though not the bone deformities. But calcium and butterfat together promoted more or less normal growth. Cod liver oil, on the other hand, warded off both the xerophthalmia and the rickets, so calcium could not itself be the anti-ricketsial factor, even if it was utilized by the body in the suppression of rickets. There had already been indications of a link between rickets and dietary phosphorus, and the protective action of phosphorus and calcium (in their proper proportion) was clear.

The next useful observation came from Gowland Hopkins, who discovered that vitamin A was easily oxidized by atmospheric oxygen, and its activity thereby annihilated. When he bubbled oxygen through butterfat its vitamin A activity was similarly lost. Cod liver oil thus treated, on the other hand, retained activity. It was left to McCollum to show that the oxidized cod liver oil did not cure or protect against xerophthalmia, but was effective against rickets. There were, then, two active substances in cod liver oil and in dairy products, one of which was a specific against rickets. McCollum called this vitamin D.

It was known that sunlight, and especially ultraviolet light, had an anti-ricketsial action and babies indeed were sometimes treated by irradiation (Chapter 1). All this was confirmed by McCollum and others, but McCollum then did a characteristically shrewd experiment: he irradiated animals with ultra-violet light, and on examining their livers found that vitamin D had accrued there. The body thus synthesizes its own vitamin D when properly stimulated. (In excessive amount vitamin D is dangerously toxic. It is a familiar tale that the Inuit and other peoples of the far north always knew not to eat the liver of polar bears, because it is especially rich in the vitamin D.) Further researches along the same lines showed that vitamin D also appears when cholesterol, a fatty, waxy body constituent, is irradiated with ultraviolet light. This implied that the vitamin was chemically related to cholesterol, one of a class of compounds called sterols, which are precursors of many hormones. Eventually the vitamin was isolated in pure, crystalline state and its structure was determined through the work of several laboratories, notably that of Windaus in Germany.

A plethora of factors: vitamin B_2

McCollum, as we have seen, was not infallible. He had insisted that his 'water-soluble factor B' was a single substance, a specific against polyneuritis in birds and required for growth by rats. It was this supposition, indeed, that had led to the isolation of this so-called 'vitamin B'. But others began to articulate their doubts. There were, to be sure, certain compelling reasons for McCollum's conclusion, especially that the two activities accompanied each other through all the various steps of extraction, but this was still no proof. An American biochemist, H. H. Mitchell, was one of the doubters, and he managed to show that the relative efficacies, with respect to polyneuritis and growth, of preparations from different vegetable sources differed quite widely. It was Osborne who in 1919 discovered that one of the activities in a preparation from yeast was inactivated by heat, while the other was unaffected. He suggested the designation vitamin B_1 for the anti-polyneuritis factor and B_2 for the growth-promoting factor. In 1928 the redoubtable Harriette Chick (Chapter 1) at the Lister Institute in London hit upon a restricted diet that generated a range of dire symptoms in rats, all of them reversed when the animals were dosed with a preparation of the vitamin B_2. Two years earlier Joseph Goldberger, in his pursuit of the pellagra-protective substance (the P-P factor) had induced what he supposed was pellagra in rats. It proved to be no such thing, or at least was unrelated to human pellagra or to black-tongue in dogs (p. 149), and Goldberger's P-P preparations, which cured the rats, did not contain vitamin B_2. But in 1933 a biochemist, Lela Booher at Columbia University in New York, noticed that a yellow substance was present in whey, and she contrived to extract and purify it. This yellow material had actually been remarked on in a German publication of 1879 and was there termed 'lactochrome'. Lela Booher's preparation was vitamin B_2 (called vitamin G in some of the early publications) and a German chemist, Richard Kuhn, extracted the same substance from spinach leaves, from liver, and from kidneys. He and Paul Karrer in Zurich simultaneously effected its synthesis. It became, and still is, known as *riboflavin*. The curious sequel to the story is that riboflavin turned out neither to cure nor prevent common human pellagra, but it does act on what Goldberger and Wheeler thought was a variant of the disease, which they had called *pellagra sine pellagra*.

The pellagra factor at last

Around 1914 Goldberger and another researcher obtained, as already related, a heat-resistant preparation from liver that cured black-tongue in dogs, and was somewhat effective against pellagra in people. It turned out that the active constituent in the mixture was a quite simple compound, nicotinic acid, familiar to organic chemists since 1867, when it was prepared by oxidation of nicotine from tobacco plants. Casimir Funk had found large amounts of nicotinic acid (then called niacin) in rice polishings and was disappointed that it did not cure beriberi, although it seemed to have a generally beneficial effect when mixed in with anti-polyneuritis preparations. Suzuki, too, had purified and identified nicotinic acid in rice polishings and found it to have no effect on avian polyneuritis. Later, in 1935, the German Colossus of biochemistry, Otto Warburg, found that nicotinic acid occurs throughout animal life, in association with an important enzyme. This revelation caused the vitamin-hunters to take the substance seriously, and that same year Conrad Elvehjem and his colleagues in America found that it rapidly cured dogs with black-tongue. Success in trials of its effect on pellagra in the southern states quickly followed. The conquest of pellagra was now completed, for nicotinic acid was easily produced in the laboratory, cheap and readily administered.

The greenish-yellow disease and a sentence of death

It had been known since the mid-18th century that blood contains iron, and in 1838 the great Swedish chemist Jöns Jacob Berzelius (Chapter 5) discovered that the red pigment haemoglobin, isolated from the blood, could absorb large volumes of oxygen. He surmised that this was the chemical basis of respiration. Physicians also became aware in the course of the 19th century that certain sicknesses were linked to a deficiency of the red blood cells, in which the haemoglobin is packaged. Chlorosis, from the Greek for green-yellow, was described in the 16th century by a German physician who called it the *morbus virgineus*, because he believed that it afflicted only virgins, and arose from failure to eliminate menstrual blood. (He urged wedlock as the sovereign remedy.) The name chlorosis, given to it in the 17th century, derived from the greenish pallor of sufferers, mostly indeed young women. It is said that its association with virginity lies behind the green hue that Siennese and other early European painters imparted to the countenance of the Virgin. In England the condition was termed the 'green sickness'. Its symptoms were weakness, lassitude, and palpitations, and breathlessness after the least exertion. Thomas Sydenham (p. 115), pre-eminent among the English physicians of his time, classified it as a 'hysterical' disease, and the view that it arose from nervous degeneration persisted through the 19th century. A French doctor noted that the red blood cells were unusually small in this state. Then in 1892 Sir William Osler, then at Johns Hopkins University (and later Regius Professor of Medicine in Oxford), who was ever well ahead of his time, wrote that the complaint was most common among 'ill-fed, overworked girls in large towns, confined all day in close badly-lighted rooms'.

If chlorosis was the result of too little haemoglobin, then perhaps the fundamental cause might be too little iron? In the 18th century Sydenham had in fact proposed the 'steel tonic' to treat the disease, consisting of steel filings steeped in Rhenish wine – and a little iron might in fact have dissolved in the wine if it was sufficiently acidic. In the 19th century iron therapy became fashionable: iron (ferrous) sulphate tablets were prescribed and often worked. Chlorosis, however, was a catch-all term for all forms of anaemia, many of which would have been secondary symptoms of other diseases, and the term fell out of use around the first decade of the 20th century. Meanwhile, in the mid-19th century, it had become plain that some patients with severe anaemia did not respond to iron therapy. Some had remissions but most died and there was nothing doctors could do. One of the leaders among the English clinicians, Thomas Addison (who gave his name to Addison's disease), studied the condition and gave it the name *pernicious anaemia*, because it was so intractable, and generally nothing less than a sentence of death once diagnosed.

The breakthrough in tackling pernicious anaemia came in 1927. William P. Murphy (1892–1987) was a doctor, ministering largely to the poor Irish community in Boston, Massachusetts, when he noted that one of his patients with pernicious anaemia was showing no signs of dying. Murphy inquired into his diet, and the man owned to a great liking for liver. Murphy accordingly tried prescribing liver for his other patients with the same disease, but they showed little improvement. On further interrogation his mystery patient admitted to a singular preference: he ate his liver raw. This was the key, but talk of liver had been in the air. George Hoyt Whipple (1878–1976) was a biochemist at the University of Rochester. The son and grandson of doctors, he was especially interested in how the synthesis of haemoglobin was regulated. He made dogs anaemic by bleeding, and tested a variety of foods for their capacity to restore the red cell concentration in the blood. Liver, kidneys, and chicken gizzards were pre-eminently effective.

Dr Murphy, wanting to follow up his discovery, approached George Minot, a doctor and researcher in one of the Boston teaching hospitals affiliated to Harvard University. Together they examined the bone-marrow from pernicious anaemia sufferers who had died of the disease, and found them choked with immature red blood cells. Evidently the disease was caused by the lack of a factor that normally stimulated the release of the nascent cells into the circulation. Then they observed, when patients were given raw liver, a surge of immature red cells into the blood. The point was proved. Liver became the life-saving therapy, but to ward off anaemia an intake of a half kilogram of raw liver was required per day, and some patients preferred to settle for an early death. Minot and Murphy thereupon set out to make a liver extract that retained the activity, and in this they succeeded. In 1934 Murphy, Minot, and Whipple shared the Nobel Prize for the discovery. The active factor from liver was run to earth by teams in England and the United States, and its chemical structure was determined by a combination of organic chemistry in Alexander Todd's laboratory in Cambridge and crystallography by the remarkable Dorothy Crowfoot Hodgkin in Oxford. It is a complex molecule, with a cobalt atom at its core, and its solutions are bright pink. The vitamin was eventually synthesized by chemists at the Glaxo pharmaceutical company in England, and thereafter there was no further need for liver extracts. The factor, which has the chemical name *cyanocobalamine*, was designated vitamin B_{12}.

The fertility factor

Continuing restricted diet experiments on rats threw up yet another vital trace substance. Adult rats did not seem to need it to survive, but without it they were sterile, or at least gave birth prematurely to stillborn offspring. Sometimes the embryos were reabsorbed, while baby rats deprived of the factor were partially paralysed. Herbert McLean Evans (1882–1971) at the University of California in Berkeley, who had discovered the existence of the new factor in 1922, called this condition 'nutritional muscular dystrophy'. He found that fertility could be re-established by adding lettuce, wheat germ, or alfalfa leaves to the rats' diet, and that the activity was strong in wheatgerm oil and other vegetable oils. Attempts to repeat his work in other laboratories led to a long controversy. Results were erratic and some researchers reported that Evans's observations were irreproducible and specious. The reason, it turned out, was that the oils went quickly rancid, and the oxidized products thereby generated destroyed the vitamin. Many clinicians and researchers also looked askance at what the popular press was describing as 'the fertility vitamin', because it was becoming conflated in the public mind with the wild claims of sexual rejuvenation engendered by implants of monkey testes or injections of extracts prepared from them. The epicentre of this cult was Vienna, where Eugen Steinach and Serge Voronoff were performing testis implants in many ageing voluptuaries, all of whom seemed well satisfied with the outcome. This profitable practice was taken up on a commercial scale by an unqualified quack in Kansas, known to the world as 'Doc' Brinkley. (The procedures were also reputed to enhance health and longevity in general; they continued on the less reputable fringes of medicine for many years after they had been discredited, and numbered even such notables as Pope Pius XII among the takers. The benefits were of course illusory, for all implants would have been quickly rejected.[8])

But in the face of much discouragement, Evans persisted in his efforts to prepare the factor and by 1936 had succeeded, and had identified it as an ester (p. 253) of a complicated alcohol. The structural formula was established by the combined work of several laboratories, and in 1938 Karrer synthesized this α-tocopherol, also known as vitamin E. (Later it emerged that there are two other chemically closely related compounds, called β- and γ-tocopherol, which co-exist with the primal vitamin and are also active.) In 1923 Frederick Gowland Hopkins had found that a high-fat diet, with plenty of cod liver oil, inflicted injuries on the heart, the muscles, and other tissues of rats. This damage was later shown to be caused by free radicals. (These are chemical species formed by oxidation of fats, characterized by an extra electron, which makes them highly reactive chemically; they are much feared today in the popular imagination.) The vitamin E annihilates these free radicals – it acts as a 'scavenger' (and is itself thereby destroyed, as in the rancid vegetable oils) – though this is not its only function. Another acrimonious dispute broke out around the claims by two Canadian doctors, brothers, who reported that vitamin E could prevent and even reverse the effects of coronary heart disease, but in the end the claim was not sustained.

Koagulation

A chance observation in 1930 of a relation between diet and the rate of coagulation of blood drawn from chickens led to the identification of yet another dietary factor. The previous year a student in Copenhagen, Henrik Dam, seeking to discover whether chickens synthesize cholesterol or obtain it only from their food, observed what seemed to be a new disease. The chickens on his diet developed haemorrhages under the skin, which were not prevented by any of the dietary additives that would have ensured against scurvy, or for that matter rickets. He found that the factor was fat-soluble, but distinct from the known fat-soluble vitamins A, D, and E. He proposed the name vitamin K, from the German *Koagulation*. Dam isolated the substance, and found that it comprised two very similar active variants, which he called K_1 and K_2. It was Paul Karrer yet again who determined their structures. The use of vitamin K as a blood-clot inhibitor in patients with heart disease was explored, but proved hazardous. Today coumadin (warfarin) is the usual drug of choice for this purpose.

Yet more soluble factors B

This was by no means the end of the litany of vitamins. In particular, the 'water-soluble factor B' of McCollum yielded more surprises. Paul György, born in 1893, was a Hungarian biochemist, trained as a doctor. After the First World War he moved to Heidelberg University to work with Richard Kuhn on the chemistry of vitamins and related compounds. By 1927 he was a professor, but in 1933 the menacing political situation drove him to England. In Hopkins's Nutritional Laboratory he discovered the substance responsible for the so-called 'rat pellagra', which Goldberger had erroneously identified with the human disease. It is involved in the synthesis of proteins and called vitamin B_6, or more generally *pyridoxine*. György continued to work on this and related substances after moving to America in 1935, and he had a part in the determination of the structure of *biotin*, another trace substance essential for health. Biotin was extracted from egg yolks, but is also to be found in yeast, liver, and elsewhere. A mysterious disease of rats, induced by large quantities of egg white added to the diet, and called 'egg white injury syndrome', was found to originate from a protein in the egg white (avidin) that absorbed biotin. György established that the 'anti-egg white injury factor' in egg yolks, liver, and yeast was indeed biotin. György died, still in harness, in 1976.

Pantothenic acid is another of the B vitamins, vitamin B_5. Its existence was reported in 1933 by a leading American organic chemist, Roger J. Williams, the brother of Robert R. Williams, who played such a notable part in the identification of the anti-beriberi factor. Rats deprived of pantothenic acid suffer from a variety of disorders, ending in death, and it is presumed to be equally inseparable from human life.

Finally, mention must be made of *folic acid*, known at one time as vitamin B_c or vitamin B_9. Its existence was discovered by a British pathologist, Lucy Wills.[9] Born in 1887, a child of privilege, educated at Cheltenham Ladies' College and Cambridge, where she read biological sciences and geology, she started, as a postgraduate, on a geological research project. When The Great War began she was in South Africa. She hastened back to England, resolved now to qualify in medicine. Her interests shifted to science rather than clinical practice, and she took a position in chemical pathology at the Royal Free Hospital in London (then a teaching hospital for women). Her reputation grew, and in 1930 she received an overture from the head of the Indian Medical Service, Dr Margaret Balfour. Dr Balfour was searching for someone willing to undertake a study of a severe anaemia that was afflicting pregnant textile workers in India. Lucy Wills took up the challenge, and on arriving in India was quick to establish that the condition in question was not pernicious anaemia, for she already knew about the work of Minot and Murphy in Boston. Most important was her finding that whole liver, though not a purified liver extract, was a moderately effective therapy, but that Marmite, the spread made from digested yeast, totally cured the anaemia. Since Marmite was cheap and liver expensive, Marmite was adopted as the remedy, and the disease soon disappeared. In her publication Lucy Wills suggested that Wills's Anaemia, as it was henceforth called, was occasioned by the lack of a nutritional factor. She worked at two Indian research institutes, preparing extracts, and proving that they reversed the effects on rats of a restricted diet. It was another 10 years before the pure compound was prepared from spinach leaves (whence its name from the Latin for leaf), and its structure determined by Robert Stokstad at the laboratories of the Lederle pharmaceutical company. An American clinician fed himself on a folic acid-free diet, and showed that the characteristic anaemia developed after about four months. A lack of folic acid is now recognized as perhaps the commonest vitamin deficiency in the developed world, and particularly affects alcoholics. An insufficiency of the factor in pregnant women leads to neurological birth defects and other problems. Trace amounts of the substance are now added to bread and to some breakfast foods as a prophylactic. Lucy Wills, autocratic in manner and highly respected, continued to work on folic acid and to teach at the Royal Free Hospital until 1946, when she retired to travel the world. She finally settled down in London, represented the Labour Party as Councillor for Chelsea, and was active in local affairs. She died in 1964.

Fatty acids and more

Max Rubner (Chapter 5) showed in the late 19th century that we synthesize fats from carbohydrates. In 1920 Thomas Osborne and Lafayette Mendel reported that rats could thrive on a diet with minimal fat content, and Jack Drummond later reported that young rats grew to healthy maturity on no fat at all. He was mistaken, for in 1929 a husband-and-wife team, the Burrs at the University of Minnesota, used ether to extract all traces of fats from the diet. The effect was startling: even with a supplement of cod liver oil to supply fat-soluble vitamins, their rats fell sick. The skin scaled, necrosis developed, blood appeared in the urine, changes occurred in the kidneys, and death followed. The Burrs went on to investigate whether particular fatty acids, the building blocks of fats, could prevent deterioration or might allow recovery (as a little lard could do), and they found that one sufficed. This was linoleic acid, and its close relative linolenic acid also served the purpose. The fatty acids can be divided into the saturated and unsaturated classes (p. 85). The latter have double bonds between some of the carbon atoms in place of attached hydrogen atoms, and in general there is only one double bond in the carbon chain. But linoleic acid has two and linolenic acid three double bonds (see Appendix). The animal's metabolism does not allow for the insertion of more than one double bond into the long fatty acid chain, and linoleic and linolenic acids cannot therefore be produced in the body. Later work added a third unsaturated fatty acid, arachidonic acid, with four double bonds, to the list. Linoleic acid occurs in many foods, and is an *essential fatty acid*. The amount required in the diet is very small, but indispensable, and some baby formulas of the past, containing too little, brought grievous results. There are also some rare hereditary diseases in which the metabolic processing of linoleic acid is defective, and sufferers from these need much larger quantities in their diets to stay healthy. Linoleic acid and its relatives are incorporated into the membranes that surround the cells of the body, and they are needed for the synthesis of a large group of highly active substances called eicosanoids, the most famous of which are the prostaglandins. They are involved in inflammatory responses, in blood clotting, in the function of the arteries, and much else.

Not only are there essential fatty acids, but there are also essential amino acids, needed to build proteins. As we have seen, Frederick Gowland Hopkins first perceived that there was at least one such amino acid, tryptophan. It is now known that 10 of the 20 amino acids found in proteins are essential, that is to say cannot be synthesized in the body. Most of them are abundant in the common proteins of foods, but others – tryptophan especially – are less so, and this can pose problems to those on vegan diets.

The first indication that inorganic substances other than the abundant salts of sodium, potassium, and calcium might play a part in the life of plants seems to have emerged in 1849 when one of the great European landowners, the Prince of Salm-Horstmar, concocted a range of artificial soils based on charred sugar (which contains no inorganic matter whatever), and found that added silica, alumina, and manganese were needed for growth of healthy plants. His chemicals would have been impure, and we now know that other elements also feature in the lives of plants and animals.

Which elements, then, besides sodium, potassium, calcium, magnesium, and phosphorus, which we have already encountered and all of which are present in bulk, are essential for the processes of life? Iodine for one. Arnoldus de Villa Nova, writing in 1280, recorded that the ash of burnt sponges had long been used to treat goitre – an enlargement of the thyroid gland. The element, iodine, was discovered by a French chemist, Bernard Courtois, in 1811 in seaweed, and in 1820 Jean François Coindet, a Swiss doctor, found that the element was also present in the burnt sponge ash. In 1896 the thyroid gland was found to be exceptionally rich in iodine, and it was then an obvious question: might not disorders of the thyroid, in particular the endemic goitre seen in some localities of the world, be linked to an insufficiency of the element? So indeed it turned out. In excess amounts, iodine in some of its forms is toxic, and so its use in treating goitre fell out of favour, with unfortunate consequences for the sufferers. It was later found that iodine is contained in the compound thyroxine, which is central to thyroid activity. Goitre was a common disease until late in the 19th century: a French government report in 1864 referred to a half million cases in the country, among them 120,000 'cretinoid idiots'. Endemic goitre more or less vanished from the world when small amounts of sodium iodide were added to drinking water in the affected areas in America and elsewhere, and when 'iodized' table-salt became a common commodity. Chlorine is another element of the halogen group, to which iodine belongs. Sodium chloride is common salt, and the chloride ion (p. 172) is an essential constituent of body fluids. The other halogen, fluorine, is needed by some animals, though not humans. It does, on the other hand, give protection against tooth decay and osteoporosis. Fluoridation of water and the addition of fluorides to toothpaste reduced the incidence of dental caries in Western countries by a half.

Sulphur is an indispensable, if minor, constituent of proteins, and its many other biological compounds pervade the metabolism. Even the toxic sulphur-like element selenium is an integral part of at least one enzyme that acts to eliminate free radicals and in the synthesis of the anti-inflammatory agents, the prostaglandins. Selenium compounds are added to animal feed, and there is evidence for rare selenium deficiency conditions in humans. Iron, of course, has been known since the 19th century to be implicated in respiration, as part of the body's oxygen-carrying proteins, and it was found that copper plays a part in the synthesis of these molecules. Copper, zinc, cobalt, and manganese also occur in various enzymes that play an essential role in metabolism. Animals deprived of any of them do not thrive and show symptoms resembling those of some vitamin deficiencies. Only minute traces of each element are needed to nullify these effects. Even the quite rare metal molybdenum occurs in nature. Evolution has seized any opportunity to make use of whatever elements were presented to it, and now we can dispense with none of them.

10

Fads and Quacks

Counting the ways

There is nothing new about dietary fads. One might argue that the Galenic theories about the kinds of foods suitable for the young, the old, or the sick (Chapter 3) amounted to little else. Nor for that matter do the dietary rules prescribed by religions, which have (and had) little to do with the preservation of health. But the 19th century saw the flowering of science, and with it of pseudo-science, and this, allied to the growth of prosperity, at least in Europe and North America, diverted people's minds from the problems of staying alive to the allure of staying alive for ever. So, expedited by an irrepressible advertising industry, the golden age of fads and quackery was ushered in, and it envelops us still. One fad that pre-dates all is vegetarianism in its extreme form. Wild theories, centred on bowel movements, have also been in circulation since ancient times. In the early years of the 20th century there arose a near-religious cult of vitamin worship. Many less durable, often outlandish dietary cults have sprung up, flourished, and as quickly withered away. I shall discuss here only some of the most celebrated and potent.

Vegetarians and vegans

'Vegetarianism is harmless enough, although it is apt to fill a man with wind and self-righteousness.' So said Sir Robert Hutchison, the London doctor, who himself favoured consumption of milk on a frightening scale. But there are of course defensible arguments for eschewing meat. One is that the cultivation of livestock is expensive and wasteful, inasmuch as the nutritional yield from arable land is vastly greater if it is cultivated for crops than when it is given over to animal husbandry. So if food is scarce then meat-eating serves only to deplete limited resources. The second argument is quite reasonably based on a humane consideration for the animals, especially in our age of intensive farming. The origins of vegetarianism, which go back to ancient times, usually sprang from no such motives, although religious sects devoted to the sanctity of animal life have from time to time sprung up. But past vegetarians most often believed that ingestion of meat engendered carnal passions and a violent disposition, and impeded the path to spiritual fulfilment.

Pythagoras, the shadowy mathematician and philosopher of whose life little is known, was said to have advocated a meat-free diet to promote harmony with all of the earth's creatures and sharpen the powers of thought. Vegetarians used in fact, especially in the 19th century, to call themselves Pythagoreans (with no reference to the square on the hypotenuse). Hindus, Jains, and Zoroastrians embraced vegetarianism, and so also did the Manichean heretics in the early centuries of the first millennium. Later still, many pacific sects arose, for whom vegetarian teaching became identified with abstinence and self-denial.

In England the movement grew in numbers during the 17th century, but of course its principles ran counter to the inclinations of the majority, who enjoyed, or at least yearned for, a meat-rich diet, as sanctioned by the Church. For had not God in his infinite bounty given man dominance over the beasts of the earth and the fowls of the air? This view was robustly represented by, among many others, Henry More, who wrote that divine providence had endowed sheep and cattle with life for the sole purpose of keeping their bodies fresh 'till we shall have need to eat them'. In much the same spirit a 19th-century country cleric in England had declared that God had attached white tails to rabbits to make them an easier target for the guns. The vegetarians responded by conjuring up evidence of a sort, that the Bible had been misinterpreted. Among their leaders at the time was Thomas Tryon, whose books most eloquently expounded the vegetarian credo, and made a convert of even the young Benjamin Franklin in Philadelphia. (Franklin later lapsed when he saw a gutted cod yield up a smaller fish; this convinced him that animals, human included, were designed to predate on lesser creatures.) A near-contemporary of Tryon's was the fashionable London doctor George Cheyne, in his heyday a formidable gourmand and carnivore, who was supposed to have achieved a weight of 32 stone (450 pounds, nearly 100 kg). He became almost immobile and tormented by many afflictions. At this stage he saw the light and converted to a punishing and entirely vegetarian diet, and began to campaign for a vegetarian lifestyle.* The Society of Bible Christians, or Cowherdites (after their leader, the Reverend William Cowherd), followers of the Swedish mystic, Emanuel Swedenborg, were also vociferous vegetarians. More important were the Humanists, who numbered many prominent figures among

* Among Cheyne's best-known tracts was his *Essay on Regimen*, which contained a series of 'aphorisms'. The most famous (aphorism 13) stated that 'Every *wise* man, after *Fifty*, ought to begin to lessen at least the quantity of his *Aliment*, and if he would continue free of great and dangerous Distempers and preserve his Senses and Faculties clear to the last he ought every seven years go on abating gradually and sensibly, and at last *descend* out of Life as he ascended into it, even into the Child's Diet.' That is to say, we should all end up on nothing but milk. For, 'Milk and sweet sound Blood differ in nothing but in Colour: *Milk is Blood*.' That belief was the orthodoxy of earlier ages (Chapter 3).

their ranks, such as Robert Owen, founder of the Trade Union movement. The faithful drew strength from such powerful tracts as *An Essay on Abstinence from Animal Food as a Moral Duty* by Joseph Ritson, which appeared in 1802. Dr William Lambe was perhaps the most eloquent proselytizer for the cause, at least in England, in the decades that followed. He proclaimed the virtues of a vegetarian diet for the prevention or cure of almost every known disease, most explicitly for cancer. Another passionate advocate was none other than Percy Bysshe Shelley, who in 1813 published a polemic entitled *A Vindication of Natural Diet*. Consumption of meat, he instructed his readers, led to 'disease, superstition and crime'; human carnivores brought down on humanity all known evils; even 'the slave trade, that abominable violation of the rights of nature, is, most probably, owing to the same cause'. A host of other wicked practices were rooted in the same cultural aberration, and surfaced several times in Shelley's verse, most notably in *Queen Mab*:

> Immortal upon earth: no longer now
> He slays the lamb that looks him in the face,
> And horribly devours its mangled flesh,
> Which, still avenging Nature's broken law,
> Kindled all putrid humours in his frame,
> All evil passions, and all vain beliefs,
> Hatred, despair, and loathing in his mind,
> The germs of misery, death, disease, and crime.

Shelley had already met his watery death by 1847, when the first Vegetarian Society was established at a meeting in Ramsgate on the English south coast, Joseph Brotherton, MP in the chair. (Mrs Brotherton did her bit by publishing a cookery book for vegetarians, in which she also grappled with the problems of Biblical evidence. The story of the miracle of the loaves and fishes was based, she averred, on a mistranslation: the fishes were in reality melons or lotus plants.) The vegetarians of the United States, encouraged by two immigrant Cowherdite clergymen, formed their own society in 1850. Much the most prominent figures in the movement were Sylvester Graham, who deserves a section to himself (see below), and William Alcott, a doctor and cousin of the writer Louisa May Alcott. Starting from the proposition that there was no roast beef in the Garden of Eden, and that man's teeth differed from that of carnivores, they denounced meat-eating as barbarous and ungodly. Moreover, physiologists had established that man could get along perfectly well without meat (though not necessarily without animal products); and were gorillas and orang-utans not vegetarians, put into the world by divine providence to serve as models for mankind? In any case all bodily tissues derived ultimately from plants, either directly or by way of those that the vegetarian creatures at the far end of the food chain had eaten. Therefore, Alcott declared, it was a citizen's *duty* to abstain from meat. For not only was it the humane course, but the consumption of dead animals – 'corpses' – meant taking decaying matter, 'germs of poison', into our bodies. Sylvester Graham expressed it more earthily: 'How can a man serve God with a stomach full of grease?' Meat was also seen as a source of 'irritation', a kind of nervous agitation first defined by the famous French doctor François Broussais (who gave his name to one of the great Parisian teaching hospitals). This Alcott and Graham and their like associated with priapic excess, and indeed vegetarianism, then as before, was strongly linked to a virtuous and abstemious lifestyle, fear of God, exacting personal hygiene, and cold baths. Some thought that indulging children in their taste for sweets was a sure way to start them on the road to perdition. There were moves to extend the Cowherdite principles to schools with the aim of extirpating 'vice'. (In England, at least, several educational institutions, most famously Abbotsford, were founded with this in mind.)

Vegetarians were mocked in France and elsewhere in Europe, but societies sprang up in most countries in the second half of the 19th century. The medical profession disapproved, in part because a meatless diet was inconsistent with the prevailing view of nutrition – more or less as enunciated by Liebig (Chapter 5). In Russia Tolstoy, who earlier in life had liked to hunt on his estate, declared himself a vegetarian by reason of his new-found love of man's fellow-creatures, and he gave encouragement to an austere vegetarian religious order, the Dukhobars. They were nevertheless subjected to much persecution and eventually emigrated in large numbers to Canada, where they set up a community that still exists.

In England vegetarianism grew in strength in the late 19th century, becoming increasingly conflated with the anti-vivisection crusade. By the 1880s there were some 30 vegetarian restaurants in London alone. Many vegetarians also allied themselves with the temperance zealots (and tended additionally to eschew tobacco, tea, coffee, and even salt and spices), and they also formed a nucleus of the pacifist movement. Probably for this reason vegetarian organizations were outlawed in Russia in 1917. George Orwell thought the whole business a strange aberration. In his book *The Road to Wigan Pier* can be found the following passage: 'One sometimes gets the impression that the mere words "Socialism" and "Communism" draw towards them with magnetic force every fruit-juice drinker, nudist, sandal-wearer, sex-maniac, Quaker, "Nature Cure" quack, pacifist, and feminist in England.' He might have mentioned also the more extreme Christian sects, such as the Seventh-day Adventists. Strange beliefs about medicine and health abounded in Germany during the Third Reich.[1] The regime demanded health and fitness of all German citizens as a patriotic duty, and attempted to inculcate diets that would restrict alcohol and coffee, and avoid all forms of excess. They were to be based on 'natural foods', especially fresh vegetables and fruits, with minimal consumption of sugar and protein. Periodic fasting was also recommended. Diseases, the officially endorsed medical authorities declared, were based on faulty diets. Several of the leaders of the Nazi regime were vegetarians (like their idol, Richard Wagner), although Hitler apparently would take modest amounts of meat. The proclamations seem to have done little to change the eating habits of most Germans, whatever their political leanings. Diets deteriorated only with the shortages following on the defeats in the East, when substitute foods were introduced and hunger ensued.

Vegetarians now come in several varieties. Ovolacto-vegetarians eat eggs and dairy foods (although until recently it was difficult to find cheese made without rennet from sheep stomachs. Now much of the rennet, or rather its active enzyme, rennin, is prepared in genetically modified bacteria.) Lacto-vegetarians eat dairy products but no eggs, fruitarians rely primarily on raw fruit, and vegans eat no animal products at all, or indeed use them for clothing or any other purpose. Theirs is a dangerous diet, which can lead to malnutrition, especially in infants, because of a dearth of vitamin D, riboflavin, and, most of all, vitamin B_{12}, without which anaemia will supervene. (Vegans are urged now to take vitamin supplements.)* Some of the more extreme zealots insist that omnivores are all debilitated by toxins generated from putrefying meat, and that their bodies seethe with uric acid (p. 115) and mysterious entities called 'necrones'. But, despite the strange images that the mention of vegetarianism brings to mind, it has to be reiterated that there is much to be said for limiting (or eliminating) meat in the diet, and there is evidence that vegetarians are somewhat less prone to some cancers, and on average live longer.

The illustrious Doctor Sawdust

The most durable food fads sprang up in the United States, especially in the wake of industrialization in the 19th century. As lifestyles changed, a sense of lost innocence began to afflict the pious, who saw in the sophistication of food a particular target for their disapproval. The 'back-to-nature' movement thus became identified with revivalist religion. The most prominent of its leaders was a New England Presbyterian clergyman, Sylvester Graham. Born in 1795, Graham came of an English family that had established itself in Boston in the previous century.

* George Bernard Shaw was one of the 20th century's more prominent vegetarians (although he accepted the necessity for killing pests, such as mice and squirrels, as long as it was done humanely). He suffered from pernicious anaemia and was kept alive by liver extracts, the source of vitamin B_{12}. He was fiercely castigated for this by other vegetarians, but responded that their position was absurd. Should diabetics also sacrifice their lives by refusing insulin from animal pancreas? In one of his letters (to the writer, Hal Caine) he recounts the story of a lunch at the house of his friend William Morris: 'Mrs Morris did not conceal her contempt for my folly. At last pudding came; and as the pudding was a particularly nice one, my abstinence vanished and I showed signs of a healthy appetite. Mrs Morris pressed a second helping on me, which I consumed to her entire satisfaction. Then she said, "That will do you good, there is suet in it." And that is the only remark, so far as I remember, that was ever addressed to me by this beautiful and stately woman, whom the Brotherhood of Rossetti had succeeded in consecrating.'[2] Mrs Morris was not alone in deploring Shaw's aversion to meat. His friend, and perhaps mistress, the actress Mrs Patrick Campbell, was supposed to have addressed him thus: 'One day, Shaw, you'll eat a beefsteak, and then God help all women.'

He advocated a vegetarian diet, comparing the anatomy and physiology of man to those of the orang-utan, and a return to the fare that Adam and Eve enjoyed in the Garden of Eden, based on 'fruits, nuts, farinaceous seeds, and roots, with perhaps some milk, and it may be honey'. All this was set out in merciless detail in a magnum opus, under the name of Sylvester Graham, MD (which he was not), the *Lectures on the Science of Human Life*. Graham's doctrines excited the ire of butchers and bakers, who felt their livelihoods threatened by the growth of his movement, and he was several times set on by angry representatives of both trades.

The Grahamites, as they were called, also favoured hard mattresses, open windows, cold baths, and (within limits) chastity (for the emission of sperm was, they believed, debilitating), but a healthy diet and good husbandry were their prophet's main preoccupations. Most famously, bread had to be made in the home, and of wholemeal, unsieved flour (Chapter 6). It was this that earned Graham the sobriquet of Doctor Sawdust. Sylvester Graham, who promised his adherents the health, energy (by implication, also virility), and longevity of the biblical prophets, died at the age of 57 after years of poor health. His followers are still numerous and vocal. The editor of the 1980 edition of his great opus tells us that 'Modern dietary science (trophology) may be said to have had its beginning with Sylvester Graham, and his "Lectures on the Science of Human Life" is still abreast of our time in most particulars. If you want the "newer knowledge of nutrition", you'll find it in this book.' He goes on to offer comfort to 'the sincere seeker after truth, who comes to the hygiene classics after years of wandering in the wilderness of false, misleading literature ...'[3] In truth, Graham's book is a grandiloquent mishmash of half-digested and usually misrepresented gobbets of 18th- and 19th-century physiology. The only coherent element in his philosophy is an obdurate vitalism, which caused him to reject all explanations of biological processes in terms of chemistry. William Beaumont's experiments on the digestion of meat in his 'walking stomach', Alexis Saint-Martin (Chapter 5), were futile, Graham insisted, and he tried without success to gain access to Saint-Martin himself. His deepest scorn was reserved for Liebig: 'It is not in the power of chemistry in the least possible degree', he declared in *Lectures on the Science of Human Life*,[3] 'to ascertain what substances the alimentary organs of the living animal body require for nourishment of the body, nor from what chemical elements the organic elements are formed.'

Graham was much sought after on the lecture circuit. He held courses at the illustrious Franklin Institute in Philadelphia and elsewhere up and down the east coast. He circulated huge collections of testimonials from grateful adherents, who purported to have gained health and well-being by following his dictates. He urged the dissemination of his version of physiological wisdom, and, in response, Ladies' Physiological Reform Societies were founded to discuss these matters, and a chain of Graham boarding houses sprang into being. The *Graham Journal of Health and Longevity* was also launched during this period, and, when

Graham widened his interests to embrace education in general, books and pamphlets appeared with such titles as *What a Young Man Should Know*. His thoroughgoing scientific obscurantism notwithstanding, Graham's advice on diet and plain living almost certainly did some good. Many latter-day dietary systems differ remarkably little from his. Graham's name lives on in the flour that he championed, and in Graham crackers, which approximate to the British 'digestive biscuits', but are today made with refined white flour.

Roughage to riches

Sylvester Graham's prescriptions did not make him hugely rich. It was James Caleb Jackson who made his fortune out of Graham's name. First he marketed Graham's flour, and then a series of other profitable products, including the first 'breakfast cereal', concocted on Graham's principles, which he called Granula. This set off what became known as the 'corn flake crusades'. Graham's prescriptions for a healthy life were embraced by the recently founded Seventh-day Adventist sect. One of its elders, Ellen G. White, was inspired to create a prototypic health farm in Michigan. This was the celebrated Battle Creek Sanitarium (not Sanatorium), known from coast to coast as 'the San'. It imposed a regime of abstinence, healthy and frugal eating, a water cure, and exercise. Miss White had a protégé, a young man called John Harvey Kellogg, whom she enabled to train at the Hygeio-Therapeutic College in New Jersey, and then as a real doctor at the University of Michigan and Bellevue Hospital in New York. Kellogg returned to the San as Director, and developed a diet of his own. It was based on an idea which had taken root among sections of the medical profession that the causes of disease lurked in the colon, and that constipation spelled death,* for the fermentation of its putrefied contents generated toxins ('ptomaines') that seeped into the bloodstream. This was the dreaded 'autointoxication' that the Victorians so feared, and about which Kellogg preached. In its worst and most dangerous form this primitive belief gave birth to the laxative regimes, and to the cult of 'colonic irrigation'. But Kellogg's remedy was dietary – and of course vegetarian, for in Shelley's portentous words, meat-eaters' vitals were 'devoured by the vulture of disease' – and it relied on roughage. His first attempt was a 'Granula', like

* This doctrine originated in Europe, where spas with their aperient waters were all the rage. The most influential of the medical luminaries who grappled with the problems of constipation was the English surgeon Sir William Arbuthnot Lane. According to Lane's gospel, it was the effects of our unnatural upright posture allied to social inhibition that prevented us from voiding our waste with the frequency permitted to animals. Because of its supposed ill-consequences, extending from halitosis to cancer, Lane believed that drastic measures were needed to ease the costive bowel. He first tried a series of lubricants, such as pints of cream taken at frequent intervals, and then the more effective castor oil. This left his patients with bloated livers from which, at autopsy, the oil could be wrung like water from a saturated towel. For the most severe cases he resorted to surgery. He was a skilful operator and most of his patients who submitted to extirpation of their colons survived. Lane died in 1945.

Jackson's, made of ground dried bread and biscuits, but then in 1894 he made a chance discovery that was to change the pattern of American, and in some measure European, eating habits: Kellogg and his younger brother, William Keith, who managed the San's finances, had prepared some boiled wheat to be squashed in a roller mill. The wheat was apparently forgotten for some days, and it dried. When passed through the mill the grains were flattened into flakes, rather than emerging as the usual thin leaf of dough. William thought the flakes palatable and offered them to the residents as 'Granose'. His brother took out a patent on 'flaked cereals' of wheat, barley, oats, maize, or other grains. The Kelloggs' Shredded Wheat became hugely popular around this time. It was the younger brother, however, whose business acumen led him to establish the Sanitas Food Company to market these products, and later, in 1899, the Sanitas Nut Food Company, which also produced peanut butter (likewise invented by John Kellogg). All this helped the San to prosper mightily, and the rich of both America and Europe, including many of its aristocrats, flocked to it.

Self-denial was another of Kellogg's prescriptions. He believed in the benefits of a severe restriction of calorie intake, and he ran a successful campaign to persuade the State Government of Michigan to reduce the diet in its institutions, including hospitals, from an average of 5000 calories per day to a mere 2000. This must have caused much misery and perhaps shortened quite a number of lives.

Diet was not John Kellogg's only obsession. He enlarged the water-cure facilities at the San, with medicinal baths and purges along European lines, and expounded his philosophy in a book, *Rational Hydrotherapy*. He inveighed against masturbation and denounced other lapses, especially over-indulgence and its outcome, obesity. For this – the condition of most of the visitors to the San – he offered (besides the meagre diet) treatments with cold douches, icy baths, sweating, and a series of devices of his own invention, such as abdominal rollers and chest- and stomach-pounding machines. Vibrating chairs were designed to stimulate the internal organs and 'electrotherapy' with alternating current was also on offer. Instruments, likewise devised by Kellogg, with names like plethysmograph, ergograph, and pneumograph, monitored the clients' physical state and progress. In all, some 300,000 people subjected themselves to the San's exacting regime.

The Kelloggs were only the first in a mighty movement, and next on the scene was an untutored visionary by the name of Charles W. Post, who sought relief from physical and mental ills at the San, which he endured for 10 months. The stay did him little good, and he was cured of his ailments only when he discovered Christian Science. He thereupon resolved to set up his own establishment in competition with the San, and on the adjoining plot of land he built the La Vita Inn. Post's regime forbade alcohol and stimulants such as tea and coffee, but it was not strictly vegetarian, and he experimented with wholesome alternatives. One that he brought to fruition was Post Toasties, a breakfast food, and another was Postum Cereal Coffee Food, a concoction made of toasted grains, not very different from the brew served up at the San. It instantly caught on, and astutely advertised, made Post very rich. Next he invented a new breakfast cereal made of a baked and crushed mixture of wheat with malted barley flour. This was called Grape Nuts, and with each packet came a copy of a pamphlet, penned by Post, with the title, *The Road to Wellville*. It outlined its author's views on how to achieve bodily and spiritual well-being, and assured the purchaser that the contents of the package would cure appendicitis, and alleviate tuberculosis, malaria, and loose teeth. Grape Nuts and Instant Postum still survive as a monument to Post's vision and commercial acuity.

John Harvey Kellogg, meanwhile, observed with dismay the growth of the monster to which he had given birth. From about 1902 the breakfast cereal boom spread across America, and Battle Creek was transformed into an industrial centre. Entrepreneurs of all stripes, including many of Kellogg's own employees, raised money to set up companies and build factories. New products promising health, happiness, and longevity engulfed the market. Some succeeded, many were ill-conceived, unattractive, unpalatable, or downright offensive, and failed. The advertisements became increasingly extravagant in their claims, and engaged the skills of artists and poetasters. In 1903 the number of patent breakfast foods on sale in the United States exceeded 100. The cereal craze also spread to Britain, where Quaker Oats and Hornby's Steamed Oats, both invented in America, had made a modest splash at the start of the 20th century, and was just as ferociously promoted. Here is one of a series of doggerel advertising jingles, composed to promote a cereal product called Force:

> Jim Dumps was a most unfriendly man,
> Who lived his life on the hermit plan.
> In his gloomy way he'd gone through life
> And made the most of woe and strife,
> Till Force one day was served to him.
> Since then they've called him Sunny Jim.[4]

Kellogg, a man of principle, did not approve of all the hysterical commercialization, and soon fell out with his brother who, he felt, was debasing the austere virtues on which the San had been founded. William Kellogg created a new company in 1906, the Battle Creek Toasted Corn Flake Company, printed his and not his brother's signature on each packet of Corn Flakes, and embarked on a campaign of ever more aggressive and inventive advertising. The Kellogg booklet, *Health from Day to Day*, sold in vast numbers. One hundred years later the company is still growing.

The heirs of Graham

Sylvester Graham cast a long shadow, far into the 20th century. There were many followers who disseminated and embroidered his message, but none more successful than Bernarr Macfadden (*né* Bernard Adolphus McFadden in rural Missouri in 1868). Orphaned at an early age and with minimal education, Macfadden somehow found the means to start *Physical Culture*, a magazine dedicated to promoting a life of exercise, healthy diet, and fasting. He laid down separate prescriptions for fitness for women, which included the renunciation of corsets and high-heeled shoes. *Physical Culture* was a huge success, and with the profits Macfadden expanded his publishing operation into a substantial empire, putting out magazines dedicated to such subjects as romance and detective fiction. Among the most popular were *True Story* and *Dance Lovers Magazine*. At the same time Macfadden had been establishing one health farm after another, such as the International Healthatorium near Chicago and the Physical Culture Training School. They did well, but megabuck success came only in 1929, when he bought an institution originally built by James Caleb Jackson – he of the Granula – in upper New York State. Macfadden transformed it into the Physical Culture Hotel, where his system of exercise and a rigorous vegetarian diet, interspersed with fasting, on which he laid much stress, was endured by many affluent clients, some of them consumptives – a condition that he claimed his methods would cure. Macfadden was already the object of critical scrutiny by official bodies when he tried to persuade President Roosevelt to appoint him Secretary of Health. Thwarted, he tried to run for Governor of Florida and then for President. But his position, undermined by his persistent claims of unlikely cures, was further weakened by a strange episode that occurred in 1928: in that year Johanna Brandt published a book under the title *The Grape Cure*. In it she related that she had shaken off terminal stomach cancer by subsisting on nothing but grapes for a period of years. The therapeutic merits of grapes, she claimed, had been acclaimed in European works since the 16th century, and she was in possession of other proofs of their efficacy. Macfadden printed the story in one of his organs, the *Evening Graphic*, and continued to advert to it in the years that followed. There was indignant criticism, and in 1950 Macfadden hit back with an offer of $10,000 to anyone who could prove that grapes did *not* cure cancer. The respectable medical profession was, of course, appalled that such irresponsible nonsense should be foisted on the public, and Macfadden's reputation suffered. As to the grape cure, Johanna Brandt's book was, startling to discover, relaunched in 1989 under a new title, *How to Conquer Cancer, Naturally*, with commendations from two doctors and personal reports of miraculous cures from 22 patients. As for Macfadden, he remained a good advertisement for his diet. He married his fourth wife – a woman little more than half his age – and made a well-publicized parachute descent in France on his 84th birthday. He died at 87, apparently from the consequences of an enlarged prostate of which he tried to cure himself by fasting. His name is still revered by diet cranks in the United States and his books remain in print.

A variant of Sylvester Graham's system of diet arose in Europe at the beginning of the 20th century. It was started by a Swiss doctor, Maximilian

Bircher-Benner, who conceived the notion that cooking deprived foods of their nutritious virtues, their 'vital substance'. He broadened this primitive vitalism by the corollary that cooked foods generate elements of decay in the digestive tract, which lead to the feared autointoxication. Man should eat – and this included babies – only raw food, which had received all the cooking it needed by exposure of the plants to the sun. Bircher-Benner's followers called themselves apyrotrophes. One of the cornerstones of the diet offered in Bircher-Benner's clinic in Zurich was a mixture of fruit, nuts, and grains, which he called by a Swiss diminutive, muesli. The packets of some makes of this elixir still carry Bircher-Benner's commendation. This, at least, is more wholesome than the Max Gerson diet, the product of the imagination of a German doctor born in 1881. It consists of an average of 13 glasses (some 20 pounds) each day of fresh, liquidized carrots, apples, broccoli, and other fruits and vegetables, allied to weekly injections of liver extract and five coffee or camomile tea enemas. This treatment is still dispensed by the Gerson Institute, and is said to be a sure cure for even advanced cancer.

Another diet evangelist of the early 20th century was William Howard Hay (1866–1940), an American doctor, who overcame a formidable variety of medical misfortunes in his early life by dint of diet. Late in life he articulated his philosophy in a widely disseminated book, *Health via Food*, and applied it in his private health resort in Pennsylvania. His particular fixation was pH, the measure of acidity: low pH signifies acid, high pH alkali, and in between, at pH 7, lies neutrality. Disease, Hay taught, was caused by excess acidity, resulting from the digestion of meat and of refined carbohydrate. Moreover, meat to be digested required an acid, and carbohydrate an alkaline milieu. It followed that proteinaceous foods must never be combined with starches, for then, by tortuous reasoning, acidity would be generated and constipation ensue. A delicate adjustment was therefore called for. Hay's diet comprised mainly salads, vegetables, and fruit, with very little protein. The mayhem wrought by excess acid in the human entrails was indeed a recurring theme in diet advertising of the time. Sarah Tyson Rorer, whose influence peaked in the 1890s, was a prominent spiritualist and Principal of the Philadelphia School of Cookery, 'the Queen of the Kitchen' and author of many books on cooking, housekeeping, and diet. She inveighed against excess, against alcohol, against starchy, sweet, and fatty foods (leaving relatively little that her devotees were permitted to eat), but especially against mustard, vinegar, and pickles. 'If salt and vinegar will eat away copper, what must it do to the delicate mucous lining of the stomach?' This, and the advertisement that shrieked, 'the acid in your stomach would burn a hole in the carpet', disregarded, of course, the fact, already known to all who cared to look into the matter, that the normal stomach contains a concentrated solution of hydrochloric acid.

Hay's beliefs (which still command a following) may have been inspired by the theories of a British doctor, prominent and vocal in Victorian London,

Alexander Haig. Haig was another monomaniac, whose particular obsession was uric acid. This is a normal product of metabolism, and a small amount is always found in the blood, but when a disturbance in the metabolism allows it to accumulate, gout (p. 115) is apt to result. Short of gout, though, uric acid in the blood caused, according to Haig, a type of poisoning for which he invented the term collaemia, and from this all manner of ills would surely flow. Chemically speaking, uric acid is derived from a class of biologically important substances called purines (they occur, for example, in DNA), and Haig therefore proscribed all foods known at the time to be rich in purines. Out therefore went meat, all but a few fishes, beer, coffee, tea (as bad as opium), and chocolate. As Haig's obsession grew, more and more foods were added to the list, and in the end what was left would barely sustain life, and assuredly not without anaemia. Haig's most famous work, *Uric Acid as a Factor in the Causation of Disease*, was published in 1892 and ran into seven editions. Diet faddists still refer reverentially to Haig's ravings. Regulation of acidity (pH) has also become a mantra for contemporary faddist sects.

In the wake of Dr Hay came a procession of other clamorous soothsayers. Dr George A. Harrop, at one stage of the illustrious Johns Hopkins University, seeking a healthy low-calorie regime, settled in 1938 on bananas and skimmed milk, which he thought would be filling and provide everything the human constitution might require. An altogether more masterful figure, with no medical education but a great deal of commercial acumen and charisma, was the egregious Gayelord (formerly Helmut Eugen Benjamin Gellert) Hauser. A formidable publicist and virtuoso of the lecture platform, Hauser wrote hugely successful books, especially *Better Sight Without Glasses* and *Look Younger, Live Longer*, which sold nearly half a million copies within a year of its first appearance in 1950. His diet emphasized a black treacle, called blackstrap molasses, yeast, wheatgerm and other seeds, yoghurt, and dried skimmed milk. Hauser attracted many famous people eager to live longer, and ageing film-stars hoping to look younger. His company ramified into cosmetics and other products and his name still lives.

The way of the carnivore

There is a Newtonian principle that to every fad there is an equal and opposite fad. So the wisdom of John Kellogg and Sylvester Graham is cancelled out by that of Dr James H. Salisbury (1823–1905). He was born in New York and spent his early years after university in the study of plant physiology. He seems to have been a skilful microscopist and published several papers. But alas, Salisbury was a fantasist, whose weakness first revealed itself in an article published in a French journal claiming that fresh lumps of earth, placed in an open bedroom window, could occasion fever. Next he asserted that syphilis and gonorrhoea were caused by the spores of certain plants. Later he identified these spores in blood cells of patients with the same diseases and also in cases of measles, malaria, and rheumatism. His interest in diet developed during his service as a medical officer in the Civil War. He thought that the endemic condition of 'camp diarrhoea' was 'consumption of the bowels', and he treated the soldiers with a diet of grilled steak and coffee. He proceeded to convince himself that all disease stemmed from bad diet and enunciated his theories in a book, *The Relation of Alimentation and Disease*. He revealed to his readers that man was constituted as two-thirds carnivore and one-third herbivore, for he had, amongst other anatomical features, the teeth of a carnivore (which is to an extent true; Salisbury might also have mentioned that our gut is certainly not that of a herbivore). His conclusion had been strengthened by dietary experiments that he had performed on pigs and also on himself. They involved living for periods on only a single food. He had tried beans, for instance, and proved by examination in the microscope that they were not properly digested in the alimentary tract. He had made assurance doubly sure by a similar study on six human subjects whom he paid to subsist on baked beans for several days.

The diet that Salisbury devised for his patients, and urged on the world, was based primarily on grilled chopped lean steak, formed into a patty – in essence a hamburger, prepared according to a minutely specified procedure – and hot water. Salisbury believed that disorders were caused by something he called 'fibrous tissue', and the hot water served to flush this out of the body. The hamburgers, weighing about one pound each, should be consumed for breakfast, lunch, and dinner, and a pint of hot water should be sipped slowly one hour before each such meal to cleanse 'the system'. Vegetables and starches, and even the connective tissue found in meat, on the other hand, would accumulate in the intestines, secreting products that poisoned and paralysed the tissues, and a dire litany of maladies could follow, notably heart disease, tumours, mental illness, and tuberculosis. His prescriptions brought Salisbury into conflict with John Kellogg, who accused him of visiting on the population everything from rheumatism to 'nervous prostration'. A follower of Salisbury's introduced the diet to the English, where, despite its huge popularity in the United States, it never caught on, perhaps because of the cost of beef at the time. Salisbury steak is still, of course, to be found on restaurant menus, as a hamburger with social pretensions. Salisbury's regime is a precursor of today's high-protein slimming diets, about which more in the next chapter.

Salisbury was by no means the first to decry carbohydrate-rich diets. That may have been the admired gourmet Jean Anthelme Brillat-Savarin (p. vii), who lent his name to many extravagant confections (not least the rum-soaked savarin). He nevertheless persuaded himself that the delicious Parisian bread-rolls and cakes, not to say pasta and potatoes, engendered obesity and ill-health, and he proclaimed as much in his famous book, *La Physiologie du Goût*, of 1825. Then in 1863 the carbohydrate-deficient diet surfaced once more in the insistent affirmations of a London undertaker to the quality, by the name of William Banting.

He was short and globular, and an object of 'cruel and injudicious' comment. On medical advice he gave up starchy foods and adopted a diet of lean meat, green vegetables, unbuttered toast, and soft-boiled eggs. As a result his girth soon shrank to respectable proportions. Delighted, he resolved to spread the good news and composed a tract with the title *Letter on Corpulence*. He caused 2500 copies to be printed, which he distributed free. His prescription was widely touted on both sides of the Atlantic, and dieting became known in the argot as 'banting'.

Health through starvation

Without doubt the cheapest of the many prescriptions for dietary health was fasting. The cult has been around since ancient times, but enjoyed a wide resurgence towards the beginning of the 20th century. The most influential proponent was probably one Hereward Carrington (1880–1958), an aficionado of the occult and intimate of 'The Great Beast', the 'magician' Aleister Crowley (who advertised himself as the wickedest man on Earth). Carrington delivered himself in 1908 of a mighty volume, entitled *Vitality, Fasting, and Nutrition*, which promised that fasting would ensure a long life and freedom from disease. A fervent apostle of the creed was the inimitable Upton Sinclair, author of (among many other novels) *The Jungle* (p. 133), and the most credulous of faddists. He published in 1911 his book *The Fasting Cure*, in which he assured his readers that a strict regime of deprivation would cure any of a long inventory of diseases, including cancer, tuberculosis, asthma, syphilis, locomotor ataxia, and, to cap it all, the common cold. In what passes for a caveat he remarks: 'I have known two or three cases of people dying while they were fasting, but I feel quite certain that the fast did not cause their death.' The irony in all this farrago is that we now have good evidence for an increased life-span in rodents kept in laboratory conditions on a very low-calorie diet (p. 245).

Eternal life through yoghurt

The conjecture that soured milk might be the elixir of youth – the reason for the legendary longevity of the mountain-dwellers of Asia minor and the Caucasus – was first heard in Western Europe around the middle of the 19th century, but was given scientific expression by a Russian biologist, Élie (Ilya Ilitch) Metchnikoff. He was born in 1845 in a village near Kharkov in the Ukraine and graduated in zoology at the local university. Metchnikoff was a scientist of exceptional insight and imagination and made a series of important discoveries, chief of which was that of phagocytosis – the capacity of certain specialized cells to engulf and eliminate intruders, such as bacteria. Phagocytosis is one of the cornerstones of the body's system of immune defence against infection, and for this work Metchnikoff received the Nobel Prize in 1908.

By that time he had settled in Paris and was working at the Pasteur Institute, where he made the observation, important at the time, that syphilis could be transmitted to monkeys. But his preoccupation in his later years was with the mechanism of ageing. He developed the idea that phagocytic cells rampaged through the body, consuming the moribund tissues, although the theory that caused so much excitement in Europe, and soon thereafter in America, was that we were all being gradually poisoned by toxins secreted by our gut bacteria. This calamitous process, Metchnikoff had no doubt, could be prevented by replacing the noxious by benign bacteria. An abundant source of the latter was *kephir*, a sort of buttermilk preparation, popular in Eastern Europe and the Middle East, and closely related to the now ubiquitous, but then exotic, yoghurt. The bacteria in question were *Lactobacillus bulgaricus*, or *Lactobacillus acidophilus*, together with several species of the so-called *Bifidobacteria*. The digestion of milk by *Lactobacillus* produced lactic acid, which is responsible for the sour taste. These microbes, Metchnikoff thought, would multiply rapidly in the gut, and would drive out the resident *Escherichia coli*. On a diet that included an abundance of yoghurt one might then live a healthy life into a prodigious age, perhaps even without limit. Metchnikoff tried the diet himself and owned to feeling much the better for it. His book, *The Prolongation of Life* in its English version, appeared in 1907 and was rapturously received and translated into many languages. Pamphlets on how to prepare fermented milk soon appeared, and circulated in their millions. Metchnikoff, a depressive who was much affected by The Great War, nevertheless died in 1916 at the age of 71. His influence endured, and inspired such seekers after spotless intestines as the visionary surgeon Sir William Arbuthnot Lane (p. 194 f.n.). Yoghurt worshippers are, of course, with us still.

Horace the chew-chew man

Among the many outlandish doctrines that sprang from this fertile soil, blossomed, and suddenly faded, none was odder than Fletcherism.[5] Horace Fletcher, the Great Masticator, was a prosperous man of business before he started masticating, but his life was, by his own account at least, a burden to him, for his waist was distended and his hair had turned white. He was prematurely aged and a martyr to dyspepsia (a great American preoccupation at the time) and catarrh. Then in 1898, shortly before his 50th birthday, a great dietary vision was vouchsafed to him. He was directed down the path of enlightenment by a friend who had taken to heart the advice of the ascetic British Prime Minister William Ewart Gladstone, that one should chew each mouthful of food 32 times – once for each tooth. Fletcher began to chew, and the more he chewed the better he felt. Soon he had lost 60 pounds and felt rejuvenated (though whether his hair regained its pristine colour was not revealed). The process of digestion, Fletcher asserted in his many articles, began in the mouth (true), and it was only by chewing each mouthful until it had been reduced to a liquefied, saliva-impregnated sludge that the first stage of digestion could be properly accomplished. Any residual refractory fibrous remnants (dangerous to the health if ingested) could then be discretely spat out, while the remainder passed into the gastrointestinal tract, where, unlike the coarsely divided chunks swallowed by the ignorant, it would be effortlessly digested to the limit. Here Fletcher made his most important anatomical discovery: there was a kind of valve, the 'food gate', at the back of the throat, which opened to receive only the fully liquidized mass, while everything else took a different route.

At the outset Fletcher's pronouncements provoked mainly mockery, but by 1906 the tide had turned. For one thing, he did indeed appear lean and fit on a diet that, he kept repeating, contained only two-thirds of the calorific value then regarded as necessary to support the life of a grown man; nor had he lost more weight in the years since his girth had stabilized. This could only be because the interminable chewing released the nutritional essence that was otherwise lost, although self-control and rejection of all excess (not, by all accounts, very assiduously practised by Fletcher himself) was also part of the formula. The nature of what one ate was immaterial, for the digestive capacity released by chewing promoted total absorption of everything of nutritive value, ensured that none of it was turned into body fat, and thus eliminated any tendency towards obesity, and the dreaded autointoxication.

But in addition, Fletcher was an accomplished liar. He had managed to bring himself to the attention of many prominent nutritional scientists. Some showed interest, and one remained a keen advocate. He now styled himself Dr Fletcher and claimed to have studied with Frederick Gowland Hopkins in Cambridge and performed Hopkins's most important laboratory analyses. Moreover, he had taken a course of study with Arthur Gamgee, nutritionist and sometime Dean of the medical school of Victoria University in Manchester, and graduated with acclaim. And finally he had spent seven years in Wilbur Atwater's laboratory (p. 111) at Wesleyan University. Fletcher had indeed come to know Hopkins – an acquaintanceship that Hopkins does not mention in his memoirs – but the extent of his research seems to have been to collect and weigh his own waste products. Perhaps it was on these studies that he based his insistence that faeces of a healthy human 'have no more odor than a hot biscuit'. As to his brush with Atwater's laboratory at Wesleyan University, calorimetry experiments there had shown only that he could not have maintained his body weight on the purported intake of two-thirds of the minimum requirement. Nevertheless, Fletcher had turned the full force of his personality on Sir Michael Foster, head of the Cambridge physiology department and the doyen of British physiologists, and persuaded him that there might be something to his theories. Foster had introduced him to Hopkins, who agreed to conduct some experiments. *The Lancet* conjectured that Fletcher might be on to something. An editorial commended the virtues of self-restraint and also observed that the 'commissariat problem' of the army might be alleviated by his methods: 'Napoleon's dictum that an army "moves on its belly" is to be altered and the instructed army will hardly need a belly to move on.'

Fletcher's articles proliferated in newspapers and magazines. He was suddenly in great demand on the lecture circuit, and for a time shared the platform with John H. Kellogg, attracting audiences of a thousand and more. Fletcher's was a captivating presence on the podium, with his baby face and bright eyes behind steel-rimmed lenses, and his ready humour. He had also been a notable athlete in his youth at Dartmouth College in New Hampshire (which later bestowed an honorary degree on him), and was still able to impress his audiences with remarkable physical feats for a man of his age, putting down his prowess, of course, to his diet. His fame spread to Europe and especially Britain, and chewing parties became popular in fashionable circles. These 'muncheons', in which the participants were enjoined to chew with their heads low over the plate so that the tongue could hang down, were often coordinated by a conductor, who timed the mastication of each mouthful and rang a bell or struck a gong when the moment came to swallow. Many famous names were to be found among Fletcher's followers. Henry James was one who spread the word, referring to 'the divine Fletcher', and in England the royal physician became a devotee, and presumably urged the system on the King and his family.

In 1900 Fletcher bought a palazzo on the Grand Canal in Venice, where he housed his wife and step-daughter. He was fortunate in the step-daughter's suitor and eventual husband, an English hotel doctor, resident in Venice, Eric van Someren. This man was won over by his step-father-in-law's powers of persuasion, and did much to bring the chewing principle to the attention of prominent people. Among them, after van Someren presented the theory of Fletcherism (at Foster's instigation) at the International Congress of Physiology in Turin, were several scientists. Fletcher's ambitions by this time knew no limits, and with Foster's encouragement he tried to raise money from philanthropists to found an institute of nutritional research, first in Venice, later in the United States. Such a centre was indeed established a little later at Harvard University with support from the Carnegie Institution, but Fletcher was excluded, and the first director was one of Atwater's associates at Wesleyan University, who had examined Fletcher's claims some years earlier and found them spurious.

Fletcher was chagrined, but did not repine, and continued to proselytize in America and Europe. He paid down-and-outs to chew in a Chicago restaurant, and reported improvements in their condition. Anaemia, appendicitis, colitis, alcoholism, insanity, and sudden death were added to the catalogue of conditions infallibly cured by chewing or precipitated by the want of it. To these Van Someren, apparently on the basis of personal experience, added diabetes, and 'morbid sexual cravings'. Moreover, Fletcher now had a patron. At Foster's suggestion he had approached Russell Chittenden at Yale (p. 89) as the authority best placed to set up a rigorous programme for testing the efficacy of the mastication technique. Chittenden had already dissociated himself from the calorie requirements laid down by Atwater, which differed little from those of Voit a few years earlier (Chapter 6). Where Atwater had recommended a daily intake of 2200 calories for an average man in a sedentary occupation, and 4500 calories for one engaged in work demanding continuing muscular exertion, Chittenden insisted that a daily intake of 2500 calories would suffice for a regime of heavy labour or for soldiers in the field. Perhaps the increased food value liberated by chewing could explain the deficit? To what extent Chittenden's arguments for a low calorie intake were in fact based on his contacts with Fletcher, and Fletcher's claim to have maintained his weight on 1600 calories per day, is unclear, but he was almost certainly eager for support, both moral and financial, and Fletcher had also offered him a significant sum to help his research.

By this time the chewing fashion had waned, at least in Europe, although its begetter continued to travel, write, and lecture, flamboyant still in manner and dress. Van Someren had jumped ship: his espousal of Fletcher's teaching had made him unpopular among the restaurateurs and hoteliers of *La Serenissima*, who did not want their visitors to stint on food and wine. His hotel patients, frightened away by the rumours assiduously spread, deserted him, and he and his wife departed for friendlier climes. Undeterred, Fletcher strove to revive interest in Fletcherism by publishing a book with the title *Fletcherism: What Is It; or How I Became Young at Sixty*, along with many articles. In the United States he launched a campaign to inculcate the chewing habit in young children, and even for a time ran a kindergarten in New York. Then, with the start of The Great War in 1914, he enjoyed one more hour of fame. Long before, in 1898, at the height of the Boer War, he had heard that the British Army was being incapacitated by typhoid fever, and had sought to persuade the Army medical service that Fletcherism would avert the putrefaction of the ingested food and eliminate the problem. He failed, but a few years later, probably as a result of Chittenden's professed successes in his experiments on soldiers (p. 89), the Royal Army Medical Corps did institute some trials. The results were a fiasco: the men could not function on a diet of 3500 calories per day. But in 1914 Chittenden tried again. The Royal Society committee, which the Government had asked to look into nutritional requirements of the soldiery and those engaged in war work, angrily rebuffed his overtures. But Fletcher had stated that Britain could be self-sufficient in food (thereby defeating the Atlantic blockade) if only the whole population could be induced to chew. There was support for this notion in the press and in political circles, but in the end it came to nothing.

At the outbreak of war Fletcher was in Brussels and approached the United States ambassador, whom he had met on the lecture circuit, to offer his help to the American Commission for Relief in Belgium. The ambassador was probably having a hard time of it, as the German army swept into the country, and accepted Fletcher's offer. The head of the Commission was a future President,

Herbert Hoover, who would have been only too glad to be told that the food he was able to provide could be made to go further, if only Fletcher's prescriptions were followed. Fletcher returned home, and seems to have done good work agitating for help for Belgium, once more on his lecture tours. But in time he seems to have become a thorough nuisance to Hoover, and in 1919 when the War was over he tried again to reach Belgium, was marooned in Copenhagen, and there died. The residual popularity of the chew-chew diet did not long survive his demise.

The wilder shores of quackery

Quacks, or quacksalvers as they were originally called, have always been with us, but diet quacks appeared in unprecedented profusion in the latter half of the 19th century, and received a further boost when vitamin became a catch-word early in the 20th century. It would be tedious to go into more than a few of the many infamous examples of charlatanry, but one that deserves mention is Lydia Pinkham's Vegetable Compound. Mrs Pinkham was born Lydia Estes in rural Massachusetts in 1819, and in her early years was much devoted to good works. In 1843 she married a prosperous landowner, and only began to produce her tonic when his fortunes declined some years later. Exactly what the brew contained is unclear except that alcohol formed a large part, but the liquid and pills were advertised as curing 'constipation, biliousness and torpidity of the liver', at 25 cents the pack. The virtues did not end there, for Mrs Pinkham was especially eloquent on feminine complaints. 'All Ovarian troubles' would yield to the potion, whether 'Inflammation, Ulceration, Falling and Displacements of the Womb', along with tumours, which it would 'dissolve and expel' from the uterus. There followed Lydia Pinkham's Blood Purifier, which would 'eradicate every vestige of Humors from the Blood, and at the same time will give tone and strength to the system. It is far superior to any other known remedy for the cure of all diseases arising from impurities of the blood, such as Scrofula, Rheumatism, Cancerous Humors, Erysipelas, Canker, Salt Rheum and Skin Diseases.' Lydia Pinkham did not live to see the full consecration of her products, for she died in 1883, but at its zenith in 1925 her company's annual income was $3.8 million (equivalent to more than 10 times that today).

With the passage into law in the United States of the Pure Food and Drugs Act in 1906, fraudulent claims for the efficacy of crank diets and supplements abated. In most European countries legislation and enforcement lagged behind, but even in the United States large-scale fraud did not by any means cease, as entrepreneurs looked for loopholes in the law. So the Horlick's company (p. 114) dreamed up the new concept of 'night starvation' and was able to advertise its milky drink with the warning that 'right through the night you've been burning up reserves of energy without food to replace it. Breathing alone takes twenty thousand muscular efforts every night.' (Who would have thought it?)* Other con-men simply played cat-and-mouse with the law. Throughout the first half of the 20th century, for instance, Royal S. Lee, a qualified dentist, ran his Foundation for Nutritional Research, which distributed literature commending in fulsome terms the products of his own company. The most famous of these was Catalyn, a diet supplement consisting of milk, sugar, starch, and bran, with other uncertain vegetable matter. It purported to prevent or cure many diseases, and in the 1930s the Food and Drug Administration (FDA) impounded shipments. In 1956 Lee ran into deeper trouble for his book *Diet Prevents Polio*. He continued in and out of court until, in 1962, he was punished by a sizeable fine and a year's suspended prison sentence, for 'misbranding' no less than 115 diet products and fraudulent claims of cures for some 500 afflictions. The following year a statement from the FDA called Lee 'probably the largest publisher of unreliable and false nutritional information in the world'. Lee's remarkable career was finally terminated by his death in 1967.

Among the many miscreants who preyed on the public in the United States, and especially on the sick, was Dudley LeBlanc, state senator for Louisiana, who peddled on a nationwide scale an alcoholic nostrum called Hadacol, and got rich. (When asked by Groucho Marx what Hadacol was good for, LeBlanc was supposed to have replied, 'It was good for five and a half million for me last year.') Perhaps the most flamboyant of all was Adolphus Hohensee, whose years of ascendancy were the 1940s and 50s. All the world agreed that Hohensee was a hypnotic performer on the lecture platform. His histrionic talents drew huge audiences and his multitude of slavishly devoted followers were convinced that their hero was a genius, persecuted by a corrupt and incompetent medical profession and by pharmaceutical interests, out to suppress the discoveries that offered relief from any number of distressing conditions. Hohensee's remarkable career and his many encounters with the law (and also the story of Dudley LeBlanc) are absorbingly told by James Harvey Young in his book *The Medical Messiahs: A Social History of Health Quackery in Twentieth-Century America*.

The vitamin-mongers

What Elmer McCollum called 'the newer knowledge of nutrition' was quickly translated into medical practice and, as we have seen, saved many lives and allevi-

* Such advertising was allowed to continue in Britain until well into the 1940s, and the opportunity was nowhere better exploited than in women's magazines, such as the *Lady's Companion*. Horlick's was one of the most vigorously promoted products ('When men feel ... fagged out, dull and wretched, it's a sure sign of "Night-Starvation".') Among the 'tonics' or 'pick-me-ups' were such products as Iron Jelloids, Carter's Little Liver Pills (which turned the urine a brilliant turquoise), and Bile Beans – 'These fine vegetable pills', the ladies were instructed, 'free your system of harmful impurities that can so easily upset your health and looks. They cleanse the bloodstream, tone up the entire system. You feel new life, new vigour, new energy, and your eyes and complexion reflect the radiance of the new health this favourite family tonic-laxative has given you.'

ated much suffering. The concept of vitamins engaged the popular imagination to an astonishing and unaccountable degree,[6] nor did it take long for unscrupulous entrepreneurs to recognize an opportunity. In the United States it was only the FDA that stood between them and a gullible public, and there were ways of circumventing this obstacle, for instance by recruiting thousands of door-to-door salesmen. The advertising traded on the proposition that if vitamins were required for healthy life, then more vitamins would afford even greater health and vitality ('to give you that firm flesh pep'), not to mention sexual virility. Against such bombardment of the senses, the truth – that a normal diet supplied all the vitamins that the body requires – could not prevail. Advertisements tabulated the signs of an insufficiency of vitamins (without distinguishing between one vitamin and another) thus: you would have 'dull eyes, dull or falling hair, poor bone structure, soft dentine and enamel, tiredness and irritability, serious deficiency diseases'. Primed with vitamins, on the other hand, your eyes would be bright, your hair 'lustrous', your bones and teeth 'good'; you would, in short, enjoy a 'fit feeling' and find yourself suffused with 'buoyant vitality'. The advertisers also preyed on fears of disease and most of all on the consciences of mothers of young children. Cod liver oil, which had been in use since the 18th century, enjoyed a remarkable resurgence. The affirmation by the world's largest producer, the Squibb company, ran: 'Inside, where it can't be seen, the trouble starts! A defective development of the bone structure so insidious that it is more than likely to touch even the well-cared-for baby in intelligent modern homes!' By 1937 the consumption of cod liver oil had reached 5.8 billion gallons. The dangers of hypervitaminosis, or vitamin poisoning, were not widely recognized, nor that of ultraviolet irradiation, which Elmer McCollum and others had shown to promote the synthesis of vitamin D (p. 179). There is no record of the extent of the skin cancers that must have been occasioned by Dr Sereda's 'violet ray machine', which he claimed cured some 80 diverse diseases. Irradiated cottonseed oil (a waste product of the cotton industry) and olive oil ('bottled sunshine') were marketed at an inflated price. Milk was also touted for its vitamin D content. The popular dietician Adelle Davis observed in her best-seller *Let's Get Well* that 'I have yet to know of a single adult to develop cancer who has habitually drunk a quart of milk daily.' Even a doctor on the staff of the Bellevue Hospital in New York contended that vitamin D in milk cured such conditions as polyneuritis, which is common in alcoholics and caused by a failure of thiamine absorption (p. 174). He had encountered a barman whose polyneuritis, the result of a daily whisky intake of 15 fluid ounces, was cured by lashings of milk. But it was vitamin pills that swept the market. In Britain, as in America, 'vitaminized' foods were even advertised for dogs and cats.

The strengthened legislation of the Food, Drug and Cosmetic Act of 1938 restricted the worst abuses, and factories producing quack nostrums were regularly raided, but the vitamin industry received a major boost when, after prolonged legal wrangling, vitamins were designated foods and not drugs. That meant that preparations could be sold over the counter in pharmacies and grocery shops. This gave scope for many new fraudulent products. James Harvey Young in *American Health Quackery* gives an instance of a particularly shameless imposture, when an enterprising mountebank in the Midwest took the name of Silent George of Shawneetown, whence emanated tins of condensed milk, sprayed gold and relabelled as Swamp Rabbit Milk, 'a balanced product for unbalanced people, rich in Vitamins J, U, M and P' (none, of course, known to science, but hinted on the label to possess powerful aphrodisiac qualities). Silent George evidently did a roaring trade before he was raided by officers of the State Department of Agriculture and shut down. Royal Lee's Catalyn, too, contained every known vitamin and several more.

Medical opinion, it should be recorded, was not altogether immune to the vitamin hysteria, and many unnecessary and even dangerous products found endorsements from respected members of the profession. A curious instance was Dr Russell N. Wilder of the Mayo Clinic, one of the foremost centres of clinical science in the United States. His particular obsession was with vitamin B_1 or thiamine (p. 174), otherwise 'the morale vitamin', a deficit of which does certainly lead to neurological disease. Wilder's experimental subjects were factory workers who, on being fed a diet deficient in thiamine for some weeks, ceased to be 'sociable, contented' and were horribly transformed: they were now quarrelsome, depressed, tired, and *even* mounted strikes. Whether it was an unpalatable diet that produced this dire effect, or indeed the thiamine deficiency, is uncertain, but when the regime was discontinued, at all events, the workers recovered their biddable dispositions. Wilder also had the startling insight that 'Hitler's secret weapon' was a devilish scheme to deprive the enslaved peoples of Europe of thiamine. But vitamins, he declared, would win the war for our side.

The Laetrile scandal

Perhaps the most pernicious of the fraudulent claims centred on a purported cancer cure.[7] Laetrile was an extract of apricot kernels containing amygdalin, a substance that breaks down to generate hydrocyanic (prussic) acid. The active ingredient was purportedly synthesized in the laboratory by Dr Ernst T. Krebs, Sr, who peddled various quack remedies and whose first 'cancer cure' was something he called vitamin B_{15}, or pangamic acid. Laetrile would, according to Krebs, release its hydrocyanic acid in the body and this would kill the cancer cells (if the quantity generated was insufficient to kill the patient). Krebs may have believed his own fantasy, but of the trials with patients that he claimed to have done there is no trace. Krebs passed the torch to his son, Ernst T. Krebs, Jr, MD, as he styled himself (although he had no doctoral degree). Serious marketing began in the late 1950s. When the claims for Laetrile came under the baleful scrutiny of the FDA, its promoters changed its description to a vitamin, vitamin B_{17}. Cancer (any cancer), they now asserted, was a vitamin deficiency disease. A prolonged legal battle developed. Attempts to ban the marketing and use of Laetrile were countered by a flood of new products containing the same vegetable ingredient, and by indignant cries of suppression of citizens' freedom to swallow whatever they pleased. There were of course many testimonials to Laetrile's success, and there were in all an estimated 70,000 purchasers in the US, and no doubt many more in other countries. Laboratory trials were undertaken, and on the principle that any proposition, no matter how absurd, will find a supporting publication somewhere in the medical literature, there was indeed a positive report or two. But at last, in 1980, proper trials conducted by the National Cancer Institute laid the matter to rest. In the interim, desperate patients continued to pay large sums for shipments of Laetrile from Mexico. Laetrile has more or less vanished from the scene, although it is still allegedly in use in some clinics in Mexico, and it can still be bought from suppliers advertising on the Internet.

Linus Pauling and vitamin C

Perhaps the strangest obsession linked to vitamins was that of Linus Pauling (1901–94), arguably the outstanding chemist of the 20th century, Nobel Laureate twice over (once for chemistry and once for peace). His interests had long reached into aspects of biology, but his espousal, when he was well into his sixties, of vitamin C as a panacea is unaccountable. He convinced himself that early man had adhered to an opportunistic diet in the wild that included a huge intake of plants rich in vitamin C. Therefore evolution had ensured that human health would be safeguarded by vitamin C, ingested by the gram. Pauling indulged himself in this manner, and claimed to be immune to colds in consequence. Although his theory received no support from mainstream medical science, it was greeted with uncritical enthusiasm on the fringes and in the press. In old age Pauling cast off all restraint, and insisted that a vitamin C diet would prevent or cure cancer. This brought him into a harsher conflict with the medical establishment, which feared that sufferers from cancer would take to dosing themselves with vitamin C pills and refuse valid treatments. Pauling's wife was one who died of cancer, after he had declared that his vitamin C therapy would cure her and rejected medical intervention on her behalf. It was plausibly suggested in some quarters that the huge concentrations in her stomach of vitamin C, a strong reducing agent (p. 176), could well have generated reactive free radicals and hastened her death. Pauling's movement, which he called 'orthomolecular medicine', still thrives in the United States and elsewhere.

11

The New Millennium: Profits and the Higher Quackery

Deprivation, excess, and punishment

An Eastern European proverb has it that a man with food may have many problems, but a man without food has only one. Famines are still common in the modern world, whether through drought, erosion, or war, but in the countries of the prosperous North malnutrition is self-imposed. It has even, here and there, increased in recent decades. In the United States, in Britain, and to a varying extent in other countries of the North this is a consequence of the ubiquity of processed foods. In Britain it is estimated that some 70% or more of the food consumed is processed. The proportion is greatest by far in the diet of the poorest section of the population. In the United States the National Scurvy Institute has asserted that half a million Americans die each year from scurvy, presumably undiagnosed. This, of course, need not be taken too literally: the Institute is a private foundation which (incredible to relate) questions even now the link between scurvy and vitamin C. Yet there is evidence that scurvy does reveal itself, especially in the old and the very young, and among heavy smokers, who are thought to require a greater daily intake of the vitamin. If scurvy really is once again on the increase this can be due only to a rise in consumption of processed food at the expense of fresh fruit and vegetables. Potatoes, for example, which for centuries staved off malnutrition in many European countries, have made up a progressively smaller part of the average diet in recent decades, except in the form of crisps and other denatured products. The consequences are widespread. Studies in British prisons have revealed extensive malnutrition among the inmates, habituated to inadequate food since childhood, commonly aggravated, of course, by drugs. It has been reported that prisoners given vitamin supplements for a minimum of two weeks are less likely by a margin of 35% to re-offend.[1] Most startling perhaps are the results of recent analyses of stature in Europe and the USA.[2] Whereas a century ago white Americans were the tallest of all Caucasian populations, they are now merely the heaviest. Western Europeans, not excluding the citizens of what was formerly East Germany, are up to three inches or so taller than the Americans (recent immigrants excluded). While the inferiority of health care and other social provisions for the underprivileged may play a part in this inversion, it is hard to imagine that poor nutrition would not be a major cause.

One of the primary causes of the changes in eating habits around the world is the revolution in agriculture in the Western countries. Subsidies to farmers in the United States and the European Union (and before it the EEC) have led to massive over-production, and the accretion of 'food mountains', especially of cereals, meat, and butter. These have been unloaded below cost on overseas markets, to the great detriment of the farmers in agrarian countries in the Third World. Urbanization in the poorer countries has also brought about a gradual shift in dietary preferences towards those of the West – to processed, and fatty and sugar-rich foods. This is not to say that cultural differences in diet have vanished. Economics alone can scarcely account for the average consumption of some 300 g of meat per head per day in Argentina and Australia, compared with 10 g in India and Indonesia. Diets in Eastern Europe and notably in Russia tend still to be based to a great extent on meat.

In the prosperous West it has been the power of the supermarkets and ruthless promotion, aimed especially at the young, by Coca-Cola, McDonald's, Mars, and other food colossi that have driven up the consumption of fats and carbohydrates, particularly sugar. In Japan, for example, the incursion of these juggernauts doubled the fat consumption in the two decades after 1960. Parts of animal carcasses once thought unfit for humans to eat are now, thanks to new technologies, turned into pies, sausages, and hamburgers. This has brought down costs and made such confections increasingly appealing to the less well off. They are rendered palatable by the fruits of intensive research in the laboratories of the synthetic flavour, fragrance, and dyestuff industries. These foods are by no means cheap relative to the costs of the ingredients, and the profit margins are immense. It is estimated that the *average* American consumes three hamburgers with four loads of chips (french fries) each week. (Systematic deregulation by the Administration at the start of the millennium, aimed at increasing the industry's astronomical profits still further, has included such measures as re-classifying french fries as a natural, rather than a processed food.) McDonald's is also the largest outlet for Coca-Cola, and, according to Eric Schlosser (*Fast Food Nation*), bought in the concentrated syrup for dilution with soda water at (in 2001) $4.25 a gallon, while selling a medium Coca-Cola, with its nine-cents'-worth of syrup, for $1.29. Far more is spent on advertising than the cost of the food.

The proportion of fat, sugar, and salt in processed foods has risen steadily. Such products as jams, cakes, and ice-creams on the one hand, and pickles, sauces, and salad dressings on the other are in general now of a sweetness at which the average palate would have rebelled 50 years ago. The increased sugar intake has brought with it an epidemic of diabetes, in Britain the greatest single financial burden on the National Health Service, and it is spreading now to countries such as China, where it was formerly rare. In the United States the sugar industry is heavily subsidized and little is done to restrict the sugar content of foods. The bulk of the sugar consumed by most people is now, in fact, in processed foods. The sugar moguls, such as the executive of the US Sugar Corporation, are among the Republican Party's most generous donors. There seems little prospect, therefore, of much immediate action on obesity and health in the United States, and little more in Europe either. The fact, nevertheless, is that in the United States other calorific sweeteners have to some extent edged out sugar in recent years, notably high-fructose (fruit sugar) corn syrup, now preferred by Coca- and Pepsi-Cola (about which more shortly.) A report by the World Health Organization, urging action to combat obesity and its numerous attendant ills, was met by threats from the United States' regime of withdrawal of the country's contribution to the body's finances. Obesity has led to an increased consumption of artificial sweeteners, such as cyclamates and Aspartame (Sweet 'n Low). One can hardly doubt that these must be doing some good, but, as with all issues concerning nutrition, even that has been questioned: does the lower level of satiety that these substances engender compared to sugar result in increased craving after food and more weight gain? Perhaps, but more likely not. One may choose whose evidence to believe. It is not only sugar, but also the enormously increased salt consumption, deplored by medical authorities (if by no means unanimously, as will appear), that comes mainly (in Britain 80%) from processed foods.

The hidden ingredients

Adulteration of food, especially in the developed world, exceeds today anything that the public guardians of standards of purity in earlier times, men like Accum and Hassall (Chapter 7), could have imagined. For modern food technology relies on additives, numbered in the thousands. Most are aimed at making food more attractive, some at making it go further at less cost. A scientist in the employ of one of the largest food producers in America recalled a message from the management to the laboratory workers: the company's business is to sell air and water. Whipped cream blasted from an aerosol dispenser, for instance, is lighter and more fluffy than anything the classical recipes for *crème chantilly* can hope to achieve. It is also stabilized by added colloids,* and has the virtue of low calorie content. It has, on the other hand, almost no discernible taste. Cellulosic

* Colloids are particles or, as in foods, giant molecules (polymers), such as proteins or natural or synthetic polysaccharides, that will disperse in a solvent. In the present context, the solvent is water. Polysaccharides (which include starches and glycogen) are made up of long chains of chemically linked sugar molecules. They may be insoluble, like the woody cellulose of plant cell walls, or soluble (colloidal). Synthetic polymers used in the food industry are often derived from cellulose by chemical modification.

polymers give texture to mousses and ice-creams, and manifest themselves in the ghostly skeleton that remains standing in the dish when a block of a typical commercial ice-cream is allowed to melt. Of greater commercial importance are the polyphosphates – so far as is known, harmless – universally deployed to increase the bulk of meat and fish. The solution is injected into the meat (muscle) on the abattoir production line before rigor mortis has set in, and it is then most often frozen. The natural fluid in the meat is thereby acidified, which causes the muscle protein to absorb more water, and also produces a change in texture. The water gain can be very substantial, and the process is commended in advertisements to producers and butchers on the grounds that it allows them to sell water in the guise of meat.

The polyphosphates also function as stabilizers of the so-called 'meat emulsions', which are not true emulsions (fine dispersions of fats in water, like milk, or of water in fats, like butter), but watery suspensions of meat, connective tissue (cartilage, membranes, and so on), and fat particles. Most often these are produced by grinding or chopping whatever remains on stripped carcasses, or by generating a slurry of meat and other bodily residues with jets of high-pressure water. The ooze of fat particles and fibrous matter from connective tissue and muscle, all dispersed in water, is mixed with polyphosphates and 'fillers', such as cereals, starch, dried skimmed milk, eggs, or blood plasma, and can then be cooked and fashioned into sausages, chicken 'nuggets', or other processed delicacies. (Skin seems actually to be the only chicken-derived ingredient in some nuggets.) The least attractive parts of the animal, such as the lungs and stomach, may be extracted with alkali to yield soluble protein. This is then recovered in solid form by neutralizing with acid, and is often extruded through fine tubes, emerging as fibres suitable for spinning and incorporating into many types of meat product, with an admixture of a suitable filler. Fish for such confections as fish-fingers is treated in much the same way as meat, and textured vegetable protein also serves as the basis for many varieties of snacks especially. 'Reformed' ham, which can be sliced like the real article, is another triumph of colloid chemistry. New uses for the traditionally inedible elements of animals and plants are referred to as 'added value' and are perpetually sought after in a hugely competitive industry.

Another consideration is the texture of fatty processed foods, which relies often on the hardness of the fats that are added to them. Hardening of fats implies the elimination of the much more fluid unsaturated fats, leaving only the saturated variety (p. 85 f.n.). The main source is palm oil, now produced in vast quantities. When cooled it throws down a crystalline deposit, in more innocent days a waste product, of the saturated fraction. This 'tree lard', as it is sometimes called, or 'fractionated fat', is collected and used as a stiffener. These are the 'unhealthy fats' (about which more later) that often make up the bulk of the fat intake of those who rely on convenience foods. Saturated fats are widely distributed in today's edibles; they give resilience, for instance, to white bread and other products of the bakeries. Much the worst are the fats made up of *trans* fatty acids, which, because of their chemical structure (see Appendix), are the most solid of all. They are formed by the chemical process of hydrogenation* – addition of hydrogen – to the more fluid unsaturated kinds of fatty acids of vegetable oils. They were the basis of margarines and found wide use as stiffening agents, but now they have for the most part been supplanted by 'fractionated fats'.

The finely divided nature of the material that goes to make up processed foods renders them particularly susceptible to deterioration during storage, and so preservatives, especially antioxidants, are added to prevent fats from turning rancid. Such substances as anisole and butylated hydroxytoluene (BHT) are safe, at least in small amounts, although there has been much debate about possible carcinogenicity when they are absorbed in quantity. Sodium sulphite is another ubiquitous preservative the safety of which has been questioned. Humectants – substances that retain water – are also incorporated in most processed foods to maintain their moisture content. Such aliments as sauces, mousses, custards, and ice-creams, and especially baby foods, contain large amounts of emulsifiers, commonly modified starches. Also generally present are processing aids, most importantly anti-caking agents, especially sodium aluminium silicate, and lubricants. Consider, for illustration, the British Imitation Cream, which comprises refined vegetable oil, sugar, methyl ethyl cellulose, polyoxyethylene, sorbitan stearate, salt, sodium alginate, flavouring, colouring.

It is not unknown for a chemical reaction to occur between a pair of unrelated additives or an additive and a component of the food, to form a toxic compound. When a solvent, trichloroethylene, was used to extract fats from soya bean meal destined for cattle feed, the cattle developed an anaemia caused by destruction of the bone marrow. A toxin, it turned out, had been generated from the soya meal by a complex enzymic process, initiated by the solvent. Other examples are on the record. Enzymes (the natural proteins that bring about chemical reactions) are also often deliberately added to modify the consistency of ingredients, notably flours or starches, which can be turned into glues, pastes, or creams by enzymic treatments. A proteolytic enzyme – one that digests proteins – finds employment as a meat tenderizer. Papain, from the papaya plant, is the one generally preferred. One way in which it is used is by injection into the jugular vein of cattle immediately before slaughter. The bloodstream conveys the papain into the small blood vessels that permeate the muscle, but it remains latent because it is activated only by heat. The papain-impregnated meat (marketed originally as Protem meat), has to be roasted according to a quite precise schedule, for when the enzymic digestion has gone far enough to produce a tender

* So as not to offend anyone with a knowledge of organic chemistry, it ought to be stated that, strictly speaking, the process should be referred to as 'partial hydrogenation'. Complete hydrogenation would lead to saturated fatty acids, lacking the double bonds needed to form a *cis* or *trans* configuration. This is explained in the Appendix.

joint the heat at the roasting temperature destroys the papain. If the oven warms too slowly and the meat remains for too long at a temperature in the range of about 35–50°C the enzyme will convert the meat into a puddle of brown slush.

But it is when we consider flavouring and colouring substances that chemistry really comes to the fore. In 1986 Erik Millstone, surveying the products on the shelves of English supermarkets, found that Bird's Raspberry Flavour Trifle was totally innocent of raspberries. The giveaway here is the designation: if it were a Raspberry Flavour*ed* Trifle it would have had at least a passing acquaintance with a raspberry. It contained, at all events, the largest number – 22 – of additives in the various foods that Millstone surveyed.[3] The permissible food colourings are now everywhere tightly controlled, and the hideous inorganic and other poisons that did such mischief during the 19th century (Chapter 7) have no counterpart in today's gastrochemical banquet. Because of the intensity of modern dyes the quantities that are needed to give allure to something that would otherwise be an unappetizing grey are generally small. There have nevertheless been occasional alarms, and also concern about the rare individuals who develop an attachment to a particular product, which they then consume in excessive amounts as their staple meal.

In Britain no artificial colouring may be added to unprocessed foods, but a resourceful producer can sometimes find ways around this obstacle. So, for example, cows can be fed dyes to impart a more creamy colour to their milk and butter in the winter, when these are normally paler. Consumers in Britain also prefer brown to white eggs, and so the hens are given colorants in their feed. According to Erik Millstone, orange growers in some countries inject dye into the trees, which then finds its way into the rind of the fruit (imparting a pleasing tinge to under-ripe oranges) and thence into the cartons of juice. A particularly unpleasing instance of such practices is the addition of pigments to the feed of the farmed salmon that now make up so large a proportion of the total fish consumption of Europeans and Americans. These salmon, fatty (though with little of the omega-3 unsaturated fatty acids (p. 255) that so commend fish to nutritionists) and infested with sea lice, are reared on 'salmon chow', which consists of pulverized mackerel, herring, sardines, and other fish. Because they are denied the pink krill on which they feed in the wild, their flesh, in the absence of dyes, is a dispiriting grey. The salmon farmers therefore add a synthetic chemical, canthaxanthin, to the feed. This comes in various concentrations, allowing the farmer to select the depth of colour he thinks will best please his customers. Canthaxanthin has a questionable reputation: it has been used as a tanning pill and is suspected of causing retinal degeneration. Its use in tanning is now forbidden in Britain, though not yet in the United States or Canada. (There are, as we shall see, other grounds for unease about the farmed salmon.)

The chemical garden of flavourings is richer and more subtle than the dyes. Until about the middle of the last century some simple organic esters (p. 253) were added to boiled sweets, but it was only the 'flavour enhancers', common salt and monosodium glutamate (now added in awesome quantities to processed foods, and physiologically equivalent to salt), that were widely used. Now the numbers of permitted artificial flavourings is estimated as about 5000. In the United States, where the technology is most highly developed, disclosure of additives is not required, but Eric Schlosser in his remarkable investigation of the American food industry has sought out an example – a typical strawberry flavour cocktail, such as goes into a strawberry milkshake in a hamburger chain. He reproduces a depressing inventory of organic chemicals: it begins with amyl acetate, amyl butyrate, amyl valerate, anethol, anisyl formate, benzyl acetate, benzyl *iso*butyrate, butyric acid, cinnamyl butyrate, cognac essential oil, and diacetyl dipropyl ketone, and goes on to name 34 other substances. The quantities of all the ingredients of this chemical feast are, to be sure, very small and almost certainly harmless (or as the Food and Drug Administration (FDA) terms them, 'Generally Regarded as Safe'). In many cases the flavouring materials are found in nature, but most often they are based on something discovered in the laboratory. The taste of the french fries that drew so many of the American (and later the world's) population into McDonald's came from the beef fat in which they were cooked (giving them a higher content per ounce of saturated animal fat than the hamburgers they accompanied). But in 1990 a wave of cholesterol panic swept America and McDonald's were forced to resort to vegetable oil and chemistry. International Flavors and Fragrances of New Jersey were entrusted with the challenge and it did not take them long to find a solution. The smell and taste of beef fat was perfectly reproduced by synthetic chemicals added to the vegetable oil. Eric Schlosser confessed that when blindfolded and exposed to the vapour of the flavouring agent deposited on a piece of filter paper, he could swear that he was in close proximity to a portion of french fries. The amount of all flavouring agents ingested each year by the average consumer is hard to estimate; Erik Millstone puts the nationwide total for Britain in 1985 at 1000 tons. According to Felicity Lawrence[4] the annual cost of food additives worldwide is $20 billion, of which $1 billion goes on colourings. It is worth reiterating, though, that natural flavours of foods are just as much chemicals as are synthetic additives, and few of them do us any harm.

Undesirable, undesired

Today's food also contains unintended additives. Cured meats (notably bacon) and fish were traditionally, and still are, prepared by treating with salt and potassium nitrite or nitrate (saltpetre), or a mixture of the two. But today curing is a much more rapid process than of old because the meat is injected with the solution, rather than merely immersed in it. It is known that the nitrite–nitrate mixture gives rise in time to a class of chemicals called nitrosamines, which have been linked to stomach cancer. At least, though, the risk (not great) is known, and has been for many years. Of far greater concern are the trace substances that modern intensive agriculture has put in our food. Pesticides, weed-killers, and other chemicals such as the notorious PCBs (polychlorinated biphenyls) and dioxin, often find their way into the food-chain. Pesticide-resistant insects and fungi have become more prevalent, partly because of intensive farming methods that exclude crop rotation. Traces of unpleasant substances, such as dithiocarbamates and neurotoxic organophosphorus compounds, in fruits and vegetables are hard to eliminate, and commonly exceed the (admittedly rather arbitrary) permitted levels in produce in the shops. The best hope is that genetically modified crops will in time reverse these trends.

Farmed fish offer up their own soupçons of the unexpected. Farmed salmon ingest greater quantities of PCBs than fish in the wild. They get them from the oils concentrated in the salmon chow, for the fatty flesh of the fish from which it is prepared soaks up the oceans' chemical pollutants. The extent of the danger to consumers remains a matter of debate, especially when set against the benefits of the high content in the flesh of some farmed fish (though less so in salmon) of the healthy omega-3 fatty acids. The hormones and antibiotics found in fish or in the meat of factory-farmed cattle, sheep, pigs, and chickens are another matter.[5] Since 1989 the European Union (EU) has maintained a ban on beef imports from the United States and Canada, to the enormous displeasure of successive American administrations. Hormones accelerate growth and increase the weight of the adult animals – the factors that determine profits. Even in America the primal hormone distilboestrol was withdrawn in 1989 from use in animals – some years after it had become clear that it was carcinogenic, and that considerable concentrations passed into the meat. It was replaced by other sex hormones, especially synthetic progestins (related to the natural progesterone) and oestradiol, whereas in the countries of the EU no hormones are permitted. The recommended means of administering the hormones is by way of an implant in the head of the animal, but most producers simply inject them into the muscle. The quantities of hormone in the meat are in effect unregulated, and the average American consumes each year 100 kg – one-tenth of a ton – of meat. What the consequences might be of the doses of hormone swallowed year after year by steak-eaters, and especially by children and pregnant women, is by no means clear, but, depending obviously on quantity, these substances cause cancers of the breast and reproductive organs and feminization of the human male.* Protests by the US Department of Agriculture and the FDA that the concentrations of hormones in American beef are negligible have been rejected by European scientists on the grounds that no record of testing exists. (The World

* Some of the effects that extraneous hormones may exert can be gleaned from the ailments that beset East German athletes, given steroidal hormones in the guise of vitamins. It was discovered after the reunification of Germany that some 10,000 young people had been treated in this manner, including 14-year-old girl swimmers. Depression, heart failure, and tumours were among the afflictions uncovered in an investigation.

Trade Organization sided with the Americans in the dispute, but American endocrinologists have supported their European confrères.) A hormone, bovine growth hormone, or BGH (otherwise recombinant bovine somatotrophic hormone, rBST), is also administered to dairy cattle (again not in the EU) to accelerate maturation and increase milk yields,* and traces appear in the milk.

Cows treated in this manner additionally secrete other substances in their milk, notably a growth factor which, like the BGH, is a carcinogen. The FDA nevertheless allows the use of BGH to the despair of many American doctors. The carcinogens appear also to survive in the millions of hamburgers made from the meat of worn-out dairy cattle, supplemented by the PCBs and dioxin fed to the animals in fat recycled from carcasses.

A much greater abuse, the consequences of which are already upon us, is the profligate and reckless use of antibiotics in farming generally. The overwhelming proportion of antibiotics now produced is fed to livestock. They serve primarily as 'growth promoters', acting, it is thought, by killing the gut bacteria that participate in normal metabolism. Without the bacteria the animal feed is more efficiently assimilated. A secondary function is to limit the spread of diseases in the hideously crowded and insanitary conditions of factory farms. The antibiotics kill most *Salmonella*, *Campylobacter*, and other pathogens, but spare the resistant strains, which then grow and reproduce. The manure from the farms contaminates earth and ground water, and is spread through rivers into drinking water. Multiple Drug Resistance of bacteria is now a burgeoning danger in hospitals and claiming many lives.[6] The best hope of combating a lethal drug-resistant staphylococcal infection is the doctors' 'drug of last resort', vancomycin, which is reserved for use only *in extremis*. But vancomycin-resistant strains of bacteria have emerged, and the reason is surmised to be the use of a chemically related antibiotic, avoparcin, as a growth promoter in poultry factory farms. Vancomycin-resistant bacteria were discovered by Danish veterinarians in chickens that had been fed avoparcin. The Scandinavians at once banned its use, but other countries have followed suit only slowly or not at all. The decision of the EU was delayed by vociferous protests from Roche, the manufacturers of avoparcin, which predicted disaster for the agricultural industry and mass bankruptcies of farmers. Eventually a ban was implemented, nobody went bankrupt, and the incidence of vancomycin-resistant bacteria in poultry meat diminished. Other new antibiotics (such as the fluroquinolones, once the great hope of the medical

* The modern, hormone-primed, dairy cow is not a happy or a healthy animal. It produces about 20 litres of milk per day, whereas 2 litres would suffice to feed its calf. The cow is repeatedly milked until it runs dry and has to be inseminated once more. By then it will have delivered some 10,000 litres of milk (more than 2000 imperial gallons), contaminated to a greater or lesser degree with blood and pus. If, as often happens, the cow is afflicted with painful mastitis, it will be treated with massive doses of antibiotics, a measure of which will pass into the milk. Instead of living [illegible] years or so, as it would if left to itself, today's dairy cow lasts on average [illegible] years before it reaches the end of its useful life and is turned into hamburgers.

microbiologists) suffered the same fate as their predecessors. A report in the United States in 2001 described the isolation of Multiple Drug Resistant *Salmonella* from many samples of minced meat in supermarkets. American children now swallow some 50 antibiotics and hormones with their milk. And much of their, and our, meat and eggs are contaminated with resistant pathogens – *Salmonella*, *Campylobacter*, and dangerous strains of the gut bacterium *Escherichia coli*, which, if the food is not properly cooked, can cause distressing illness, even death. Another pernicious practice that has come to light is the addition of antiparasitic drugs to chickens, and also to farmed fish. Permitted levels seem often to be exceeded, and quantities of the drugs turn up not only in the meat but also in the eggs. The toxicities of these substances are still uncertain. In chicken-feed the antibiotics are supplemented with arsenic compounds that also inhibit bacterial growth and are certainly toxic, and there have been reports of arsenic in the drinking water. This effect is aggravated by the use of the chicken manure to fertilize fields of maize and soya beans, which are then used to feed more chickens, thereby concentrating the additives.

The disaster in Britain of mad-cow disease (bovine spongiform encephalopathy, or BSE) sent shock-waves through the cattle industry in other countries as well. It impelled the FDA to reconsider the practice of feeding animal waste to cattle. Until that time most livestock in the United States had been routinely fed a protein-rich diet of rendered carcasses of sheep and cattle, together with dead dogs and cats supplied by animal shelters, and treated chicken manure. Now only dead pigs, horses, and chickens (along with their waste products, rich in antibiotics and pathogenic bacteria) are permitted. Chickens, conversely, are fed the remains of cattle. The outbreak of BSE in Britain was caused apparently by a change in the temperature at which the animal wastes were treated to produce cattle-feed, but lesser outbreaks occurred in other countries. In Switzerland, for instance, cattle were allowed to feast on their own species, on dead pets, and on human placentas collected from hospitals. At least the organs in which the agent of BSE (prion) is concentrated – brain, spinal cord, blood, eyeballs – are now at last excluded from use in processed foods and animal feeds in the United States (despite strenuous objections from the industry), as well as in Europe. Blood, guts, and other parts, however, are still permitted. The link between BSE and similar diseases in sheep and the human fatal degenerative conditions, new-variant Creutzfeldt–Jakob disease (vCJD), is not yet established for certain, but the capacity of the BSE agent to pass from one species to another has been demonstrated. It is hard to imagine that we have seen the last of prion disease in humans.

Fruit and vegetables are not proof either against all hazards, although safe threshold levels of pesticides remain largely a matter of guesswork. Tests in 2004 have, though, disclosed the presence in some produce of pesticide levels far above the statutory limits, as well as traces of banned substances. Apples are commonly sprayed 16 times before they are harvested, with pesticide mixtures containing up to 36 chemicals.[7] This may all be harmless, though one cannot be sure, and 'organic' produce is not necessarily any more reassuring, for it is usually treated with the conventional pesticide spray of toxic copper sulphate, and nourished with manure, which probably contains pathogens and chemicals fed to the animals. Again, the best hope of greater safety lies in genetically modified strains.

Imposing controls on the food industry has been difficult for governments committed to exports and the free market. Making good the deficiencies brought about by the processing of foods is an altogether simpler matter. Micronutrients have been added to foods since early in the 20th century, starting with bread and also margarine, which was enriched with vitamins A and D (thus allowing the manufacturers to claim for it health-giving properties superior to those of butter). Today the practice is of wider importance. Calcium – essentially chalk – is still added to white bread to make up for this common deficiency, and combat osteoporosis. Vitamins, especially A, C, D, the B vitamins, and folic acid, which are often destroyed during processing, are added to bread, milk, and other products, and so also is iron. There have been some objections from the fringes, just as addition of fluoride to drinking water to inhibit tooth decay was opposed with extraordinary passion, but there can be little doubt that public health, and especially that of children, has greatly benefited (even if the virtues of milk itself have become a matter of debate (p. 102)).

Modern bugbears: health and happiness

Advances in medical science are news as never before, and the relentless emphasis on health and longevity has created an illusion that death has become more or less optional. But medical science has grown into a vast and boundlessly competitive undertaking. Publications are the coin in which researchers pay their debt to the bodies – public, charitable, or, too often of late, commercial – on which they depend for their academic survival. At times the coinage has been debased. The pressures to publish, or worse, too great an appetite for public attention, has led to the appearance of spurious or worthless factoids in the guise of validated truths. To these human failings must be added the intrinsic difficulty of designing valid trials of health effects in heterogeneous populations. The weaker the evidence the greater usually is the fervour with which a position is defended, and the denser the fog that envelops the debate. And it is of course always easier to expel the toothpaste from the tube than to get it back in. Propositions, then, that are in actuality conjectural or just wrong are often presented to outsiders as established facts. The links between fat-rich diets, cholesterol, and salt, and heart disease and strokes are among the most contentious of all.

Cholesterol has been a topic of dinner-table conversation in America and in at least some European countries for many years. It usually features as one of humanity's most sinister enemies. An American biochemist and authority on nutrition, David Kritchevsky, has expressed it thus: 'In America, we no longer fear God or the communists, but we fear cholesterol.'[8] Yet cholesterol is an important and ubiquitous constituent of the body's cells, and a participant in essential metabolic processes. It is transported back and forth in the blood in the form of complexes with proteins, and it is the variation in the amounts of these from one individual to the next that is the cause of all the agitation. The main cholesterol–protein complexes are of two major kinds, the high- and low-density lipoproteins (HDL and LDL). One of the few unquestioned facts is that people with high serum cholesterol concentrations, in particular when contained in LDL, are more likely to be hit by coronary heart disease. There is also evidence that all-*trans* unsaturated fatty acids (see Appendix), and perhaps also saturated fatty acids, in the diet increase the risk. This much has been common currency for more than 30 years. But we can now add something more, namely that, although the saturated fats and cholesterol are abundant in meat and dairy products, most of them, at least in the industrialized societies, and practically all of the all-*trans* fats, are consumed in processed and deep-fried foods. What is by no means clear, indeed contrary to the balance of the evidence, is that reduction in dietary saturated fats, or of fat generally, leads to any drop in serum cholesterol or greatly increases the life expectancy of healthy people.[8]

The debate has been governed more by emotion than by reason since its beginnings soon after the Second World War, when what appeared to be a steadily mounting epidemic of heart attacks was sweeping the United States. The famous Framingham Heart Study was undertaken in response, and reported its findings in 1961. It was not clear until much later that the supposed epidemic was a mirage, stemming from the reduction of deaths from other causes, especially infections, and the resulting increase in longevity of the population. There were thus many more people living into the age at which heart disease was to be expected. The epidemiology of the Framingham study was based on a comparison of diets in Japan and Crete, where the incidence of heart disease was low, with those in Finland and the United States, where it was high. Cholesterol was the word that broke for the first time on the American consciousness, for serum cholesterol levels were the big difference. The recommendation that virtuous Americans should reduce their intake of fat, and in particular of saturated fat, was endorsed by another body, the McGovern committee, but came in for no little criticism since the final link in the chain – that high blood cholesterol is the culprit, rather than reflecting a disposition to heart disease – was missing. The committee's suggestion was based in essence on the postulate that a reduced-fat diet might not do any good, but at least it could do no harm. A distinguished nutritionist from one of the country's great seats of learning, the Rockefeller Institute (now University) in New York, rejoined that this prescription amounted to a nutritional experiment on the entire American public. Scientific opinion now divided, and the beliefs of the two sides became increasingly entrenched.

More epidemiological analyses followed. In 1987 the Harvard Multiple

Sclerosis study appeared. The data it collected went beyond its remit, for they related life expectancy to diet quite generally. The outcome was that a low-fat diet would win a heavy smoker, a hypertensive, or some other poor insurance risk an extra year of life on average, while a healthy person would gain three days. A second study, undertaken on the initiative of the Surgeon General's Office, found that a drastic reduction in fat intake would prevent 42,000 deaths in the United States per year, but this corresponded to an average increase in life expectancy overall of only 3–4 months. A woman dying prematurely at 65 could expect to add two weeks to her life, assuming she had avoided saturated fats from the beginning. The scientists involved submitted a report to the *Journal of the American Medical Association* in 1991, which the Surgeon General's Office attempted to suppress by telling the editor that the study was 'deeply flawed' in its data analysis; but the independent expert reviewers to whom he sent the manuscript thought otherwise, and it was duly published. Soon thereafter, a Canadian study came to very similar conclusions.

None of this satisfied the cholesterol zealots, who made free with accusations that the authors had been nobbled by the food industry. Thereupon the National Institutes of Health, the largest centre of medical research in America, initiated yet more studies, comparing diet and heart disease in different locations around the country. Again no significant link was found. Another analysis, more comprehensive in that it considered other risk factors for heart disease, concluded that a high-fat diet actually *increased* longevity. A drug trial on one of the first statins (drugs that lower blood cholesterol) concluded that for men at highest risk, with cholesterol levels in the top 5% of the range, cholesterol reduction effected a modest gain in life expectancy. The reduction in the probability of a heart attack over seven years was 1.6–2%. This was sufficient to excite the media and provoke reports that cholesterol had indeed been proved a killer.

The Americans, terrorized by the intimations of premature demise, have reduced their average fat intake over the last 30 or so years by nearly 10%. This has occasioned a drop in average blood cholesterol, but no decline in heart disease. There has, to be sure, been some increase in longevity, but this can be accommodated by improvements in diagnosis and treatment of heart conditions once they appear. The evidence that a low-fat diet and reduced serum cholesterol does marginally improve the prospects of those most at risk of death from heart disease caused a boom in sales of the new generation of statins. These are now being liberally prescribed throughout the Western world. At the last count the market was worth $4 billion a year in the United States alone. It seems in fact to be well established now that the most recently developed statins do work, if not in the manner foreseen: what they appear to do, though nobody has yet explained how, is to cause the substance of the damaging arterial plaques to fragment and thereby restrict its accretion on the artery walls. But the distressing fact is inescapable, that in America obesity, the pre-eminent predictor of heart disease (not to say elevated blood pressure, diabetes, and kidney failure), rose sharply at about the time the warnings against the dangers of cholesterol-rich food took effect and the proportion of fats in Americans' diets began to fall. The same pattern is now emerging in Britain and some other countries. Social engineering is seldom simple.

Nor, it might be said, is epidemiology – the search after causes of medical conditions through the examination of social and environmental variation – for the design of trials, with elimination especially of extraneous factors, is notoriously treacherous. Over a period of about five years up to 1982 evidence was repeatedly brought forward in favour of a link between dietary fat and cancer. Nobody seemed to demur. And yet, to quote a review in the leading American journal *Science*, '15 years and hundreds of millions of research dollars later, reports by the World Cancer Research Fund and the American Institute for Cancer Research found no "convincing" or even "probable" link'.

Fats good and fats bad

Cholesterol in the serum of the blood is, as we have seen, carried by lipoproteins – proteins complexed with lipids (fats). They are, as diet-conscious citizens all know, of three kinds: the low-density lipoproteins (LDL), the very-low-density lipoproteins (VLDL), and the high-density lipoproteins (HDL). The VLDL group, which form a small proportion of the total, represent an intermediate stage in the formation of LDL. (This and the other terms in this section are explained in the Appendix.) About 70% of the serum cholesterol is normally carried in the form of its esters in LDL, and it is from LDL that the material of the arterial plaques is deposited. The LDL is therefore commonly known as the 'bad cholesterol'. The rest of the cholesterol is in HDL, which is formed in the circulation from fatty precursors issuing from the liver and intestine. It is generally accepted that HDL has a protective action on the blood vessel walls. This therefore represents the 'good cholesterol'.

Fatty acids, both free and in the chemically bound form in which they are stored and released (the triglycerides (p. 253)), are transported, like cholesterol, in the lipoproteins in the blood, and the type of fatty acid present is related to the concentration of LDL. Saturated fatty acids, which preponderate in meat and dairy foods and most processed foods, are associated with high levels of LDL, as also are the *trans* fatty acids, whereas monounsaturated fatty acids (abundant in seed oils, such as olive oil) and polyunsaturated fatty acids (in some other vegetable oils and in fish) are linked to a reduction of LDL in the blood.

From all this it would appear that for a healthier and longer life one should adhere to a diet low in cholesterol and the wrong kinds of fatty acids, avoiding therefore meat, butter, eggs, milk, and cheese, but especially processed foods. But there is good news too. In the first place, carbohydrates and proteins also have an effect on cholesterol and fatty acids levels in the blood. Furthermore, and more interestingly, saturated fats, while certainly increasing LDL, also increase the concentration of HDL, so their beneficial effect may negate or even outweigh the deleterious. It has also been found that stearic acid, a saturated fatty acid and the principal fatty component of chocolate, elevates the HDL level, while barely affecting that of LDL. It is only the *trans* fatty acids that, on current evidence, are best avoided, for they both increase the amount of LDL and reduce that of HDL. On the other hand, the HDL of people who already have cardiovascular disease or diabetes has been found to be chemically modified in such a way as actually to increase the formation of fatty deposits in the arteries.

The article in the journal *Science*, cited above,[8] concludes with an analysis of the likely dietary effects of consuming an average American steak, which consists typically of equal amounts of protein and fat. After grilling, 50% of the fat is monounsaturated and 90% of that is oleic acid (the fatty acid of olive oil, therefore healthy), so 45% of the fat is saturated. But one-third of that is stearic acid, apparently health-giving. The remaining 5% is polyunsaturated, and even better. Seventy per cent, then, of the fat should lower the 'bad' serum cholesterol, and as for the rest, it will raise HDL as much as it does LDL. A comparable quantity of carbohydrate-rich food (pasta or potatoes, for instance) would be expected to cause a greater increase in LDL.*

And finally it is worth noting that studies in Japan have concluded that low cholesterol levels are bad for men, being strongly correlated with strokes. Measures to raise the blood cholesterol of Japanese men have been proposed. Even in the United States there is now some evidence that, while very high blood cholesterol levels (a hereditary condition in a small percentage of the population) are linked to premature death, the longevity of men (though not, it seems, women) with low blood cholesterol is also reduced. The debate will no doubt continue, and the participants will belabour each other with whichever item of evidence feeds their preconception. That it is prudent to avoid excess of any type of food seems now to be the clearest message.

Below the salt

The hopelessly obfuscated question of salt intake encapsulates perfectly the difficulties that beset the social and epidemiological investigation of diet. From official pronouncements and numberless dramatic effusions in newspapers you might conclude that the large amounts of salt considered necessary to make manufactured foods palatable are the greatest present threat to our health and survival. The price of continuing indulgence will be an epidemic of strokes. But what is the evidence? Elevated blood pressure (hypertension) is undoubtedly a

* But the last word has not been spoken, and few generalizations in this area seem to survive for long intact. There is evidence that people who already have heart disease respond differently from healthy subjects to the [illegible] of LDL and HDL, as also do men from women, and women who take hormone supplements from those who do not. So the picture is still blurred.

cause of strokes, of coronary and congestive heart disease, and of kidney failure. The evidence linking hypertension to salt intake in human subjects, though, does not bear close scrutiny, except in the case of the small minority of genetically predisposed individuals. A former director of the Center for Food Safety and Applied Nutrition at the FDA opined that the salt debate was the 'number one perfect example of why science is a destabilizing force in public policy'. At the outset, he explained, the evidence for a link between salt intake and hypertension was weak, though tending towards positive. With the passage of time the quality of the accumulated data has greatly improved and seems now no longer to support the case for reducing salt intake.[9] But erosion of the evidence has not put a stop to the argument. An entrenched section of the medical profession remains obdurate. The debate revolves around the design and analysis of epidemiological studies. The diet on which the bushmen of the rain forests survive contains very little salt, and hypertension among them is rare or unknown. This has been taken as evidence for a causal link between the two, but what of other differences between the lifestyles, let alone genetics, of bushmen and Wall Street traders, say? They differ in the quantities of fruit and meat that they eat, the alcohol that they drink, the amount of exercise they take, their average age, the occupational stresses they have to endure, their hereditary tendency to obesity, and endless other factors. The matching of samples, then, is critical. As one of the researchers into the salt question has put it, if everyone in our population smoked a packet of cigarettes every day and those of another smoked none, you would conclude beyond a doubt that lung cancer is a genetic disease. There is a further limitation to the epidemiological approach: if foreswearing salt caused an average drop in blood pressure of 1 mm of mercury (out of say 150), this would be below anything a statistical survey could detect, and it might diminish the likelihood of a stroke by perhaps one-hundredth of a per cent. This is far less than would matter to us as individuals, but worldwide it would save many lives. Animal studies offer a more direct approach to an understanding of diseases, but those of course also have their limits, especially when small effects, like that of salt, are involved. An American researcher announced that in a susceptible strain of rats a high salt intake caused hypertension; but the level at which this effect revealed itself was equivalent to more than one pound of salt per day for a man.

In the end, then, it has all come down to epidemiology, and there have been many studies and meta-analyses (analyses of the combined results of all the relevant and properly designed single studies). The debate has now taken on a theological character: there are believers and disbelievers, and passions have often supplanted reason. Yet, there seems little doubt that the sceptics have won the argument. They include the majority of the most respected epidemiologists and statisticians, and they have enumerated the many flaws in design – in matching groups of subjects with high and low salt intakes, while disregarding other differences in diet – and in statistical analysis of the results, on which the inference of a positive (and causative) relation between salt and hypertension rests. A survey of 56 clinical trials in which groups of healthy people were deprived of all salt beyond that in the natural foods that they ate, in comparison with a control group fed a normal salty diet, is shown below (from reference 9). The bars show the statistical confidence limits. From this meta-analysis one can only infer that there is no evidence of any effect of salt intake on hypertension. But the controversy still rages notwithstanding, and such questions as whether a lifetime's high-salt diet leads to earlier onset of hypertension in the later years are hotly debated. Uncertainty surrounds in particular people with rather rare forms of hereditary hypertension, linked especially to mutations in the proteins that convey sodium across the cell membranes (commoner in black and Asian than in Caucasian populations).[10]

It seems that facts supporting a general link between salt intake and hypertension are, at best, elusive. But there *is* compelling evidence that diets rich in synthetic and processed foods and deficient in fresh fruit and vegetables, and in calcium and other minerals, do promote hypertension, and so of course does obesity. Certainly hypertension is prevalent in the developed world, and diet is taken to be a determining factor. There are grounds for believing that insufficient calcium and potassium may contribute.

Fear of fibres

Dietary fibre, also known as roughage, is the part of vegetables and fruits that passes through the digestive tract unchanged. It comprises the woody polymers, cellulose, hemicellulose, and lignin, and also pectin. The roughage craze began with Kellogg and his coevals (Chapter 10) and took wing in 1932 with the rise of the New Health movement, but it received an enormous boost in the 1950s from the utterances of Dr Dennis Burkitt, which were eagerly taken up by the press. Burkitt (1911–93) was an Anglo-Irish missionary doctor, working in Africa, who gave his name to a virus-induced cancer, Burkitt's lymphoma. He was struck by the rarity of bowel (colorectal) cancers among the rural Africans, common as they were among the whites, and he developed the theory that infrequent evacuation was the cause of the disease. It was in essence a resurrection of the eccentric views of Sir William Arbuthnot Lane more than a half century before (p. 194 f.n.). The African diet contained a high proportion of fibre, and this, Burkitt knew, had a laxative action. He devotedly weighed and studied African and European faeces and formulated the theory that the white man's waste fermented longer in the gut and gave rise to carcinogenic products. Burkitt's conclusion stimulated several studies, which appeared to support his beliefs. But epidemiology of this kind is, as we have seen, a trap for the biased seeker after revelation, and Burkitt was naïve; he did not consider all the other differences in diet, lifestyle, and genetics between his Africans and the Caucasians.

Burkitt's dietary prescriptions were of course espoused by the manufacturers of breakfast cereals, and a new cult of dietary fibre consumption was launched, with bran the supreme exemplar. But there was never any evidence of a link between constipation and cancer, nor yet that fermentation of foods in the gut gives rise to carcinogens. Later studies arrived, for the most part, at a contrary conclusion: fibre intake either had no effect on cancer incidence, or tended perhaps even to increase the risk. As with salt, agreement is not unanimous, but the best evidence is that fibre does nothing to prevent bowel cancer. Moreover, it is established that fibre has a high avidity for calcium, and so exacerbates the risks of conditions that result from calcium deficiency, particularly osteoporosis. What, on the other hand, it does achieve is a reduction in serum cholesterol, and that perhaps is to the good. One takes one's choice of life's hazards.

Sugar, excess, and its nemesis

Carbohydrates, which come mainly in the form of sugars and starches, are broken down by metabolic reactions to the simple sugar glucose. Glucose is the essential nutriment of the brain and of tissue and blood cells generally. Its level in the blood has therefore to be constantly maintained. Glucose enters the cells according to need, and the excess is turned into a polymer (p. 213 f.n.), glycogen, each molecule of which consists of perhaps 100,000 glucose units, chemically linked together. The glycogen is stored in the liver and muscles and acts as a reservoir from which glucose can be rapidly mobilized, as required. This is the preferred mechanism for generating muscular energy, being much faster than the degradation of fats (which exist in muscle, and not only in the adipose tissue – body fat).

The agent that controls the entry of glucose into cells is the hormone insulin, a small protein. Insulin is formed in specialized cells, clustered together in the so-called islets of Langerhans in the pancreas, and a rise in blood glucose is the signal for its release into the bloodstream. On meeting its target cells it causes channels to open and admit the glucose, thereby again diminishing its concentration in the blood. Insulin production thereupon ceases, to complete the feedback loop. Insulin has the additional function of shutting down lipolysis – the breakdown of fats – after a meal. Therefore when carbohydrates are being digested to produce more glucose the attendant rise in insulin concentration ensures that lipolysis is suppressed. This affords a justification for vigorous exercise, accompanied of course by the additional sensory mortification of fasting, for when glucose is being consumed the insulin level remains low and fats are mobilized and 'burned' for energy.

Diabetes is the failure of the regulatory mechanism. It can arise in two ways. In Type 1 diabetes the pancreatic cells fail to produce insulin. This condition is treated with insulin injections. Type 2 diabetes is both more common and more complex, for it is caused by the failure of cells in the tissues to respond to insulin. In the relatively rare, purely hereditary form, the insulin receptors – the molecules on the cell surface to which the insulin must attach itself before the channels that admit glucose can open – are missing or defective, but more often Type 2 diabetes is linked to diet. The consequences of untreated or refractory diabetes are well known. Blood glucose runs out of control, and at high concentrations it inflicts damage on nerve endings, which are in time destroyed. This in turn interrupts blood flow to the extremities, and numbness and eventually gangrene can supervene, as well as leakage from the blood vessels in the retina at the back of the eye, ending often in blindness. In an attempt to subdue the high glucose concentration in the blood, the pancreas in Type 2 diabetes goes on secreting insulin, and the persistent high concentration (hyperinsulinaemia) plays havoc of its own in the kidneys, the arteries, and elsewhere.

The way in which excess and the modern Western style of eating cause diabetes is, in outline at least, well enough understood. It is linked to a phenomenon, previously unknown, that has asserted itself in recent years.[11] In earlier times it was the common custom in most societies to sit down to meals three times a day. In between, nothing or little would pass the lips. This style of sustenance has all too often now given way to frequent nibbling of packaged snacks and especially the consumption of sugary drinks. The pattern is commonly established in childhood. In the United States Coca-Cola and Pepsi-Cola machines are even installed in many schools, in return for a substantial bribe paid directly to the school.[12] The result is that the normal sequence of metabolic events, in particular a sharp rise in blood insulin after a meal, followed immediately by a decline (the so-called insulin spike), breaks down. Instead the insulin is maintained at a high level throughout the day, and metabolic mayhem ensues. Animal studies indicate, and observations in obese human subjects also imply, that under constant stimulation the insulin receptors on muscles and fat cells progressively atrophy. So also does an insulin-activated switch on the liver, which normally limits the secretion of fats, in the form of triglycerides (the state in which they are mostly ingested and stored). The triglycerides pumped out in mass by the liver assist in the continuing annihilation of receptors. The loss of insulin receptors thus leads to *insulin resistance*, culminating in Type 2 diabetes. Fats deposit in the muscle, and the disordered system tries to derive energy from proteins, so that the substance of the muscle itself is degraded.

But there is worse to follow. The adipose (fat) cells in the fat deposits, assailed by glucose and triglycerides, become refractory to insulin like everything else. Part of the excess triglycerides, packaged with cholesterol in VLDL and LDL, deposit plaques in the arteries. Another part surges into the fat cells (the normal triglyceride storage depots), but when overloaded like this the cells react by breaking down triglycerides into free fatty acids (see Appendix) and expelling them into the bloodstream. The free fatty acids are bad news, for in high concentration they are toxic and cause widespread destruction, most notably to the insulin-secreting cells in the pancreas. Insulin production eventually shuts down altogether, and now there is the full weight of Type 1 insulin-dependent diabetes. This scenario has been called Syndrome X by one of the leading researchers in the field,[13] and also appears in reports as 'metabolic syndrome' and 'insulin resistance syndrome'.

Type 2 diabetes is inextricably linked to obesity, with its attendant evils, for it is in the fat cells that the worst of the mischief brews up. Obesity, even without diabetes, carries a hugely increased risk of high blood pressure, heart disease, hormonal and joint problems, and susceptibility to some cancers, along with a strongly increased tendency to atherosclerosis, gall-stones from the excess cholesterol, and fatty degeneration of the liver. And a large measure of all the ills can be put down, as we have seen, to the huge quantities of sugar in soft drinks, processed foods, and snacks. Erik Millstone, writing in 1986, quotes the response of a British sales representative to the accusation that the amounts of sugar in almost all comestibles had been rising alarmingly: 'Yes, but you can't get the buggers to eat more than two pounds a week.'[14] Since then the industry has found new ways.

How bad became worse

In 1971 Japanese chemists made a discovery that was to enrich the food industry still more. They developed an enzymic method to prepare from maize (which was then being produced in greater quantity than the market could absorb) a syrup rich in fruit sugar, fructose. It is known as 'high fructose corn syrup', or HFCS. Fructose bears a close resemblance to glucose in its chemical structure, but is about six times as sweet. HFCS, moreover, is extremely cheap, and was therefore eagerly seized on by the industry and especially the soft-drink manufacturers. It became the sweetening agent in the colas, and so reduced the cost that the Coca-Cola company was able to advertise an increase in the size of their cans at the same price (while presumably still increasing profit margins).

Now to the less obvious consequences: despite their close chemical similarity, the human body discriminates between glucose and fructose, and responds to them in quite different ways.[14] Small quantities of fructose, such as occur in fruits, can be perfectly well tolerated, and furthermore, ordinary sugar from cane or beet, sucrose, is made up of a glucose and a fructose unit, chemically coupled together. So when sugar is broken down in the digestive system it gives rise to both constituents. (This is accomplished by the honey-bee, and accounts for the sweetness of honey, which contains a mixture of the two sugars, also known as invert sugar). But the trouble begins when the amount of fructose taken in, whether as such or derived from sucrose, becomes too large. For fructose heads directly to the liver,[15] where it enters the metabolic pathway by which lipids are formed (see Appendix). A large increase in the production of triglycerides then results. In animal studies this was found to develop into a chronic state. It inevitably leads in turn to a rise in VLDL and therefore of LDL and of blood cholesterol. Some fructose also remains in the blood, where it can cause the same mischief that is wrought in diabetes by too much glucose. Frequent fructose intake additionally induces seemingly irreversible disturbances in glucose metabolism, and, not only that, it also fails to stimulate the postprandial secretion of insulin and of the hunger-controlling hormone, leptin (see below) – all effects that glucose normally encompasses. The unchecked LDL increase is greatest immediately after a meal, when its action on the arteries is maximal. The incursion of HFCS into the modern diet must be seen, then, as an unmitigated catastrophe.

The genes bite back

In the last three or four decades Type 2 diabetes has reached truly epidemic levels in some populations, mainly of indigenous peoples. A remarkable example is the case of the Pima Indians of Arizona, half of whom are now diabetic, and nearly all those diabetics are obese. It can hardly be doubted that this is a consequence of changes in diet over the years, and indeed the Pimas' preferences have shifted from the grains and proteins of old to foods rich in fats and sugar. A century ago the Pima diet contained about 15% fat; the fat content now is 40%. But this cannot explain why the consequences have been so much more severe than in the bulk of the American population, which enjoys a similar diet. An explanation offered in 1962 by an American geneticist, James V. Neel, is now widely favoured. His theory posits that in communities inured by their environment to a hard and uncertain life, for whom a bad harvest could spell disaster, a process of evolutionary selection must operate. The genetically advantaged individuals would be those with the highest capacity to store fats and a limited response to insulin. They would accumulate fat during a time of plenty, and thus be more likely to survive a period of famine by mobilizing fat reserves for energy, and safeguarding the glucose supply to the brain. Neel termed this propensity the 'thrifty genotype'.[16] When the Pima Indians' lifestyle changed – when they ceased to rely on farming and hunting for their food and adopted the diet of the urban American population – the thrifty gene became a liability. The calorie-rich diet now overtaxed the capacity of the pancreas to produce insulin and the deposition of body fat set off the cycle that ends in obesity and diabetes. The genetically identical population of Pima Indians across the border in Mexico, who have not been granted the benisons of the American lifestyle, display no abnormal incidence of obesity or diabetes. In environments in which a state of relative abundance was more or less guaranteed, as in much of Western Europe, the thrifty gene would have been an evolutionary disadvantage, or at least of no benefit, from the beginning, and would not have emerged as the predominant genotype. A final consideration that also enters into the health of a population is prenatal influence on the metabolism of offspring. Malnutrition in early fetal life can, it may be recalled (Chapter 8), predispose to obesity and diabetes, though of course this can hardly explain the spread of obesity in the prosperous Western world.

Eat and grow thin

Two-thirds of Americans are now said to be overweight, and about a quarter clinically, life-threateningly, often grotesquely obese. Britain still lags some way behind with 21% of overweight adults (up by a third on a decade ago), but well ahead of other European countries; around 30,000 deaths a year in Britain are linked to obesity. In some vanished societies corpulence was admired as a mark of prosperity and success. A king of Rwanda in the 19th century found fat women so alluring that he forced his many wives to drink milk all day from a gourd until they grew so gross that they could only, according to the account of an English explorer, grovel on the floor like seals. In the United States the Fat Men's Club of Connecticut thrived from 1866 until fashions changed in 1903, by when an obesity syndrome, dignified by the designation *adiposis dolorosa*, had crept into the textbooks. A kind of fat liberation movement seems now again to be re-emerging in America, driven by women who have renounced the torments of slimming and resolved instead to celebrate gastronomic freedom and luxuriant flesh. But for most fat people their condition is a constant mortification, and carries always of course the threat, perpetually held up to them, of a premature demise.

How then to reverse such misfortune – but painlessly? Eating habits and fat deposition are established in early childhood, and many American infants are almost weaned on colas and other sugary drinks. Losing weight by reducing food intake is a slow and harrowing process, and the search for tolerable regimes or, best of all (setting aside corsets, 'obesity belts', or merely jackets with vertical stripes), slimming pills, has gone on for many years. In 1907 the *British Medical Journal* undertook an examination of 'obesity cures' as part of a wider investigation of 'nostrums' peddled at the time. Its report began with the following observation:

> In general, claims that are made for these articles are less extravagant than in many other cases, and a reason for this is not far to seek; it is important that the consumer of the medicine shall be encouraged to persist in its use for a considerable time, and any statement as to a rapid cure which would very soon be found to be at variance with the facts would probably only lead to the discontinuance of the medicine, and therefore defeat of the maker's object. Nevertheless, the emphatic and confident statements, backed by testimonials, which are so important a weapon to the quack-medicine maker, are by no means abandoned.[17]

There follow descriptions of the analyses performed by Dr Robert Hutchins. He found, he reported, extracts of *Fucus vesiculosus*, the plant known as bladder-wrack, which was reputed in folk medicine to be good for slimming, but most often the essential ingredient was nothing more than citric acid, the acid of citrus fruits, for, Hutchins points out, sucking a lemon was commonly held to cause weight loss. So Antipon, for example, was a solution of citric acid, coloured red with cochineal in alcohol (cost of bottle two shillings and six pence, cost of ingredients 1½ pence). Russell's Anti-Corpulent Preparation came in a plain bottle with an unlabelled package, an explanatory pamphlet sent separately by mail on account of 'many suggestions received, principally from ladies'; so 'The servants and others attached to the household may therefore be safely entrusted to open the box; inquisitiveness, if present, will not be rewarded.' The liquid was, as before, a coloured citric acid solution. Other products contained large amounts of sugar, which could hardly have assisted the slimming process, and in addition there were ointments, made up of such ingredients as beeswax, lard, and ox-bile, to be rubbed in 'where needed'.

These impostures were at least harmless, and indeed, when Dr Hutchins followed up some of the many testimonials he found most of them to be genuine, a remarkable affirmation of the powers of suggestion. Unfortunately, the more effective shortcuts to weight loss have been a great deal less benign. At times in the past pharmacists have sold, overtly or surreptitiously, tapeworm eggs as slimming-aids. Diuretics, laxatives, and emetics were frequently offered. In the USA Lucy Kimball's Powder, which contained sodium carbonate (washing soda), magnesium sulphate, commonly known as Epsom salts, and soap, a formula indicative of a powerful combined emetic and laxative action, was probably the most famous. The pokeberry, a toxic seed with emetic and laxative properties, also often featured in weight-loss pills. All these did of course induce a transient loss of weight, mainly as water, though at no little cost to health. Around the beginning of the 20th century much more dangerous concoctions appeared in the pharmacies. Some contained arsenic, some strychnine, others caffeine on the mistaken supposition that it would accelerate the metabolism and promote rapid oxidation of fat. Probably the most lethal of them all was dinitrophenol. This is an inhibitor of the synthesis of adenosine triphosphate (ATP), the fuel of muscles (see Appendix), and provokes rapid consumption of glucose to make up the energy deficit; it is also a notorious industrial toxin. It was incorporated into several proprietary products, and caused illness and, it is surmised, many deaths.

An extract of seaweed, which contains iodine, an essential element in thyroid function (p. 186), in the form of potassium iodide, enjoyed a brief vogue as Allan's Anti-Fat. ('Through the study of physiological chemistry', the advertising read, ' a *specific* has at length been discovered' and named after its discoverer, the eponymous Allan.) This would have been as harmless as it was ineffective, but it co-existed with more scientific, and at the same time far more damaging formulations that contained the thyroid hormone thyroxine. Initially these preparations were just extracts of thyroid glands collected from abattoirs. Overstimulation of the thyroid gland does increase the metabolic rate and was supposed to assist the mobilization and oxidation of fats. Thyroxine was widely used until the second half of the 20th century, when the side-effects, especially disturbances of the heartbeat, were found to be too alarming. Attempts to counteract this effect with arsenic, digitalis, or strychnine merely compounded the troubles. Another product still widely used in the United States, at least, for weight loss is ephedra, or Ma Huang, a Chinese plant, and the source of ephedrine, a close analogue of adrenaline (epinephrine). It intervenes in carbohydrate metabolism, but more than this, it accelerates heart rate and output and increases blood pressure. It can occasion a series of unpleasant and dangerous side-effects, and has been known to cause death, especially when taken as a stimulant by athletes.[18]

Amphetamines, which became available in 1938, were at one time much favoured appetite suppressants, and to offset their stimulant effect (nervous jitters) narcotics would be added. It eventually transpired that the efficacy of amphetamines for slimming was transitory, and moreover that they were addictive. An amphetamine-related compound that, despite the unease of several experts, is quite widely used is sibutramine, marketed as Meridia, or in Europe, Reductil. Its site of action is the brain, and there is concern about possible long-term side-effects; there have been reports of elevated blood pressure[19] and possible impairment of memory.[20]

Other preparations purporting to repress appetite or alter the digestive processes ('starch blockers' for instance, which were meant to impede the action of starch-digesting enzymes, but were themselves digested in the stomach before they could get to work) have come and gone, often withdrawn from the market as ineffective or dangerous. An inhibitor of pancreatic lipase, an enzyme that breaks down ingested fats, has enjoyed a little more success: this is orlistat (trade name, Xenical). It is still sometimes prescribed in cases of gross obesity, although the weight loss of patients in clinical trials was fairly modest and in general transitory. It reduces the absorption of the fat-soluble vitamins, which have to be given as a supplement. Moreover, since the drug allows (like Olestra, which we will come to) a sizeable proportion of ingested fat to pass untroubled through the alimentary tract, this can emerge as an oily discharge. Gas also forms in the gut, and when the need to evacuate comes upon the victim the urgency may be uncontrollable. The usual consequence is almost certainly a cautious reluctance to take meals at all, and in this sense the drugs may well work.[21] Perhaps the most notorious and expensive mistake in the melancholy history of slimming aids was a mixture of the sedative drugs, fenfluramine and phentermine, introduced in 1966 as Fen-phen for treatment of obesity. It was an instant success, prescribed in the United States on a vast scale. But all too soon, in fact early the next year, side-effects began to show themselves. The culprit was the fenfluramine, and equally one of its two constituents, called dexfenfluramine, separately marketed as Redux. These were the cause of a previously rare, incurable, and often fatal malfunction of the heart valves. No less than a third of patients on the drug were affected, and it was hastily withdrawn. A class action against the pharmaceutical company responsible for Fen-phen, Wyeth-Ayerst Laboratories, followed and resulted in a $3.75 billion settlement.[22]

Yet hope springs eternal, and there are today some 100,000 websites advertising slimming medicines (many now classed as foods, and therefore available in the United States, and hence by post elsewhere, without prescription).* In the US, since deregulation in 1994, there is no requirement for manufacturers to divulge the contents of their pills, and it may safely be assumed that most have no effect at all beyond giving their purchasers the reassurance that they can now eat anything they like. There is, to be sure, legislation to discourage mendacious advertising, and so a branch of the advertising industry now flourishes, dedicated to a form of psycholinguistics, the invention of locutions that will deceive the consumer without at the same time transgressing the law. Rimonabant (see below) is on the horizon and may be what everyone is looking for, but as matters stand, the respectable medical profession at least is unanimous that the only known safe way to maintain normal weight or reduce the excess is still by diet. But here the unanimity ends.

A sad fact is that it is the cheapest processed foods that are also the unhealthiest because, thanks to the relentless advance of food technology, production costs have gone down and down. It is consequently the poorest people in the prosperous countries who are the industry's chief prey. So in the United States one-quarter of the lowest-income group are clinically obese, compared to one-sixth of the rest. It is hard to imagine how this trend could be reversed.

The satiety protein

Much more interesting than all the drugs was the discovery a few years ago of a new protein hormone. Leptin was hailed as a natural means of combating obesity. It was brought to light through a genetic analysis of an obese strain of mice, and was isolated by researchers at the Rockefeller University in New York. Leptin is secreted mainly by adipocytes (fat cells) and is the natural regulator of body weight. The amount of leptin circulating in the blood is greater in corpulent than in thin people and its level drops after a period without food. Obese mouse strains have also been bred that are deficient in the leptin receptors, to which the hormone binds in order to exert its effect. A similar, and fortunately extremely rare defect has been recognized in humans, and reveals itself in early childhood. The afflicted children, like others who have the receptors but fail to produce leptin, are perpetually hungry, suffer torments if they are prevented from eating, and so become rapidly and life-threateningly obese. There is no known cure for a lack of leptin receptors, but leptin deficiency seems now to be treatable with leptin.[23]

The function of leptin is apparently to register when enough fat has been produced in the body after a meal. It does this by attaching to a leptin receptor in the hypothalamus, a structure in the brain that governs a variety of bodily functions. The receptor then sets in train the sequence of reactions that leads to

* Americans are now spending some $18 billion each year on dietary supplements of one kind or another, up from $4 billion during the previous decade.

breakdown of the fat (see Appendix), while also suppressing hunger. When the Rockefeller researchers injected leptin into genetically obese mice the animals rapidly lost weight, while normal mice were unaffected. When this result was announced in the press the Rockefeller University switchboard was blocked by calls from distraught dieters, and the pharmaceutical companies were not far behind an eager public. The rights to develop leptin were sold to one bidder for $20 million. When leptin treatment was tried out on the grossly obese leptin-deficient children there was a rapid and dramatic response. But the proportion of obese people whose bodies make no leptin is vanishingly small, and for the rest, who have quite enough leptin to saturate the receptors, injecting more of the hormone is without effect. The hope that leptin will be the answer to weight control at this moment seems remote.[24]

But science has not stood still since the discovery of leptin. The control of hunger has turned out to be a highly complex mechanism. Receptors for leptin have been identified, and in rare instances are subject to mutations that make them unresponsive to leptin. There are receptor molecules in the hypothalamus, mainly in a small area called the arcuate nucleus, for other hormones besides leptin, all of which send out hunger or satiety messages. The production of leptin is itself controlled by the *melanocortin* system in the hypothalamus, a complex feedback loop linked to other functions, including the control of hair and skin pigmentation. Failure of one or other step in the melanocortin cycle is thought to be responsible for 5% of cases of gross obesity in children. There is even evidence that insulin has a secondary effect in regulation of appetite. Another player in this intricate network is a peptide (like a very small protein) called ghrelin. There is indeed a rare condition associated with gross obesity and at the same time over-production of ghrelin. Another peptide, termed PYY, the concentration of which rises when one eats, exerts a contrary effect: in animals it was reported to reduce the inclination to eat. These compounds and their receptors present possible targets for slimming therapies, all being explored by pharmaceutical companies. But interfering with signalling mechanisms in the brain is a hazardous tactic, and the débâcle of fenfluramine is a spectre that haunts the industry.

Most recently there has been enthusiasm verging on euphoria about a drug of a new kind, called Rimonabant.[25] This emerged from the well-known circumstance that cannabis users experience hunger when they surface from their drug-induced stupor. It was found that a circuit in the brain responds in this manner to the cannabinoid class of chemicals generally. Substances that block the cannabinoid receptors were found to reduce appetite, and also, as a bonus, the craving of smokers for nicotine. Clinical trials were reportedly successful: nearly two-thirds of the obese subjects lost 5% of body weight over the period of the trial, and half of those shed 10%, without any recorded ill-effects. Even if drawbacks do arise, as history warns us they may, almost any risk appears preferable to such desperate remedies of the moment for life-threatening obesity as surgery, in particular the stomach-stapling operation, with its side-effects and uncertain outcome. More extensive trials of Rimonabant are awaited with no little interest by obese citizens and shareholders alike.

The battle of the diets

Meanwhile the Atkins diet holds sway, and it works, though at what cost is also not yet clear. Most diet regimes in past years have been based on limiting the intake of carbohydrates or more often of fats, and sometimes both. These all have a solid scientific basis, for energy intake is very strongly correlated with weight gain, and obesity develops when the first exceeds the second. The problem in evaluating the relative effects of fats and carbohydrates on increase and stability of body weight lies, as before, in the elimination of other factors. Both ways of examining the problem – epidemiological analyses of pre-existing groups (or populations) with differing kinds of diet, and extended trials with carefully selected groups of volunteers – are littered with pitfalls. People with a genetic disposition towards corpulence or skinniness may respond quite differently to a given diet. Some participants in a study will pull themselves together during the period of the trial and eat less. Others simply lie; according to the best investigation of this phenomenon, obese people confess on average to only 60–70% of their actual calorie intake, and they probably dissimulate most on fatty foods. Or subjects forced to deny themselves one kind of food may develop aberrant habits in respect of another. The duration of the trial is a major factor in determining the outcome: some diets have an appreciable short-term effect, which dissipates as time goes on. If an overweight group is chosen for study, have participants with diabetes or other conditions that alter the metabolism been identified? And so on. It is not surprising, therefore, that studies done at different times in different countries and by different researchers, not all of them competent in the craft of epidemiology and some blinkered by preconceptions, have produced divergent results. Meta-analyses (as explained above) are considered the most reliable indicators. The most extensive of these concludes that a low-fat diet leads to weight loss in the obese and weight stability in those of normal proportions.

The once popular Pritikin and Ornish diets operate on a simple principle. They work, for those who can endure them, because they specify foods low in calories – mainly cereals, vegetables, and fruit, and very little fat. Robert Pritikin, son and heir of the originator, Nathan, has refined the formula by introducing the concept of 'calorie density' – that what matters is not how many calories you consume, only how many lurk in each pound of the food that you eat. But to get your energy requirements solely from apples, celery, and bran, say, must make demands on the jaw muscles, taste-buds, and the labouring gut that few could sustain, and so the calorie intake must surely drop and the waistline accordingly shrink.

An innovation in recent years has been the introduction of polyols as sweeteners. These, more properly known as sugar alcohols, are derivatives of sugars, and are added, not in tiny quantities like aspartame and other artificial sweeteners (p. 213), but in the same sort of amounts as ordinary sugar. Many different polyols – sorbitol, mannitol, xylitol, and others – are used in processed foods. They have a lower calorific value than sucrose (cane or beet sugar), are more slowly absorbed and therefore cause less of a rise in blood glucose, and have few side-effects, other than a laxative action when eaten in large quantity.

But where diet is concerned, fashion rules. Thirty years ago Dr Walter Kampner of Duke University in Durham, North Carolina, developed the gruesome 'rice diet', which he administered to more than 18,000 obese patients in his private institution at extortionate cost. Kampner's activities made Durham famous, as 'Fat City'. The diet was based on unsalted, unflavoured rice from which starch had been extracted by repeated washing, both before and after cooking, until, in the words of one lapsed sufferer, it tasted like damp Kleenex. After a time scraps of real food were added. Kampner's regime worked because of his intimidating presence, which cowed his patients into persisting. He was even accused of having used corporal punishment on at least one recalcitrant patient. A book extolling the virtues of the rice diet, written by a beneficiary, appeared in 1986, long after Kampner's death and remained for some time on the best-sellers' list. Next, 25 years ago, the Scarsdale diet (high protein, low calories, mountains of grapefruit, and a medicine-chest of appetite suppressants) swept the market. It was conceived by Hermann Tarnower, a fashionable New York physician (the centre of a *cause célèbre* when he turned up as a corpse, murdered by a disaffected mistress), and was probably dangerous for it was deficient in some vitamins and in minerals, and is thought to have occasioned some deaths. Then came Pritikin's low-fat diet and soon thereafter the fruit-based Beverly Hills diet. Other formulae came, tarried briefly, and passed into oblivion, some of them sufficiently unbalanced to cause malnutrition diseases and a number of deaths.

Today, fashion has moved (if transiently) from low fat to the low carbohydrate in the celebrated Atkins diet, first introduced in 1972 and latterly enjoying a huge resurgence: 30–40 million people are said to be or to have been Atkins devotees. Robert Atkins was a New York doctor who propounded the attractive view that a rich regime of meat, butter, cheese, and eggs was the perfect slimming prescription, as long as carbohydrates were eliminated. It must be said that at the time of his demise from a fall on an icy street in New York in April 2004 Atkins was, at 18 stone (250 pounds), grossly overweight. Nevertheless, his diet, at least on the evidence of the limited studies that have been published, does seem to work about as well as the low-fat diet. Yet the most recent controlled trial on obese subjects found that an initial weight loss was not maintained in the longer run. The Atkins rationale is that without carbohydrate the levels of glucose and of insulin ('the fattening hormone', as he called it) would not rise after meals and thus fat deposition would be prevented. It does appear that slimmers tolerate the low-carbohydrate better than the low-fat regime and persist with it for longer, nor is there evidence so far of any rise in blood cholesterol. Even so, it has often been reported that the diet comes, after a time, to appear monotonous and unappealing, so that appetite and therefore food intake diminish, and that this accounts for the weight loss. The insulin theory is weak, since insulin causes fat deposition only when the calorie intake is excessive. Other authorities see risks in the Atkins method, and regard it as 'a dangerous experiment' that could lead to kidney damage, due to the huge nitrogen load imposed by the digestion of so much protein, loss of bone density (osteoporosis) as a result of increased calcium excretion, heart disease, and a return of the ancient scourge of gout. A respected organization, the Physicians' Committee for Responsible Medicine in the US, and bodies in Britain and elsewhere in Europe have voiced their concern that doctors prescribing the Atkins diet may be putting their patients' lives at risk. Long-term hazards that may follow from the grossly unbalanced intake of nutrients implied by Atkins's formulation have yet to be determined. One common manifestation is certainly ketosis. In carbohydrate deficiency, as in starvation, diabetes, or the Atkins diet, body protein is mobilized as an energy source. This results, by a well-understood mechanism, in accumulation of acetone in the blood. Acetone, in the normal way a rapidly processed metabolic intermediate, is a highly volatile substance and some of it therefore vaporizes in the lungs, and its sweetish smell, resembling that of nail varnish, becomes apparent on the breath.* As Atkins's description of his diet has passed from one edition to the next its tenets have actually been progressively modified, and at the last count the permitted quantities of butter, cream, and cheese, and of *trans*fats (p. 224) had decreased.

The Atkins diet may soon have had its day. It may be supplanted by Dr Steven Pratt's formulation for weight loss, as set out in his book, *Super Foods: Fourteen Foods That Will Change Your Life*, that is to say, beans, blueberries (or in Pratt's vocabulary, brainberries), broccoli, oats, oranges, pumpkins, salmon (but see p. 216), soya, spinach, tea, tomatoes, turkey, walnuts, and yoghurt. Pratt lists the micronutrients that these comestibles will provide, among them many antioxidants. Pratt's diet has received some endorsement from the medical profession in the United States.

The South Beach Diet appears to be an offspring of something called *The Zone*, the title of a book by two American zealots, Barry Sears and Bill Lawren. Published in 1995, this sold 1.5 million copies and was translated into many languages. The 'zone', within which you must remain to be certain of losing weight, is defined by the proportions of protein, carbohydrate, and fat in your diet (the ratio is set at 3:3:4). This is supposed to maintain an optimal metabolic

* The great British biologist J. B. S. Haldane, whose regular weekly articles, published between the wars in the *Daily Worker* – for he was a dedicated communist – are still read for pleasure, propounded a theory, probably with frivolous intent, about ketosis. The acetone, he suggested, might have been the original 'odour of sanctity', diffused by the saints, who habitually mortified their flesh by starvation.

balance by controlling your insulin levels. This simplistic proposition is supported by reference to a forgotten publication of 1956. There have been other dietary prescriptions without number (among them tracts with such irresistible titles as *What would Jesus Eat?* mostly fruit, seeds and nuts, it seems), and they keep appearing, often in the best-sellers' lists.

Another strong contender in the fashion stakes is the South Beach Diet, formulated by Dr Arthur Agetson, a cardiologist in Miami. This is a carbohydrate-based regime, and Agetson's book in which he first set it out sold 7 million copies (but bear in mind that the sales of diet books in the US amount to $2 billion a year – as though the very possession of a library of such volumes would of itself lead to weight loss). All you have to do, says Agetson, is to keep away from the 'bad carbs', those with a high *glycaemic index* (GI). Bran cereals, wholemeal bread, baked beans, and pulses in general, are all good, while bread baked with refined flour and patisserie are bad. Saturated fats are to be avoided in favour of mono- and polyunsaturated, which are permitted in moderation.

What then is the glycaemic index? It is certainly now a term implanted in the minds of all dieters. The proponents of high-carbohydrate regimes allocate glycaemic index points to each item of food expressed relative to glucose, with GI of 100. The GI measures the rate with which a given carbohydrate is digested and converted into blood glucose. Potatoes, most kinds of rice, and white bread, for example, have a high, and pulses and breakfast cereals a low GI. Foods in the first category elicit a sharp transient increase in insulin after a meal, resulting in the conversion of the blood sugar into energy stores and deposition of fat. Foods with a low glycaemic index exert a more gradual effect, thereby moderating insulin release, and appear also to induce satiety to a greater degree and so diminish food intake overall. There is evidence that proteins in the diet also help in this regard. Animal studies have provided strong support for this scheme, which emerged originally from a search for the best diets to meet the needs of diabetics.[26] The virtues of a high-carbohydrate, low-glycaemic-index diet for both normal and fat people have been promoted with evangelical fervour by many highly reputable clinicians and dieticians, mainly in Europe, Canada, and Australia. Scepticism is based mainly in the US, where the Agetson's prescription, among others, has come in for criticism from the medical profession. Many still question whether a high insulin concentration in the blood comes from high-glycaemic-index carbohydrates. Moreover, while the 'insulin spike' after meals is a feature of the normal metabolism, insulin concentration remains high (hyperinsulinaemia) in obese people, whose metabolism is chronically disturbed (see above). Nevertheless, the balance of evidence can be said to favour the glycaemic index concept.

The fatless fat

One of the most original endeavours to develop a painless dieting regime has been the invention of sugar polyesters. Sugars are molecules that bristle with reactive (hydroxyl) groups, which can be coupled by a skilful organic chemist to fatty appendages, like long whiskers. The result is a form of fat that is resistant to the enzymes of the digestive system and passes through unchanged, yielding up no calories. Such a substance is Olestra, devised by the Proctor and Gamble Company. It can be used in cooking, and served initially to prepare perfectly palatable and non-fattening crisps (American potato chips). But there is no rose without a thorn, and it quickly emerged that Olestra had its drawbacks. It has, in particular, a great capacity for absorbing fat-soluble substances, such as the fat-soluble vitamins. So the carotenes – vitamin A and its precursors, which in addition to their metabolic functions are thought to protect against many cancers – are quickly depleted from the blood by Olestra snacks, and so also are vitamins D, E, and K. Olestra tends, moreover, to cause gastrointestinal upsets, including diarrhoea and, worse, the feared 'anal seepage'. (The assay for this effect is to take a ruler to the oil patch on volunteers' underwear.) Olestra has been on the American market since 1996, but does not seem to have caught on to any great degree.[27]

The lunatic fringe

The lunatic fringe of dietary doctrine has encroached steadily on the mainstream in recent years. One large group of fanatics are the vitamin fanciers. The belief that if vitamins are good for health then more vitamins must be even better dates from the dawn of the vitamin age (Chapter 10). The style of the advertising has changed, but not its substance, for the public is still assured that a mega-dose each day will alleviate tiredness, lassitude, and stress. Vitamins were touted as slimming aids, and (before a skinny figure became the fashion) a help in building up the body. In the 1970s vitamin E was singled out as a novelty. With no evidence of deficiency in people on a normal diet, pills of the substance were proclaimed as cures for ageing, acne, asthma, atherosclerosis, cancer, diabetes, and so on. The elderly, especially, were enjoined to take huge doses. As we have seen, vitamin deficiencies occasionally reappear in the developed world among people living on processed foods, and it has been conjectured that the elderly may indeed benefit from a little extra vitamin E in such circumstances. Pregnant women are advised to take supplementary folic acid, which is also commonly added to bread and some other foods (p. 185). But the huge quantities of multivitamin pills consumed today are more likely to do harm than good. Certainly vitamins A and D in large enough doses are toxic, can cause kidney or liver damage, and hypertension, and even kill. It is surmised that vitamin A poisoning may not be rare in the countries of the West.[28] Excessive nicotinic acid and vitamin B_6 can also engender unpleasant symptoms, in the case of the latter a state resembling multiple sclerosis.

Minerals, being inherently cheap, are less ferociously touted, but it is still possible to buy a penny's worth of mineral supplement for several pounds or dollars.

Calcium has a secure place in the public consciousness, especially that of women, because of the prevalence of osteoporosis. Its calcium content is considered to rank high among the virtues of milk. Yet American women, who consume the largest amounts of calcium (often 1 g or even more each day), also have the highest incidence of osteoporosis in the world. In Asian and African societies, where the calcium intake is minute by comparison, the condition is rare, if not unknown. There is an epidemiological correlation between the incidence of osteoporosis and protein consumption, and it has been conjectured that a high-protein diet, which is accompanied by increased kidney activity (see above), leads for that reason to elimination of calcium from the blood. But the most likely major cause of the rise in osteoporosis is thought by many authorities to be a lack of exercise. The dairy industry, of course, will have none of that and has tried to revive its flagging fortunes (following the cholesterol panic) with a range of often oddly flavoured milk and yoghurt drinks.

Red meat now has a bad name in fundamentalist circles, and extracts are no longer widely regarded as all-purpose tonics. Yet Bovril still thrives in Britain – once advertised in incandescent prose like the following: 'The secret of Napoleon's power was his immense vitality. The same is true of most great men – Julius Caesar, Michelangelo, Gladstone, Cecil Rhodes – they were successful because they were never tired. Don't get tired, drink Bovril.'[29] This is the kind of language that can still be found in advertisements for patent medicines, vitamins, and now the preposterously priced minerals – the latest additions to the quack dispensary. Gullible buyers are generally informed that a correct balance of the 'natural' minerals will give protection against a range of ailments and will alleviate symptoms such as headaches, fatigue, sinusitis, and respiratory infections. Such mumbo-jumbo as 'iridology' or analysis of the hair are peddled as the ways to determine which minerals in particular are deficient. Dried digestive enzymes are also often on offer. A growing market deals in such herbal nostrums as St John's Wort. These trade on the proposition that 'natural' means safe, whereas the most virulent known poisons are nearly all derived from natural sources. The World Health Organization has issued warnings against the dangers immanent in 'natural' medicines and food additives available without prescription. In China, where many of these products originate, untold thousands of cases of serious illness and death resulting from their use are reported each year.

That people will believe anything is apparent from the rise and survival of the 'macrobiotic' diet cult, a throw-off of New Age mysticism. The movement was started, and the name Zen macrobiotics was coined, by a Japanese doctor, Yukikazu Sakurazawa (though one Christoph Wilhelm Hufeland, who lived from 1762 to 1836, ministered to Goethe, Schiller, Herder, and other luminaries of the age, and wrote profusely, developed a catch-all regime for a long and healthy life, and called it *Makrobiotik*). Sakurazawa reinvented himself in America as George Ohsawa. His inspiration derived from the notion of a 19th-century Japanese doctor that diets should balance the elements inhabiting, according to traditional Chinese philosophy, all natural objects, the *yin* and the *yang*. The *yin* is fluid and cool, the *yang* strong and hot. Foods partake of these qualities in different admixtures. Ohsawa instructed his clients that a correct balance, suited to the needs of the individual and his disorders, would effect a sure cure of just about everything, from schizophrenia (caused by excess of *yin* and requiring therefore masses of *yang*), syphilis, and retinal detachments to air-sickness and bed-wetting. He offered diagnoses, even for those in perfect health; dandruff, for instance, presaged mental illness. But the diet was severe, consisting mainly of cereals, cooked vegetables (each mouthful to be chewed 50 times), and soup, and notably deficient in several vitamins. Ohsawa died in 1966, having previously been arraigned for malpractice and raided by the FDA. The Ohsawa Institute was closed down, but the baton passed to another Japanese doctor, Michio Kushi. Kushi broadened the diagnostic repertoire to embrace diagnosis by voice, by 'meridian' (the supposed direction of 'energy' flow in the body), by astrology, and so on. One's bed must also be correctly aligned with the earth's axis, and so on. The Kushi Institute flourishes in a small town in Massachusetts, and has attracted to itself a 'macrobiotic community'. There have been reports of scurvy and other diseases of malnutrition among the followers of the macrobiotic diet, but this seems not to have discouraged the believers.[30]

An odd survival from the earlier part of the 20th century is the Edgar Cayce system. Cayce was a deranged clairvoyant, whose foundation is still extant and dispenses advice on most aspects of human life. Cayce believed, like William Hay (p. 198) before him, that some foods generate alkali in the stomach, and some acid. (Never mind that the stomach contents are bathed in a solution of hydrochloric acid.) A diet that will prevent you from falling into the pit of infirmity must comprise 80% alkali-producing and 20% acid-producing constituents. What these are can be discovered from Cayce's charts. But on no account must incompatible foods be mixed, for the results will be too terrible to contemplate. So milk must never be drunk with citrus fruits, nor anything sugary with something starchy; coffee must not be taken with milk or cream, nor apples (or green peppers) with anything else at all. But that is not the end of it, for 'even the most nutritious foods can turn to poison within the system while a person is in a negative frame of mind', so if you are under the weather just don't eat. The Foundation advertises its wares and prescriptions quite widely, and presumably still has its followers.

One of the oddest recent manifestations is the Blood Group Diet. This is sold to the public on the bizarre grounds that peoples' nutritional requirements are related to their blood groups. It takes into account the four major blood groups, O, A, AB, and B (though not the many minor blood groups), which the sales pitch categorizes as the Hunter, the Cultivator, the Enigma, and the Nomad (respectively). The evidence for any influence of blood groups on metabolism exists of course only in the imagination of the originators. So the possessors of blood group O, they assure us, are by evolutionary origin carnivores; they possess high levels of stomach acid (which will be news to science), and that somehow favours digestion of meat. Blood group A belongs to farmers and predisposes to a part-vegetarian diet, and so on. The literature of the proponents of the theory is rich in unassimilated scientific concepts, in which lectins, for instance, feature strongly. Lectins are proteins that occur in many seeds, and especially cereals and pulses. Some lectins, of which ricin is the most notorious, are toxic to everyone (regardless of blood group), and red kidney beans also contain a noxious lectin, which, if the beans are eaten raw, can occasion severe, even life-threatening, unpleasantness. Certain pure lectins, if injected into the blood, would cause the cells to stick to one another, but there are many substances in the food we eat that would have similarly lethal consequences when injected. To assert that lectins in the diet can 'cause the blood to clump', as the literature of the cult proclaims, is of course absurd. Were this the case there would be little life left on earth, human or animal. 'Metabolic typing' goes further in purporting to divine the diet best suited to the individual from analysis of the hair, as well as the blood. One must suppose, since such fads remain in circulation, that the pseudoscientific jargon is sufficient to send the consumers reaching for their cheque books.

Guilt and innocence

There is still no established reliable formula for losing weight and avoiding obesity, other than limiting the intake of calories – a painful proceeding. The old distinction between endogenous and exogenous obesity – a genetic form that you cannot help, and the kind brought on by greed – went out of fashion some decades ago, but it is recognized that there is quite wide variation, part-hereditary and part-acquired, between the response of individuals to diet. The notion that fat people have a slower metabolic rate and therefore do not 'burn' off fat like lean people does not hold water, and the many dietary prescriptions based on this notion are specious. It is in fact now firmly established that obese people have a faster basal (fasting) metabolic rate than thin people and the more obese they become the more rapid their metabolism.[31] The resting metabolic rate (lowest during sleep), when the body is performing no external work, is a reflection of the biochemical processes that constitute life. These reactions produce heat that can be measured in a calorimeter (p. 111). A part of the heat production is *adaptive thermogenesis*, responsive, that is, to external temperature and especially to diet. With a high calorie intake the heat output goes up and during starvation it diminishes – undoubtedly a survival characteristic. The level of the adaptive response to calorie intake is controlled by complex biochemical processes in the brain that send out signals to the body's cells to speed up or slow down the heat-generating reactions. These responses are disturbed in obese people, but are also influenced by genetics. One strand of obesity research is aimed at finding agents that will modulate the level of adaptive thermogenesis. The health and obesity crisis continues to stimulate much nutritional research, and perhaps eventually there will be safe drugs to solve all problems. But the great molecular biologist Sydney Brenner has told how he reacted to the euphoria of a colleague who claimed to have discovered 'the obesity gene'. He, too, Brenner replied, had long ago discovered the obesity gene: it is the gene that opens the mouth.

Live for ever

The primal search for the Elixir of Youth has never ceased. Just as Ponce de Leon's fountain of eternal youth proved an illusion, so too have all the many nostrums advertised over the years to ward off old age. Longevity is at least in part hereditary, and the affirmation by an American jazz musician, that he had achieved extreme old age thanks to never eating anything that wasn't fried, should be treated with reserve. Nor can one be too sure about the 150-year-old yoghurt eaters of Asia Minor (p. 201). Perhaps the most rational of all the measures that have been tried to prolong life is based on the theory that the agents of decrepitude are free radicals, formed as by-products of some metabolic reactions. These are highly active molecules (p. 183) and can be destroyed by 'scavengers', prominent among which are vitamin E and the carotenes, related to vitamin A. But, while these may certainly do some good, it has never been established that they really do retard the ageing process, and as we have seen, taken in the huge quantities that are not now uncommon in the developed world (and especially in the United States), they probably do harm rather than good.

The only experimental means yet discovered of prolonging life in laboratory animals is *calorie restriction*. This strikingly increases the life-span and reduces the symptoms of ageing in rodents, but a corresponding reduction in human calorie intake would mean a drop from an average 2500 to 1750 calories per day. Such a dismal regime would certainly make one's existence seem longer, but the criticism that what applies to rodents need not apply to man was countered by similar experiments on primates, which bore out the effect. One direction that the inquiry into the mechanism of ageing has taken is a search for the changes in levels of various metabolic intermediates in relation to calorie intake. All theories of ageing now embody the premise that evolution can act only in the interests of procreation, since (by definition) natural selection operates by favouring traits (genes) that promote the propagation of the species. Therefore increased survival beyond the age of reproductive activity is of no evolutionary benefit. When food is plentiful an animal can both grow and reproduce, while when it is scarce it does best to suppress both processes and switch to a mode in which the body's resources are husbanded and its reproductive capacity protected for the future. If the metabolism is boosted in times of plenty there should be more rapid utilization of the blood sugar (glucose) for the production of the body's energy source, ATP. It is the reactions by which ATP is formed in the cell's power houses, the mitochondria (p. 240), that generate free radicals. In any event, a group of researchers at the National Institutes of Health near Washington discovered that the sequence of reactions could be interrupted by a chemical analogue of glucose, which would jam the enzymic machinery. This analogue is a simple substance, 2-deoxy-D-glucose, or 2DG. When given to rats, the 2DG reduced the level of blood glucose and the amount of circulating insulin, the hormone that, as we have seen, allows the blood glucose to do its work. The metabolism slowed and the body temperature fell somewhat. The animals did not eat less, but they became leaner and apparently fitter. These results have stirred up a good deal of interest (especially, one might hazard, among older scientists), but 2DG is not yet the Elixir of Youth because, for one thing, it is toxic at only slightly elevated concentrations or when administered for an extended period. But the results do suggest that a substance may eventually be found that will simulate the effects of eternal fasting without biting back in some other way. The social consequences of course could be too alarming to contemplate.

Meanwhile, your best chance on current, and highly incomplete, evidence of dying old lies in not growing fat (for obesity, as we have seen, hugely increases the likelihood of terminal and lesser ailments); eating generous amounts of fruit and vegetables – tomatoes, which contain free radical scavengers are especially recommended; keeping dairy foods down to reasonable limits; eating fish regularly; drinking alcohol (though, sadly, only in moderation); drinking coffee, at least if your blood pressure is not too high (for this seems to diminish the incidence of bowel cancer); and possibly taking one multivitamin pill (containing folic acid) every day. And then there is the Second-World-War mantra, now seldom heard, that 'a little of what you fancy does you good'.

A bright future? The genome and all that

The sequencing of the human genome would, we were assured when this great undertaking began, presage a new age of medicine, when the diseases that had plagued the human race would finally be banished. Now the information, encoded in the order of the nucleotides (the letters in the four-letter alphabet) that are strung together to make up the long threads of DNA in our chromosomes is available to inspect and marvel at, there are still few signs of the promised dawn. But in the biotechnology industry optimism reigns, and one of the widely touted new disciplines to emerge from the genome is *nutrigenomics*.[32] Its aim is to specify diets for individuals, based on their genotypes – the nature of the relevant players in their collection of genes – that will maximize the prospects of a long and disease-free life. We know, of course, that the various elements of diet affect individuals in different ways. In the extreme case the wrong diet can be fatal, as in diabetes or phenylketonuria (p. 259) for instance, in others severely uncomfortable (as in lactose intolerance), and sometimes its effects can be insidious. So, while some of us can enjoy long life on cholesterol-rich fare, others will be felled early by coronary heart disease. In some cases the genetic basis of such variation is understood, but most often the causes are obscure. What is clear is that most of our 10,000 or so genes are subject to polymorphisms (more generally, single-molecule polymorphisms, SNPs or, colloquially, SNIPS). These arise from mutations in the long-ago and result in the substitution of one amino acid for another in the protein that the gene specifies. In most cases this is functionally neutral, so that the biological activity of the protein (most often an enzyme) is unaffected, but sometimes the activity is altered. The enzyme may become more or less susceptible to some external agency, and in fact the interaction of environmental and genetic factors is a major preoccupation of today's geneticists.

It is now a routine matter to determine an individual's genotype from a sample of DNA, taken most often from cells in a mouth swab. Companies have already sprung up offering to do this for the genes that give rise to enzymes involved in processing food, and on the basis of the results provide a table of dietary recommendations. This is nutrigenomics. An example that its champions give of the future benefits relates to an enzyme involved in the metabolism of methionine (one of the essential amino acids (p. 186), obtained from ingested proteins). There are two variants of the gene that specifies this enzyme, due to a SNP: about 1 in 10 northern Europeans and 1 in 7 southern Europeans have the rarer variant, which leads to slower processing of an intermediate in methionine breakdown, and consequently an increased concentration of this intermediate (homocysteine) in the blood. There is some evidence that high homocysteine levels are associated with coronary heart disease (but none to date that reducing it leads to a more favourable prognosis).

The problem is the complexity of interactions between different, often numerous, genes that underlie most biological processes, and also the part played by epigenetic events – enzyme-induced chemical modifications that affect the expression of a gene. There is little at this stage to encourage the expectation that a 'designer diet', constructed on the basis of the kinds of gene profiles now commercially on offer, will change anyone's life. No doubt the scope and accuracy of genotyping will improve, but the rigours of a restrictive diet will remain as objectionable as they are now. Let Mark Twain have the last word: 'There are people who strictly deprive themselves of each and every eatable, drinkable and smokable which has in any way acquired a shady reputation. They pay this price for health. And health is all they get for it. How strange it is. It is like paying out your whole future for a cow that has run dry.'

APPENDIX
The Hard Science

Nutrition today is rooted in the rigours of biochemistry. The processes by which food is utilized in the body to generate energy and replenish tissues are known in outline and for the most part in detail. There are, to be sure, still scholarly debates concerning the relation of dietary constituents to health and longevity, but for the ignorance or charlatanry that lies behind so many of the dietary prescriptions peddled in our scientific age there can be no excuse. To understand the principles of human metabolism (and thereby also recognize health quackery for what it is) requires only a modest grasp of chemistry, at about high-school level. What follows is an outline sketch of metabolic biochemistry for those who wish to sharpen their understanding of what happens in our bodies as we ingest and digest our food.

The nature of metabolism

By metabolism is meant the huge ensemble of chemical processes that go on in cells or in organisms from bacteria to man. It is described by *metabolic pathways*, the routes by which substances in the body are built up or broken down. All these processes are encompassed by enzymes, the protein catalysts that make biochemical reactions run; without their enzymes almost all these reactions would simply stall. The metabolic pathways are of two kinds, *anabolic* and *catabolic*. In anabolic reactions complex biological molecules are synthesized from simple precursors. They most often involve reduction (for practical purposes the loss of oxygen or gain of hydrogen) and require an investment of energy. Catabolic reactions break down the large organic molecules that for the most part make up food constituents into small, simple molecules. This usually involves oxidation (the reverse of reduction) and is accompanied by the release of energy, as in combustion. (Energy is measured in kilocalories, incorrectly but almost invariably called calories, or in the currently sanctioned unit of kilojoules, abbreviated kJ. One calorie is the heat energy required to raise the temperature of one gram of water by one degree Celsius, and a kilocalorie is 1000 of these. A joule derives from the discovery in the 19th century, by the Mancunian physicist James Prescott Joule, that a precise equivalence obtains between mechanical work and heat. One joule is 4.2 calories.) The so-called central pathways of metabolism have a remarkable unity, involving, as they do, only a small number of fairly simple molecules. Their elucidation was one of the great triumphs of 20th-century science. The scheme was largely complete by about 1950. The picture is complicated by hormones, some of which are small organic (that is, carbon-containing) molecules, and others, such as insulin, are proteins, which are larger. The hormones exert control over metabolic processes by regulating the activities of the enzymes in response to external or internal cues that indicate whether the supply of some essential substance should be increased or suppressed.

The Krebs cycle

The Krebs cycle, also called the tricarboxylic acid cycle or the citric acid cycle, is the energy motor of the cell, located in an organelle – one of the bodies found in its interior – called the mitochondrion. Mitochondria are autonomous entities containing their own genetic message – their deoxyribonucleic acid (DNA) – which is transmitted to her offspring by the mother: the father's genes have no part in the mitochondrion. The chemical nature of the mitochondria's DNA and the proteins that it specifies are characteristic of bacteria, and it is inferred that a primordial event caused bacteria to enter the cell and to remain and flourish there through the generations, and eventually become indispensable to the cell's function.

The Krebs cycle is named after its discoverer, H. A. (later Sir Hans) Krebs (1900–81). He was born in Germany, qualified in medicine, and joined the laboratory of the most famous German biochemist of the day, the formidable Otto Warburg. Warburg, who ran his laboratory like a military operation, conceived a low opinion of Krebs's talents and advised him to seek a career in medical practice. Krebs was not put off, and after the introduction of the racial laws in 1933, as a Jew, he sought refuge in England and began his work on metabolism in Cambridge.* Thence he moved to the University of Sheffield, and it was there that he made the great discovery for which he received the Nobel Prize in 1953. The cycle proceeds by nine steps, the 10th product being then the starting material for the next round of reactions. The cycle throws off carbon dioxide as a waste product, and, at other points, electrons, which are needed for the synthesis (by a series of reactions, termed *oxidative phosphorylation*) of the organism's principal ubiquitous energy source, adenosine triphosphate (ATP). The terminal phosphate group can be shorn off by an enzymic reaction to generate adenosine diphosphate (ADP), with release of energy to drive energy-requiring processes. The formulae of the two compounds† are:

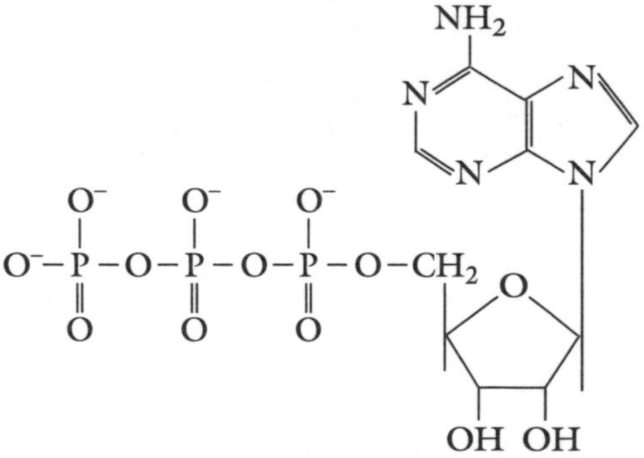

$$O^- - \overset{O^-}{\underset{O}{P}} - O - \overset{O^-}{\underset{O}{P}} - O - \overset{O^-}{\underset{O}{P}} - O - CH_2 - \text{(ribose, OH OH)} - \text{adenine} (NH_2)$$

ATP

$$O^- - \overset{O^-}{\underset{O}{P}} - O - \overset{O^-}{\underset{O}{P}} - O - CH_2 - \text{(ribose, OH OH)} - \text{adenine} (NH_2)$$

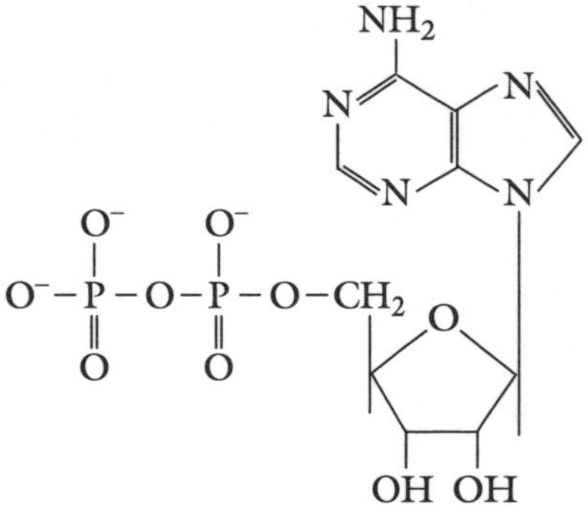

ADP

* Warburg, who was half-Jewish, remained in Germany and was not molested, indeed was allowed to continue as director of his laboratory and even to keep his stable of thoroughbred horses through the war years. The surmised reason for the regime's unaccustomed tolerance was that Warburg had developed a theory of cancer (later shown to be wrong) and a therapy to go with it. Hitler, who had a morbid dread of the disease, apparently caused Warburg to be reclassified as one-quarter Jewish and thus evade persecution under the Nazi racial laws. Warburg was notorious for his combative nature and the feuds that he entered into with his scientific contemporaries.

† *A note on chemical formulae* C represents a carbon atom, H hydrogen, N nitrogen, O oxygen, and P phosphorus, the only elements with which we shall be concerned here. Others also occur, but more sparsely, in biological substances. Carbon has a *valence* of 4, in other words, makes four bonds to other atoms, while nitrogen has a valence of 3, oxygen of 2, and hydrogen of 1, and phosphorus can have a valence of 3 or 5. In some compounds there are two bonds between a pair of adjacent atoms – a double bond. The simplest compound of carbon and hydrogen is methane,

```
     H
     |
  H— C —H
     |
     H
```

or CH_4. Ethane with two carbons is

```
     H   H
     |   |
  H— C — C —H
     |   |
     H   H
```

or C_2H_6, whereas ethylene has a double bond and the formula,

```
  H       H
   \     /
    C = C
   /     \
  H       H
```

Life depends on the ability of carbon, alone among the elements, to form long chains, and therefore compounds of great complexity. An extension of ethane is

```
     H  H  H  H       H  H  H
     |  |  |  |       |  |  |
  H— C ⁄ C \ C ⁄ C ........ C \ C ⁄ C —H
     |  |  |  |       |  |  |
     H  H  H  H       H  H  H
```

(the zigzag representation here being meant to indicate that the chain in reality is not simply straight) or for simplicity $CH_3–CH_2–CH_2–CH_2 \ldots CH_2–CH_3$. If there is one double bond in the chain, we might have $CH_3–CH_2–CH{=}CH–CH_2 \ldots CH_2–CH_3$. A common group that appears in biological compounds is the amino group,

```
      H
     /
  —N
     \
      H
```

written in short $–NH_2$; another is the hydroxyl group, –O–H, or simply –OH; yet another is the carboxyl group,

```
       O
      //
  — C
      \
       O—H
```

or more economically, COOH. Most physiologically important compounds contain rings of carbon atoms, often interspersed with nitrogen or oxygen. The atoms in the ring may be linked by single bonds or by double bonds. Rings may also be fused to one another. It is customary, for economy, to omit in writing these formulae the symbols for carbon and hydrogen, which are so prevalent in all these compounds. A corner left empty is taken to be occupied by a carbon atom, with as many hydrogens attached as will satisfy its valence of 4. Thus for example, purine, variants of which occur widely in biology – in DNA and in caffeine among many other substances – has the fused-ring structure

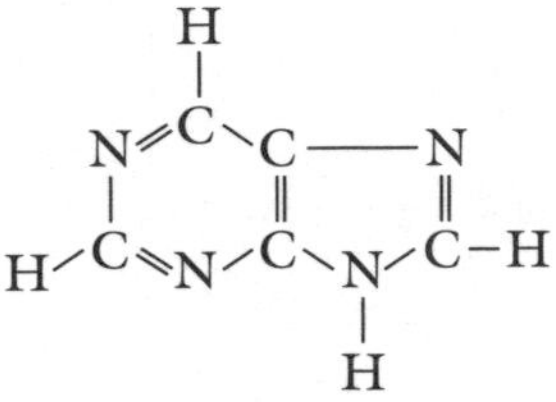

```
           H
           |
      N // C \ C ——— N
      |        ||      ||
  H / C \\ N / C \ N / C – H
                   |
                   H
```

This is always written

```
   N //\———— N
   |     ||      ||
    \\ N / \ N /
```

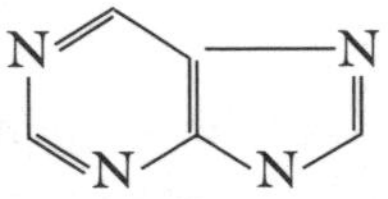

One other feature of carbon chains must be mentioned: adjacent carbon atoms, joined by a single bond, can rotate freely about that bond, whereas a double bond is rigid. (Imagine two balls, each with a hole drilled to accept a rod joining them, and then the same balls with two holes in each, joined together by two rods.) This allows two possible configurations of the chain at the double bond. These are called *cis* and *trans*. In the one case the chain must make a bend to accommodate the double-bond, in the other it remains straight:

```
 –CH2        CH2\              –CH2
     \      /                       \
      CH = CH                        CH = CH
                                             \
                                              CH2 –
```

The starting substrate (the particular species of molecule on which a given enzyme operates) is the very simple digestion product, acetate

$$CH_3-COOH$$

chemically attached to a ubiquitous carrier molecule, coenzyme A, or CoA. The acetate itself is derived from another simple substance, pyruvate

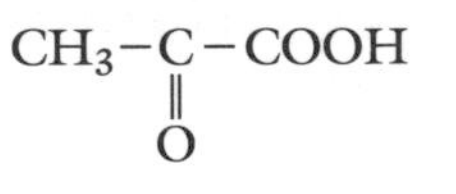

$$CH_3-\underset{\underset{O}{\|}}{C}-COOH$$

a product of sugar breakdown, as will be seen.

ATP is needed to provide energy of movement, in particular muscular contraction; it is used in a variety of synthetic processes and it fuels the pump proteins that push unwanted ions (in practice, the positively charged constituents of salts, such as the sodium in common salt, sodium chloride) out of cells. (In blood, for instance, the cells contain potassium chloride, and the serum, in which they swim, sodium chloride. The hugely important calcium ion, which is released from nerve terminals to trigger such events as contraction of muscle, must be expelled from the cell by one of the pump proteins when it has done its job.) The importance of energy processes in the function of the body can be judged by the fact that we turn over our own body-weight in ATP every day.

Glycolysis

The body's cells contain the simple sugar, glucose:

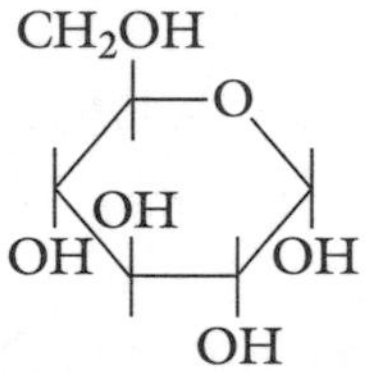

CH_2OH
O
OH
OH OH
OH

A sequence of reactions, collectively making up the process of glycolysis, consumes glucose in forming pyruvate and ATP. A feedback mechanism acts to stop glycolysis when the cell has made enough ATP, for the accumulated ATP switches off one of the critical enzymes. The pyruvate feeds into the Krebs cycle, and some of it is also used to make fatty acids. But in muscle glycolysis has a different starting-point, because it is required to generate ATP more rapidly. In the liver and muscles *glycogen* is the main store and source of glucose. It is a polymer – a giant molecule made up of as many as 120,000 glucose units chemically linked together. With the aid of an enzyme the glucose units are split from the polymer and chemically modified (by attachment of a phosphoryl group, comprising a phosphorus and three oxygen atoms) to the form in which it can feed the glycolytic reaction sequence. When the glucose supply from the diet exceeds demand, the excess is converted into glycogen in the liver, to be stored until more glucose is again needed. The formation of ATP from glycogen is accompanied by production of lactate as waste. Accumulation of too much lactate in the muscle, before it can be carried away in the bloodstream to the liver for recycling, is the cause of cramp brought on by extreme athletic exertion.

Energy supply by carbohydrates

Glucose is an essential fuel, especially for the brain, with its high metabolic rate, so the concentration of this sugar in the blood must be maintained all the time. Blood glucose normally comes primarily from the diet. Carbohydrates in the food are broken down by digestive enzymes in the saliva and small intestine to glucose and other sugars. Most of the latter are converted into forms (identical to intermediates in the processing of glucose itself) that can enter the glycolytic cascade of reactions. When the body receives no food, and the glucose cannot be replenished by breakdown of glycogen in the liver, another piece of biochemical machinery awakens to make good the deficiency. This is called *gluconeogenesis*, and involves in essence the reversal of the glycolytic series of reactions, so as to synthesize, rather than consume, glucose. Gluconeogenesis makes use of amino acids, derived from the breakdown of proteins.

Processing fats

Fats in food are predominantly in the form of ***triacylglycerides*** (or, as commonly written, triglycerides). These are esters of glycerol (popularly known as glycerine). Glycerol is an alcohol. An alcohol is loosely defined as a molecule that possesses a hydroxyl group attached to a carbon atom. Ethanol, the alcohol that we drink, has the formula.

$$CH_3{-}CH_2{-}OH$$

Glycerol has the formula

$$\begin{array}{ccc} CH_2 - & CH - & CH_2 \\ | & | & | \\ OH & OH & OH \end{array}$$

and is thus an alcohol three times over. An ester is a compound formed by reaction of an alcohol and an acid, such as acetic acid (the acid of vinegar):

$$\underset{\text{alcohol}}{CH_3{-}CH_2{-}OH} + \underset{\text{acid}}{CH_3{-}COOH} \rightarrow \underset{\text{ester}}{CH_3{-}CH_2{-}O{-}\overset{\overset{\large O}{\|}}{C}{-}CH_3} + H_2O$$

The acids that are combined with glycerol in fats have long carbon chains, with as many as 18 carbon atoms, where acetic acid has only two:

$$CH_3{\diagup}CH_2{\diagdown}CH_2{\diagup}CH_2{\diagdown}CH_2{\diagup}CH_2{\diagdown}CH_2{\diagup}CH_2{\diagdown}CH_2{\diagup}CH_2{\diagdown}CH_2{\diagup}CH_2{\diagdown}CH_2{\diagup}CH_2{\diagdown}CH_2{\diagup}CH_2{\diagdown}CH_2{\diagup}COOH$$

$$\text{or} \quad CH_3{-}(CH_2)_{16}{-}COOH$$

stearic acid

The fatty acid chains may be like the above, ***saturated***, that is, contain the maximum number of hydrogen atoms that the carbon atoms can bind, or may be ***unsaturated***, that is, contain one or more double bonds between carbon atoms, which therefore have one less attached hydrogen each (see p. 250, footnote).

Oleic acid is a *cis*-monounsaturated fatty acid (with one double bond) and the corresponding *trans*-fatty acid is called elaidic acid. As can be seen, the *cis*-double bond in oleic acid causes the chain to kink, whereas the *trans*-double bond in elaidic acid leaves it straight.

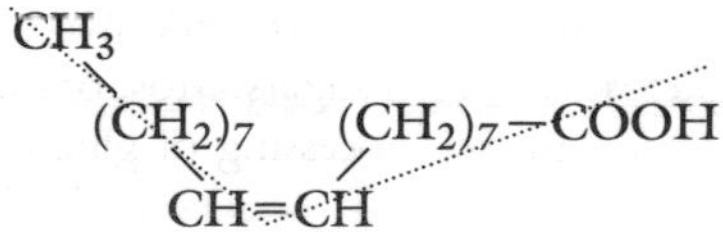

$$\begin{array}{l} CH_3 \\ \quad\diagdown \\ \quad (CH_2)_7 \quad (CH_2)_7{-}COOH \\ \qquad\diagdown \quad \diagup \\ \qquad CH{=}CH \end{array}$$

oleic acid

$$\begin{array}{r} (CH_2)_7{-}COOH \\ \diagup \qquad\quad \\ CH{=}CH \qquad\qquad \\ \diagup \qquad\qquad\qquad \\ CH_3{-}(CH_2)_7 \qquad\qquad\qquad\quad \end{array}$$

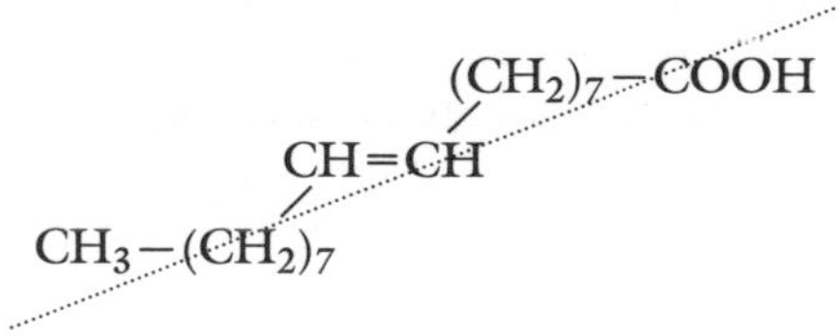

elaidic acid

The kink prevents such chains from nestling together along their whole length, and it is this that makes the *cis*-unsaturated fatty acids more liquid than the saturated (as in oils, as opposed to suet, say). A *trans*-unsaturated fatty acid (usually described just as a *trans* fatty acid) forms the hardest fat. This is the type that results from hydrogenation of plant oils (p. 214).

To be absorbed the triacylglycerides must be degraded to glycerol and free fatty acids. This *lipolysis* is mostly encompassed by enzyme molecules (lipases) in the wall of the small intestine after the fat has first been churned up with bile to produce a milky emulsion. In babies there is also an enzyme that starts the digestive process in the root of the tongue. Surplus fats go into an energy store, the *adipose tissue*, but the liberated fatty acids can be used in several ways. Some go into synthesis of phospholipids – esters containing a phosphoryl group (the parent acid here being phosphoric acid) and two fatty acid chains. These are the principal constituents of cell membranes.

Fats are oxidized in a stepwise manner to acetyl-CoA, a modified state of CoA, to provide energy by feeding into the Krebs cycle. The glycerol, meanwhile, can undergo a chemical transformation of its own to one of the compounds that likewise participate in the cycle. Some of the fatty acids are used in the synthesis of cholesterol, which, among its other functions, is an essential cell membrane constituent. Not all the body's fatty acids are supplied by the diet, and acetyl-CoA is also a precursor of more and/or different fatty acids, and especially those destined to be converted into the phospholipids of cell membranes. Some also go to make up fatty material of brain and nerves. But, as has been mentioned (p. 185), not all the fatty acids essential for life can be made in the body.

Fatty acids are continually released by enzymes from the adipose (fat) stores and transported by the blood to the sites at which they are needed. In states of starvation the body finds its energy increasingly from fat, for its glucose under these conditions has to be supplied by gluconeogenesis, a metabolically expensive process in that it consumes amino acids, and therefore protein. With glucose scarce, there is no option but to draw on the fat deposits. The breakdown of fats is accompanied by formation of compounds called *ketone bodies*, one of which is *acetone*. These intermediates of oxidative breakdown are normally present in small, transient amounts. But when there is no carbohydrate intake, as in starvation or the Atkins diet (p. 238), the concentration of ketone bodies increases, and the characteristic sweetish smell of acetone becomes perceptible on the breath. The same phenomenon (in effect starvation) shows itself in diabetes, a condition that results from failure to utilize the blood glucose, which rises to a destructively high level (p. 229); it is as though no carbohydrate had been ingested, and so the fat reserves are called on instead.

The *essential fatty acids* – those that cannot be formed by enzymic reactions in the human body and must come from food – are all unsaturated. They are called linoleic, α-linolenic, and arachidonic acid, but in fact linoleic acid suffices, because that can serve as a starting point for synthesis of the other two. The formulae for linoleic and α-linolenic acid are the following:

$$CH_3-(CH_2)_4-CH=CH-CH_2-CH=CH-(CH_2)_7-COOH$$

linoleic acid

$$CH_3-CH_2-CH=CH-CH_2-CH=CH-CH_2-CH=CH-(CH_2)_7-COOH$$

α-linolenic acid

Because the double bond comes after the sixth carbon atom in the chain, linoleic acid is designated an ω6 fatty acid (or omega-6). Omega-3 unsaturated fatty acids occur in foods too, as also do twice or even multiply unsaturated (polyunsaturated) fatty acids, pre-eminently in fish. The α-linolenic acid, as can be seen, has three double bonds. These essential fatty acids are needed for the synthesis of a group of compounds with hormone-like activity, notably on the action of smooth muscle (found especially in the uterus and blood vessel walls), the prostaglandins.

Fatty acid oxidation is one source of the ubiquitous acetyl-CoA, which among its other activities is the starting point of *cholesterol* synthesis, a lengthy process requiring the participation of a series of enzymes. Cholesterol has the structure:

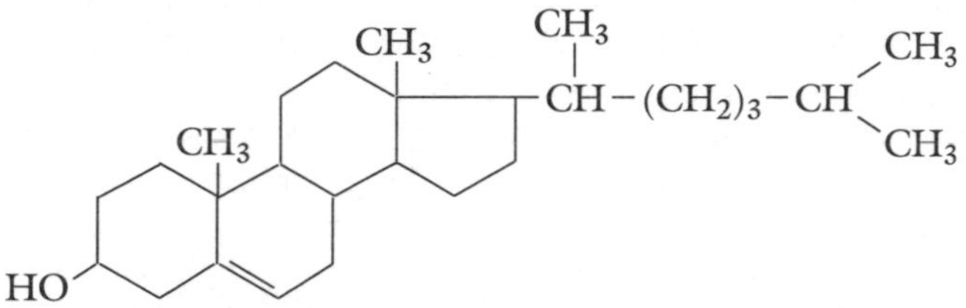

CH$_3$ CH$_3$ CH$_3$
CH$-(CH_2)_3-$CH
CH$_3$ CH$_3$
HO

It is a member of a class of substances called steroids, several of which are hormones, notably the sex hormones such as testosterone and oestrogen; another member is vitamin D. Cholesterol is not only an indispensable constituent of membranes, but also a precursor of hormones and bile acids, the substances that emulsify fats during digestion. Because it is insoluble in water cholesterol is always associated with lipids. In its free or esterified form (for its hydroxyl group makes it an alcohol) it is transported in the blood in complexes with lipids and proteins, called *lipoproteins*. The *chylomicrons*, produced from dietary fat, are one type of lipoprotein. Some are taken to the adipose tissues and deposited there, while the surplus is cleared by the liver. The more interesting lipoproteins, found in the blood serum, are derived from the body's own (endogenous) fat supplies and fall into three groups, the high-density lipoproteins (HDLs), low-density lipoproteins (LDLs), and very-low-density lipoproteins (VLDLs). These last are rich in triglycerides and represent a stage in the formation of the LDLs, which contain a large amount of cholesterol esters, along with a protein. HDL, on the other hand, is formed in the circulation from precursors that come from the liver and intestine, and one at least of its functions is to carry excess cholesterol back to the liver to be reprocessed. LDL carries the bulk (around 70%) of the cholesterol in the blood, and HDL the rest. As discussed elsewhere (p. 224), HDL has been found to exert a protective effect on blood vessel walls, whereas it is the cholesterol of LDL that is responsible for the formation of arterial plaques. Too much in the way of triglycerides in the blood is also deleterious to the arteries. In the healthy individual the blood lipids and cholesterol are maintained at a fairly constant level. If this drops below what is needed, the deficit is made good by lipid synthesized from carbohydrate.

Protein metabolism

Unlike fats and carbohydrates, proteins contain nitrogen and therefore supply that element for the manufacture of the huge number of nitrogenous compounds involved in the processes of life. Proteins are made up of amino acids, chemically joined together to form a long chain. The amino acids have the formula

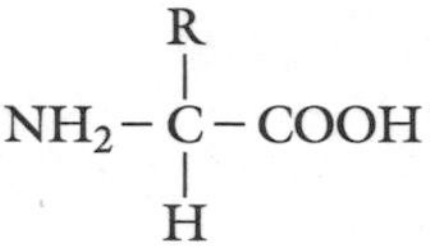

$$\begin{array}{ccc} & \mathrm{R} & \\ & | & \\ \mathrm{NH_2}- & \mathrm{C} & -\mathrm{COOH} \\ & | & \\ & \mathrm{H} & \end{array}$$

where R stands for a group, referred to as a side-chain. Each of the 20 species of amino acids that occur in proteins is characterized by a different side-chain. The protein chain is represented by the structure:

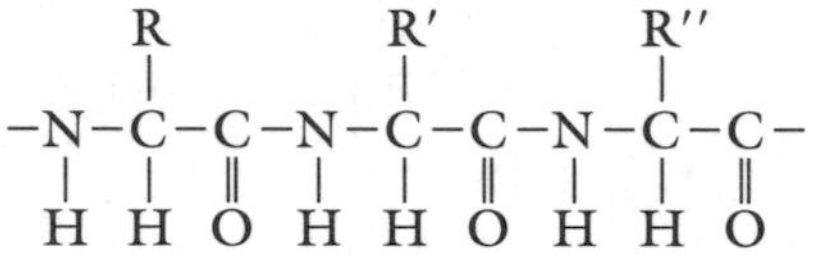

$$\begin{array}{ccccccccc} & \mathrm{R} & & & \mathrm{R'} & & & \mathrm{R''} & \\ & | & & & | & & & | & \\ -\mathrm{N}- & \mathrm{C} & -\mathrm{C}- & \mathrm{N} & -\mathrm{C}- & \mathrm{C} & -\mathrm{N}- & \mathrm{C} & -\mathrm{C}- \\ | & | & \| & | & | & \| & | & | & \| \\ \mathrm{H} & \mathrm{H} & \mathrm{O} & \mathrm{H} & \mathrm{H} & \mathrm{O} & \mathrm{H} & \mathrm{H} & \mathrm{O} \end{array}$$

in which R, R′, R″, and so on refer to the side-chains of different amino acids. A given protein is defined by the sequence in which the amino acids are arranged along the chain. Enzymes are proteins, and the solid tissues of the body (muscle, skin, hair, and connective tissue) are substantially made up of various insoluble proteins. Proteins in the diet are thus the only source of amino acids or of nitrogen for synthesis of amino acids. The amino acids are then reassembled to make new proteins and replenish the tissues. The proteins may be the insoluble structural proteins, or enzymes and other soluble, often circulating, proteins, such as haemoglobin, the red oxygen-carrying protein of blood. We use proteins to breathe, see, hear, move, and recognize and destroy alien bodies such as bacteria. Proteins are continually degraded and replaced, though at widely varying rates: the molecules of some protein types endure for months, others for minutes. Nitrogen turnover never ceases. Part of the available pool of amino acids is used to make new proteins and part as an energy source, with ammonia as the nitrogenous waste product. Ammonia is toxic and has to be converted into the simple compound urea, which is excreted in the urine. Further portions of the amino acid pool are turned into such nitrogen-containing compounds as the famous bases of DNA that encode the genetic message and those of the related RNA, ATP, adrenaline (now called epinephrine) and many others, and some go into the production of glucose by gluconeogenesis, or into the manufacture of lipids. Some are used to make acetyl-CoA, which serves so many synthetic functions. Not all the 20 amino acids can be synthesized in the human body, and for 10 of them we rely on proteins in the diet.

It will be clear by now that the metabolic reactions pertaining to carbohydrates, fats, and proteins are intimately linked. The scheme set out below summarizes the overall picture, omitting the processing of amino acids derived from dietary proteins. The diagram refers to the normal well-fed state, and does not show gluconeogenesis. Insulin, made in the pancreas in response to the level of glucose in the blood, feeds into the cycle to regulate the metabolic process (Chapter 11).

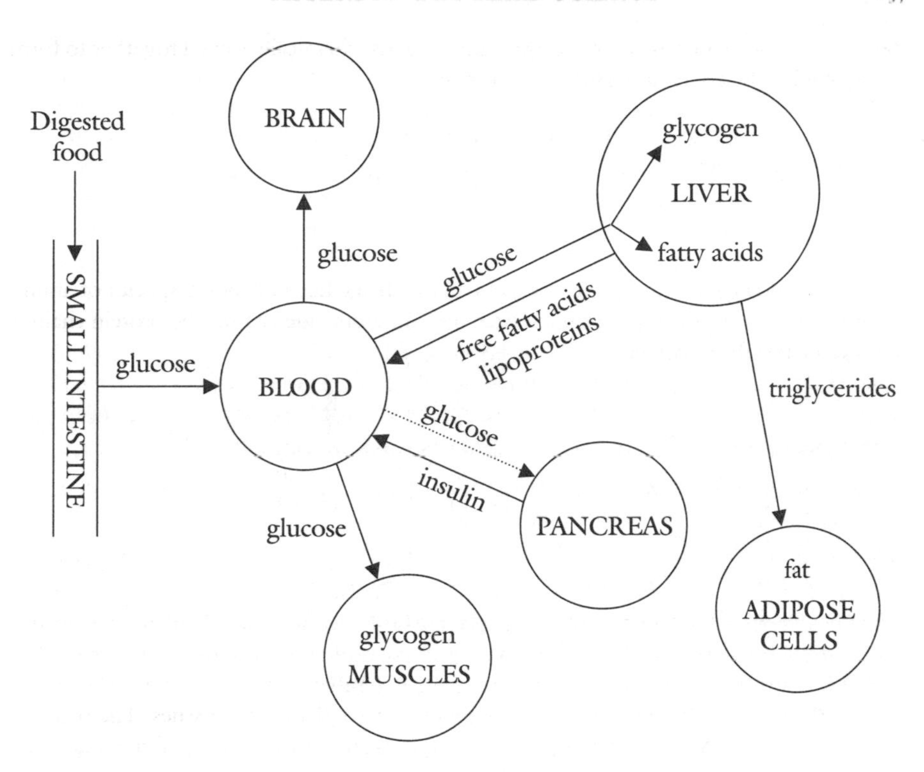

The dotted line represents the feedback step, whereby excess glucose switches off insulin formation.

Where the energy comes from

Much the largest repository of energy in the body is fat. Glycogen in the liver and muscles cannot be stored on the same scale: there is simply no space. The fat, of course, serves the additional purpose of providing heat insulation and upholstery to protect the vulnerable organs. Moreover, each gram of lipid when 'burned' yields nine calories (really kilo-calories) of energy, while carbohydrates give out only four calories and proteins much the same. The glucose in the blood needs constant replenishing: the total amount in our blood at any instant would keep us going for only a matter of minutes. The stored glycogen can put out some 600 calories, enough to sustain life unassisted by other energy sources for a day. The fat reserve of well-fed individuals can release as much as 100,000 calories of energy, which would see them through a month on a starvation diet. The wastage of muscle and some other proteins could contribute an additional 20,000 calories or so, but to call on that would be bad news (see Chapter 8).

Metabolic mishaps

Considering the complexity of the metabolic network and the number of enzymes that are implicated, it is less than surprising that a mutation will sometimes throw a spanner in the machinery. The existence of such deleterious mutations was first perceived by one of the great clinical scientists of the past, Sir Archibald Garrod (1857–1937), working in London and in Oxford (p. 168). In 1909 there appeared his seminal work, *Inborn Errors of Metabolism*. In it he set out, based on a study of four inherited metabolic disorders, evidence that the genetic rules uncovered by the Bohemian monk Gregor Mendel in the 19th century were applicable not only to the sweet peas in Mendel's garden, but also to man. Garrod formulated the principle, which became a cornerstone of 20th-century genetics, of 'one gene, one enzyme', that is to say, each enzyme is specified by its own gene, and a mutation in that gene will cause an alteration (sometimes but not necessarily deleterious), or even total absence, of the enzyme. If the enzyme is missing or is disabled by the mutation, some metabolic pathway will fail, with more or less serious effects. Since Garrod's time a large number, running into thousands, of such mutations have been detected and, with the modern methods of genetics based on DNA manipulation, they have been pinpointed in the patient's DNA. Just a few of the best-known examples follow.

Alkaptonuria was one of Garrod's four exemplars. It reveals itself in newborn infants by the colour of the urine, which goes progressively black on standing (or on the nappy). The condition leads to symptoms resembling rheumatoid arthritis, and is caused by the failure of an enzyme needed for catabolism of the amino acid tyrosine.

Phenylketonuria is caused by lack of an enzyme responsible for formation of tyrosine from another amino acid, phenylalanine. An intermediate of the reaction sequence is excreted in the urine. Mental retardation results, but today routine screening will detect the condition, which is treated by eliminating most of the phenylalanine from the diet.

Gaucher's disease, Niemann–Pick disease, and *Tay–Sachs disease* are all caused by the deficiency of enzymes involved in lipid metabolism. All give rise to distressing symptoms, and victims of either of the last two die in early childhood.

Familial hypercholesterolaemia is characterized by excessively high levels of LDL in the blood. This results in early susceptibility to heart disease. It is caused by an abnormality of the proteins (receptors) to which the LDL must attach if it is to be taken up by the liver. The condition is treated by a rigorous diet. There are several other states, caused by different mutations and collectively termed *hyperlipoproteinaemias*, with similar symptoms and outcomes if untreated.

Glycogen storage diseases are the result of inability to synthesize glycogen or to break it down to provide glucose. Some dozen types are now recognized, each due to a different enzyme defect. Some are worse than others in their consequences.

Maple sugar urine disease is detected in babies by the characteristic smell of the urine. It is caused by a failure to metabolize certain of the amino acids. If it is not treated bya rigorous diet nerve damage ensues.

Galactosaemia is failure to break down galactose, a sugar abundant in milk. It causes cataracts, enlarged liver and spleen, and mental retardation if not diagnosed.

Fructose intolerance is failure to metabolize the fruit sugar, fructose. It is life-threatening unless the sufferer is deprived of fruit and all its products.

These examples of the innumerable metabolic diseases that have come to light over the years serve only to show that metabolic reactions involving carbohydrates, fats, and proteins can all be unfavourably affected by mutations, and that the failure of one, seemingly insignificant enzyme can have catastrophic consequences. The development of large-scale screening methods will undoubtedly help in the future to avert the worst symptoms of many more of these conditions, and to warn parents about the mutant genes they may be carrying. Some of the diseases are dominant, requiring only one of the two gene copies (alleles) contributed by a child's parents to be abnormal, but many are recessive, which means that they reveal themselves only when both parents possess the mutant gene. In such instances, at least in regard to the commoner genetic maladies, prenatal diagnosis may be available and will allow the parents to decide whether the pregnancy should be terminated.

Salt and blood pressure

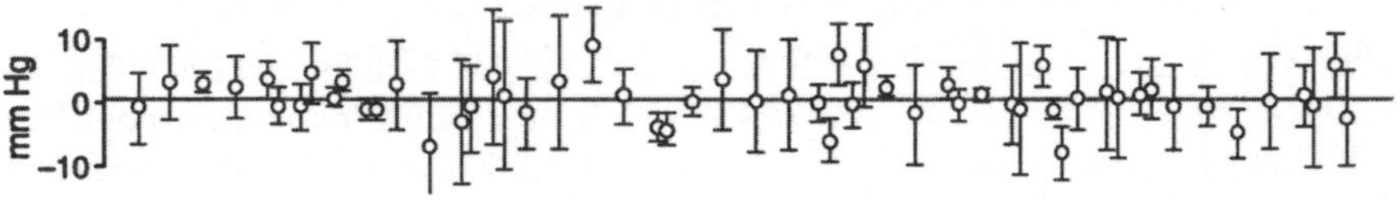

Results of 56 clinical trials, conducted between 1980 and 1998, of the effects of rigorously restricted salt intake on blood pressure in people with blood pressure in the normal range. These average out to near-zero effect.

Graudal *et al*, 'Effects of Sodium Restriction on Blood Pressure, Renin, Aldosterone, Catecholamines. *Journal of the American Medical Association*, **279** (1998), 1383

Further Reading and References

Chapter 1

Further reading

J. C. Drummond and Anne Wilbraham: *The Englishman's Food: A history of Five Centuries of English Diet* (revised edition with a new chapter by Dorothy Hollingsworth) (Cape, London, 1957; paperback edition, Pimlico, London, 1991). This is a still unparalleled work of deep, but unfailingly entertaining erudition. The revised edition regrettably omits some parts of the first edition, published in 1939.

John Burnett: *Plenty and Want: A Social History of Diet in England from 1815 to the Present Day* (Thomas Nelson, London, 1966; paperback edition, Penguin, Harmondsworth, 1968).

Reay Tannahill: *Food in History* (Eyre Methuen, London & New York, 1973; paperback editions, Penguin, London, 1988, and Random House, New York, 1995).

References

1. Thomas Fuller: *Meditation on the Times* – in *Good Thoughts in Bad Times, Together with Good Thoughts in Worse Times* (1649). Cited in Drummond and Wilbraham: *The Englishman's Food*.
2. Nicholas Culpeper: *The English Physitian Enlarged* (Peter Cole, London, 1653).
3. B. Seebohm Rowntree: *Poverty: A Study of Town Life* (Macmillan, London, 1901).
4. H. Chick (1976) Study of rickets in Vienna 1919–1922. *Medical History* **20**, 41–51.
5. Cited in note 4.
6. Medical Research Council Special Report Series No. 77 (HMSO, London, 1923).
7. John Boyd Orr, Lord: *As I Recall* (MacGibbon and Kee, London, 1966).
8. Margaret Ashwell (ed.): *McCance and Widdowson: A Scientific Partnership of 60 Years 1933 to 1993* (British Nutritional Foundation, London, 1993).
9. E. M. Widdowson (1980) Adventures in nutrition over half a century. *Proceedings of the Nutritional Society* **39**, 293–306. See also Obituary of Elsie Widdowson by Anthony Tucker, *Guardian*, 12 June 2000.
10. Ronald W. Clark: *J. B. S.: The Life and Work of J. B. S. Haldane* (Hodder and Stoughton, London, 1968).

Chapter 2

Further reading

Kenneth J. Carpenter: *The History of Scurvy and Vitamin C* (Cambridge University Press, Cambridge, 1966; revised edition, 1988).

David L. Harvie: *Limeys: The True Story of One Man's War against Ignorance, the Establishment and the Deadly Scurvy* (Sutton, Stroud, UK, 2002).

Both the above are highly readable accounts. Carpenter is a distinguished nutritionist and gives a more balanced view of Lind's achievements and shortcomings.

C. Lloyd and J. L. S. Coulter: *Medicine and the Navy 1200–1900*, Vol. IV (Livingstone, Edinburgh, 1963). This is a formidable and comprehensive work of scholarship.

References

1. Report of a Royal Society committee under Lord Lister. Cited in Lloyd and Coulter: *Medicine and the Navy*.
2. H. Chick, E. M. Hume, and R. F. Skelton (1918) The relative content of antiscorbutic principle in limes and lemons – Discussion. *Lancet* **194**, 735–736 (30 November).
3. Henderson Smith (1918) Lime juice and other antiscorbutics – Discussion. *Lancet* **194**, 737–738 (30 November).
4. A. E. Wright (1904) Discussion on the aetiology of scurvy. *Transactions of the Epidemiology Society*, 94–97. Cited in Carpenter: *The History of Scurvy and Vitamin C*.

Chapter 3

Further reading

K. Y. Guggenheim and I. Wolinsky: *Nutrition and Nutritional Diseases: The Evolution of Concepts* (Collamore Press, Lexington, MA, and Toronto, 1981).

O. Temkin: *Galenism: Rise and Decline of a Medical Philosophy* (Cornell University Press, Ithaca, 1973).

Hippocrates (trans. W. H. S. Jones): *Nature of Man, Humours, Aphorisms and Regimen* (Loeb Classical Library, Harvard University Press, Cambridge, MA, 1931).

References

1. Ralph Jackson: *Doctors and Disease in the Roman Empire* (British Museum Publishers, London, 1988).
2. R. Landau (2000) Birth of the modern diet. *Scientific American*, August, 62–67.
3. H. M. Pachter: *Paracelsus: Magic into Science* (Henry Schumann, New York, 1951).
4. Thomas Muffett (corrected and enlarged by Christopher Bennett): *Health Improvement* (1655). Cited in J. C. Drummond and Anne Wilbraham: *The Englishman's Food*.
5. Ibid.
6. Andrew Boorde: *The First Boke of the Introduction of Knowledge: A Compendyous Regyment, or a Dyetary of Helth*, 1542 (ed. F. J. Furnivall, Early English Text Society, London, 1870).
7. G. L. Carefoot and E. R. Spratt: *Famine on the Wind: Man's Battle Against Plant Disease* (Rand McNally, New York, 1967; Angus and Robertson, London, 1969).
8. Redcliffe N. Salaman: *The History and Social Impact of the Potato* (Cambridge University Press, Cambridge, 1949).

Chapter 4

Further reading

G. J. Godfield: *The Growth of Scientific Physiology* (Hutchinson, London, 1960).

Frederic L. Holmes: *Lavoisier and the Chemistry of Life: An Exploration of Scientific Creativity* (University of Wisconsin Press, Madison, 1977).

Jean-Pierre Poirier: *Lavoisier: Chemist, Biologist, Economist* (University of Pennsylvania Press, Philadelphia, 1993).

References

1. C. P. O'Malley: *Andreas Vesalius of Brussels 1514-1564* (World Publications, Cleveland, 1964).
2. M. Hunter and E. B. Davis: *The Works of Robert Boyle* (Pickering and Chatto, London, 1999).
3. G. A. Lindeboom: *Herman Boerhaave: The Man and his Work* (Methuen, London, 1968).
4. D. R. Wilkie (1969) Notice of lecture to the Institution of Electrical Engineers, 11 February.
5. Count Rumford: *Of Food and Particularly of Feeding the Poor* (Third Essay), 1796; 3rd edn (T. Cadell, Jr and W. Davies, London, 1797).
6. Sanborn C. Brown: *Count Rumford: Physicist Extraordinary* (Anchor Books, Garden City, NY, 1962).
7. Cited in reference 6.

Chapter 5

Further reading

Kenneth J. Carpenter: *Protein and Energy: A Study in the Changing Ideas in Nutrition* (Cambridge University Press, Cambridge, 1994).

Charles Tanford and Jacqueline Reynolds: *Nature's Robots: A History of Proteins* (Oxford University Press, Oxford, 2001).

Both the above are authoritative and very readable accounts of the tortured history of the field.

David M. Matthews: *Protein Absorption: Development and Present State of the Subject* (Wiley-Liss, New York, 1991). This is an academic text that enters fully into the technicalities of the subject.

William H. Brock: *Justus von Liebig: The Chemical Gatekeeper* (Cambridge University Press, Cambridge, 1997). A definitive biography of a leading figure, against the 19th-century background.

References

1. Cited in William H. Brock: *Justus von Liebig: The Chemical Gatekeeper* (Cambridge University Press, Cambridge, 1997).

2. H. Kamminga (1995) Nutrition for the people, or the fate of Jacob Moleschott's contest for humanist science. In *The Science and Culture of Nutrition 1840-1940* (ed. H. Kamminga and A. Cunningham), pp. 15–47 (Rodopi, Amsterdam).
3. Marcel Chaigneau: *J-B. Dumas, chimiste et homme politique* (Guy Le Prat, Paris, 1984).
4. F. W. J. McCosh: *Boussingault, Chemist and Agriculturalist* (Reidel, Dordrecht, 1984).
5. Cited in reference 3.
6. Cited in Brock: *Justus von Liebig*.
7. C. B. Chapman (1967) Edward Smith (?1813–1874), physiologist, human ecologist, reformer. *Journal of the History of Medicine* **22**, 1–26.
8. Colin A. Russell: *Edward Frankland: Chemistry, Controversy and Conspiracy in Victorian England* (Cambridge University Press, Cambridge, 1996).
9. Cited in reference 8.
10. W. Beneke (1851) On Extractum carnis. Recommended for admission to the London Pharmacopoeia as a most valuable remedy in the treatment of disease. *Lancet* **57** (4 January), 6–8.
11. T. Vosper (1865) On the nutritive value of 'Extractum carnis'. *Lancet* **86** (23 December), 717.
12. Frederic L. Holmes: *Claude Bernard and Animal Chemistry* (Harvard University Press, Cambridge, MA, 1974); James Olmstead: *Claude Bernard, Physiologist* (Harper, New York and London, 1938); Claude Bernard (1865): *Introduction à l'étude de la médecine expérimentale* (trans. R. C. Greene) (Dover, New York, 1957).

Chapter 6

Further reading

John Burnett: *Plenty and Want: A Social History of Diet in England from 1815 to the Present Day* (Thomas Nelson, London, 1966; paperback edition, Penguin, London, 1968; third revised edition, Methuen, London, 1989).

Roy Porter and G. S. Rousseau: *Gout: The Patrician Malady* (Yale University Press, New Haven, 1998).

References

1. Cited in J. C. Drummond and Anne Wilbraham: *The Englishman's Food: A history of Five Centuries of English Diet* (revised edition with a new chapter by Dorothy Hollingsworth) (Cape, London, 1957; paperback edition, Pimlico, London, 1991).
2. Ronald W. Clark: *J. B. S.: The Life and Work of J. B. S.* Haldane (Hodder and Stoughton, London, 1968).
3. R. Apple (1986) 'Advertised by our loving friends': The infant feeding formula industry and the creation of new pharmaceutical markets. *Journal of the History of Medicine and Allied Sciences* **41**, 3–23.
4. Cited in Roy Porter and G. S. Rousseau: *Gout: The Patrician Malady*.
5. Kenneth Dewhurst: *Dr Thomas Sydenham, 1624-1689, His Life and Original Writings* (University of California Press, Berkeley, 1966).

6. Cited in John Burnett: *Plenty and Want: A Social History of Diet in England from 1815 to the Present Day*.

Chapter 7

Further reading

Frederick A. Filby: *A History of Food Adulteration and Analysis* (Allen & Unwin, London, 1934).

H. T. McCone (1991) The history of food colorants before artificial dyes. *Bulletin of the History of Chemistry* **10**, 25–30.

References

1. Frederick A. Filby: *A History of Food Adulteration and Analysis* (Allen & Unwin, London, 1934).
2. C. A. Brown (1925) Life and Chemical Services of Frederick Accum. *Journal of Chemical Education* **2**, 829–851, 1008–1035.
3. E. G. Clayton: *A Memoir of the Late Dr A. H. Hassall: Physician and Sanitary Reformer* (Baillière, Tindall and Cox, London, 1908).
4. Arthur Hill Hassall: *The Narrative of a Busy Life: An Autobiography* (Longmans Green, London, 1893).
5. A. J. Bernays (1853) The analytical sanitary commission. *Lancet* **61**, 66 (15 January).
6. *Times*, 3 January 1852, quoted in *Lancet* **59**, 52 (10 January).

Chapter 8

Further reading

Kenneth J. Carpenter: *Beriberi, White Rice, and Vitamin B: A Disease, a Cause, and a Cure* (University of California Press, Berkeley, 2000). This is the definitive history; it is comprehensive, scholarly, and absorbing.

Alan M. Kraut: *Goldberger's War: The Life and Work of a Public Health Crusader* (Hill and Wang, New York, 2003).

Elizabeth Etheridge: *The Butterfly Caste: A Social History of Pellagra* (Greenwood, Westport, CT, 1972).

References

1. K. Takaki (1906) The preservation of health amongst the personnel of the Japanese army and navy. *Lancet* **167**, 1369–1374. (Despite the austere title this is Takaki's personal account of his work on beriberi.)
2. W. Fletcher (1907) Rice and beri-beri: preliminary report of an experiment conducted at the Kuala Lumpur lunatic asylum. *Lancet* **169**, 1776–1779; W. Fletcher (1909) Rice and beri-beri. *Journal of Tropical Medicine and Hygiene* **12**, 127–135.
3. W. L. Braddon (1911) Beri-beri. *Journal of Tropical Medicine and Hygiene* **14**, 153–154.

4. M. Glogner (1898) Onderzoek naar het verband tuschen den aard der rystvoeding in de gevangenissen op Java en Madera. *Archiven für Schiffs- und Tropenhygiene* **2** 49–56. Cited in Carpenter: *Beriberi, White Rice, and Vitamin B*.
5. Sally Craddock: *Retired Except on Demand: The Life of Dr Cicely Williams* (Green College, Oxford, 1983).
6. C. D. Williams (1935) Kwashiorkor: a nutritional disease of children associated with a maize diet. *Lancet* **226**, 1151–1152.
7. H. S. Stannus (1935) Kwashiorkor. *Lancet* **226**, 1207.
8. Ann Dally: *The Trouble with Doctors: Fashions, Motives and Mistakes* (Robson Books, London, 2003).
9. C. O'Gráda: The great famine and other famines. In *Famine* (ed. C. O'Gráda), Commemorative Lecture Series, pp. 139–157 (Teagasc, Dublin); E. C. Large: *The Advance of the Fungi* (Cape, London, 1940).
10. Robert N. Proctor: *Racial Hygiene: Medicine under the Nazis* (Harvard University Press, Cambridge, MA, 1988).
11. M. Susser and Z. Stein (1994) Timing in prenatal nutrition: a reprise of the Dutch famine study. *Nutrition Research* **52**, 84–94; L. H. Lumey and A. D. Stern (1997) *In utero* exposure to famine and subsequent fertility: the Dutch famine cohort study. *American Journal of Public Health* **87**, 1962–1966.
12. Leon Goure: *The Siege of Leningrad* (McGraw-Hill, New York, 1964); *Life and Death in Besieged Leningrad* (trans. ed. John Barber and Andrei Dzeniskevich. Wellcome Library, London, 2001).
13. F. M. Lipscomb (1948) Medical aspects of Belsen concentration camp. *Lancet* **245**, 604–605; J. Vaughan: Experience at Belsen (1945), in: Protein hydrolysates: Royal Society of Medicine Discussion, *Lancet* **245**, 723–724.
14. M. Winnick (1994) Hunger disease: studies by Jewish physicians in the Warsaw Ghetto. *Nutrition* **10**, 365–380; M. Winnick (ed.): *Hunger Disease: Studies by Jewish Physicians in the Warsaw Ghetto* (Wiley, New York, 1979).

Chapter 9

Further reading

E. V. McCollum: *A History of Nutrition* (Houghton, Mifflin, Boston, 1957).

Kenneth J. Carpenter: *Beriberi, White Rice, and Vitamin B: A Disease, a Cause, and a Cure* (University of California Press, Berkeley, 2000).

G. Wolf (1996) A history of vitamin A and retinoids. *FASEB Journal* **10**, 1102–1107.

References

1. F. G. Hopkins (1929): *The earlier history of vitamin research. Nobel Lectures, Including Presentation Speeches and Laureates' Biographies: Physiology or Medicine, 1922–1941* (Elsevier, Amsterdam, 1965).
2. Cited in reference 1.
3. E. V. McCollum: *From Kansas Farm Boy to Scientist* (University of Kansas Press, Lawrence, KS, 1964).

4. Benjamin Harrow: *Casimir Funk: Pioneer in Vitamins and Hormones* (Dodd, Mead, New York, 1955).
5. F. G. Hopkins: Autobiography. In: *Hopkins and Biochemistry* (ed. J. Needham and E. Baldwin), pp. 1–25 (Heffer, Cambridge, 1949).
6. Cited in reference 1.
7. Gerhardt Schmidt: *Das geistige Vermächtnis von Gustav von Bunge* (Verlag Zürich, Zurich, 1973).
8. David Hamilton: *The Monkey Gland Affair* (Chatto and Windus, London, 1986).
9. A. V. Hoffbrand (2001) The history of folic acid. *British Journal of Haematology* **113**, 579–589.

Chapter 10

Further reading

James H. Young: *American Health Quackery: Collected Essays* (Princeton University Press, Princeton, NJ, 1992).

James H. Young: *The Medical Messiahs: A Social History of Health Quackery in Twentieth Century America* (Princeton University Press, Princeton, NJ, 1967).

Hillel Schwartz: *Never Satisfied: A Cultural History of Diets, Fantasies and Fat* (Free Press, New York, 1986; paperback edition, Anchor Books, Doubleday, New York, 1990).

These three books are highly recommended; they are replete with curious scholarship, and entertain as well as inform. For the story of vegetarianism, see:

Colin Spencer: *The Heretic's Feast: A History of Vegetarianism* (Fourth Estate, London, 1993); also an interesting article: J. C. Whorton (1977) The formulation of a physiological rationale for vegetarianism. *Bulletin of the History of Medicine* **32**, 115–139.

References

1. Robert N. Proctor: *The Nazi War on Cancer* (Princeton University Press, Princeton, NJ, 1999).
2. Dan H. Laurence (ed.): *George Bernard Shaw, Letters*, Vol. 4 (Max Reinhardt, London, 1988).
3. T. Baker: *Sylvester Graham's Lectures on the Science of Human Life, Condensed for the Advantage of the General Reader* (Manchester Vegetarian Society, John Heywood, Manchester, 1887); Sylvester Graham: *Lectures on the Science of Human Life* (Fowler and Wells, New York, 1883).
4. Cited in E. S. Turner: *The Shocking History of Advertising* (Michael Joseph, London, 1952).
5. L. M. Barnett: Fletcherism and the chew-chew fad of the Edwardian era. In *Nutrition in Britain: Science, Scientists and Politics in the Twentieth Century* (ed. David F. Smith), pp. 6–28 (Routledge, London, 1997).
6. Rima Apple: *Vitamania: Vitamins in American Culture* (Rutgers University Press, New Brunswick, NJ, 1996).
7. T. H. Jukes: The laetrile scandal. In *Human Nutrition: Current Issues and Controversies* (ed. A. Neuberger and T. H. Jukes), pp. 233–241 (MTP Press, Lancaster, 1982); J. H.

Young: Laetrile in Historical Perspective. In *Politics, Science and Cancer: The Laetrile Phenomenon* (ed. G. E. Merkle and J. C. Petersen), pp. 11–60 (Westview Press, Boulder, CO, 1980).

Chapter 11

Further reading

Eric Schlosser: *Fast Food Nation: The Dark Side of the All-American Meal* (Houghton Mifflin, New York, 2002; paperback edition, Perennial, New York, 2002).

Ellen Ruppel Shell: *The Hungry Gene: The Science of Fat and the Future of Lean* (Atlantic Monthly Press, New York, 2002).

Greg Critser: *Fat Land: How Americans Became the Fattest People in the World* (Houghton Mifflin, New York, 2003; paperback edition, Penguin, London, 2004).

All the above are first-rate journalism, and the last two, in particular, offer good explanations of the physiological issues.

References

1. C. B. Gesch, S. M. Hammond, S. E. Hampson, A.Evers, and M. J. Crowder (2002) Influence of supplementary vitamins, minerals and essential fatty acids on the antisocial behaviour of young adult prisoners. *British Journal of Psychiatry* **181**, 22–28. See also Theodore Dalrymple: Oh to be in England: the starving criminal. *City Journal* (New York), Autumn 2002.
2. J. Komlos and M. Baur (2004) From the tallest to (one of) the fattest: the enigmatic fate of the American population in the 20th century. *Economics and Human Biology* **2**, 57–74. See also Burkhard Bilger: The height gap. *New Yorker*, 5 April 2004.
3. Erik Millstone: *Food Additives: Taking the Lid Off What We Really Eat* (Penguin, London, 1986).
4. Felicity Lawrence: *Not on the Label: What Really Goes into the Food on Your Plate* (Penguin, London, 2004).
5. Opinion of the Scientific Committee on Veterinary Measures Relating to Human Health: Assessment of potential risks to human health from hormone residue in bovine meat and meat products. European Commissions, xxiv/B3/SC4 (30 April 1999).
6. Michael Snayerson and Mark J. Plotkin: *The Killers Within: The Deadly Rise of Drug-Resistant Bacteria* (Little Brown, Boston, 2002).
7. In Felicity Lawrence: *Not on the Label*.
8. The soft science of dietary fat (2001). *Science* **291**, 2536–2543.
9. G. Taubes (1998) The (political) science of salt. *Science* **281**, 898–907; D. A. McCarron (1998) Diet and blood pressure – the paradigm shift, *Science* **281**, 933–934.
10. See, for example, F. J. He and G. A. McGregor (2003) How far should salt intake be reduced? *Hypertension* **42**, 1093–1094.
11. V. A. Zammit (2002) Insulin stimulation of triacylglycerol secretion in the insulin-replete state: implications for the etiology of peripheral insulin resistance. *Annals of the*

New York Academy of Science **967** 52–65. See also a good summary by Gail Vines (2001) Sweet but deadly. *New Scientist*, 1 September, 26–30.

12. In Critser: *Fat Land*.
13. Gerald Reaven: *Syndrome X: The Complete Nutritional Program to Prevent and Reverse Insulin Resistance* (Simon and Schuster, New York, 2000); T. J. Wilkin and L. D. Voss (2004) Metabolic syndrome: maladaptation to a modern world. *Journal of the Royal Society of Medicine* **97**, 511–520.
14. Quoted in Millstone: *Food Additives*.
15. P. A. Mayes (1993) Intermediary metabolism of fructose. *American Journal of Clinical Nutrition* **58**, S754–S765. For a clear, less technical description see also Critser: *Fat Land*.
16. I. C. Groop and J. G. Eriksson (1992) The etiology and pathogenesis of non-insulin-dependent diabetes. *Annals of Medicine* **24**, 483–489; for less technical accounts see Shell: *The Hungry Gene*, and Malcolm Gladwell: The Pima Paradox. *New Yorker*, 2 February 1998.
17. The composition of certain secret remedies. V. Obesity cures. (1907) *British Medical Journal* **3/4 3/4**, 24–25.
18. C. Haller (2000) Adverse cardiovascular and central nervous system events associated with dietary supplements containing ephedrine alkaloids. *New England Journal of Medicine* **343**, 1833–1888.
19. D. E. Arteburn, P. K. Crane, and D. L. Veenstra (2004) The efficacy and safety of sibutramine for weight loss: a systematic review. *Archives of Internal Medicine* **164**, 994–1003.
20. D. W. Clarke and M. Harrison-Woolrych (2004) Sibutramine may be associated with memory impairment. *British Medical Journal* **329**, 1316.
21. J. Garrow (1998) Flushing away fat: weight loss during trials of orlistat was significant, but over half was due to diet. *British Medical Journal* **317**, 830–831.
22. See Shell: *The Hungry Gene* for a full account.
23. J. M. Friedman (1998) Leptin, leptin receptors, and the control of body weight. *Nutritional Reviews* **56**, S38–S46; see also Shell: *The Hungry Gene*.
24. J. L. Marx (2003) Cellular warriors at the battle of the bulge. *Science* **299**, 846–849.
25. J. C. Halford (2004) Clinical pharmacotherapy for obesity: current drugs and those in advanced development. *Current Drug Targets* **5**, 637–646; T. Gura (2003) Obesity pipeline not so fat. *Science* **299**, 849–852.
26. M. E. Daly (2004) Extending the use of the glycaemic index: beyond diabetes? *Lancet* **364**, 736–737; D. B. Pawlak, J. A. Kushner, and D. S. Ludwig (2004) Effects of dietary glycaemic index on adiposity, glucose homeostasis, and plasma lipids in animals. *Science* **364**, 778–785.
27. M. F. Jacobs (1999) Olestra snacks compared with regular snacks. *Annals of Internal Medicine* **131**, 866; T. F. Barlam and E. M. McCloud (2003) Severe gastrointestinal illness associated with olestra ingestion. *Journal of Pediatric Gastroenterology and Nutrition* **37**, 95–96; but see also Food and Drug Administration (1996): Food additives permitted for direct addition to food for human consumption: olestra, final rule. *Federal Register* **61**, 3117–3173.

28. G. Bjelakovic, D. Nikolova, R. G. Simonetti, and C. Gluud (2004) Antioxidant supplements for prevention of gastrointestinal cancers: a systematic review and meta-analysis. *Lancet* **364**, 1219–1228; but see also D. Forman and D. Altman (2004) Vitamins to prevent cancer: supplementary problems, *Lancet* **364**, 1193–1194.
29. Cited in E. S. Turner: *The Shocking History of Advertising* (Michael Joseph, London, 1952).
30. Jack Raso in *The Health Robbers: A Close Look at Quackery in America* (ed. Stephen Barrett and William T. Jarvis), pp. 225–252 (Prometheus Books, Buffalo, NY, 1993).
31. Cited in Shell: *The Hungry Gene.*
32. L. Arab (2004) Individualized nutritional recommendations: do we have the measurements needed to assess risk and make dietary recommendations? *Proceedings of the Nutrition Society* **63**, 167–172; L. A. Pray (2005) Dieting for the genome generation. Nutrigenomics has yet to prove its worth. So why is it selling? *Scientist* **19**, 14–16.

Index